Linear Functional Analysis

Linear Functional Analysis

Joan Cerdà

Graduate Studies
in Mathematics

Volume 116

American Mathematical Society
Providence, Rhode Island

Real Sociedad Matemática Española
Madrid, Spain

2010 *Mathematics Subject Classification.* Primary 46–01; Secondary 46Axx, 46Bxx, 46Exx, 46Fxx, 46Jxx, 47B15.

For additional information and updates on this book, visit
www.ams.org/bookpages/gsm-116

Library of Congress Cataloging-in-Publication Data

Cerdà, Joan, 1942–
Linear functional analysis / Joan Cerdà.
p. cm. — (Graduate studies in mathematics ; v. 116)
Includes bibliographical references and index.
ISBN 978-0-8218-5115-9 (alk. paper)
1. Functional analysis. I. Title.

QA321.C47 2010
515′.7—dc22

2010006449

♾ The paper used in this book is acid-free and falls within the guidelines
established to ensure permanence and durability.
Visit the AMS home page at `http://www.ams.org/`

10 9 8 7 6 5 4 3 2 1   15 14 13 12 11 10

To Carla and Marc

Contents

Preface

The aim of this book is to present the basic facts of linear functional analysis related to applications to some fundamental aspects of mathematical analysis.

If mathematics is supposed to show common general facts and structures of particular results, functional analysis does this while dealing with classical problems, many of them related to ordinary and partial differential equations, integral equations, harmonic analysis, function theory, and the calculus of variations.

In functional analysis, individual functions satisfying specific equations are replaced by classes of functions and transforms which are determined by each particular problem. The objects of functional analysis are spaces and operators acting between them which, after systematic studies intertwining linear and topological or metric structures, appear to be behind classical problems in a kind of cleaning process.

In order to make the scope of functional analysis clearer, I have chosen to sacrifice generality for the sake of an easier understanding of its methods, and to show how they clarify what is essential in analytical problems. I have tried to avoid the introduction of cold abstractions and unnecessary terminology in further developments and, when choosing the different topics, I have included some applications that connect functional analysis with other areas.

The text is based on a graduate course taught at the Universitat de Barcelona, with some additions, mainly to make it more self-contained. The material in the first chapters could be adapted as an introductory course on functional analysis, aiming to present the role of duality in analysis, and also the spectral theory of compact linear operators in the context of Hilbert and Banach spaces.

In this first part of the book, the mutual influence between functional analysis and other areas of analysis is shown when studying duality, with von Neumann's proof of the Radon-Nikodym theorem based on the Riesz representation theorem for the dual of a Hilbert space, followed by the representations of the duals of the L^p spaces and of $\mathcal{C}(K)$, in this case by means of complex Borel measures.

The reader will also see how to deal with initial and boundary value problems in ordinary linear differential equations via the use of integral operators. Moreover examples are included that illustrate how functional analytic methods are useful in the study of Fourier series.

In the second part, distributions provide a natural framework extending some fundamental operations in analysis. Convolution and the Fourier transform are included as useful tools for dealing with partial differential operators, with basic notions such as fundamental solutions and Green's functions.

Distributions are also appropriate for the introduction of Sobolev spaces, which are very useful for the study of the solutions of partial differential equations. A clear example is provided by the resolution of the Dirichlet problem and the description of the eigenvalues of the Laplacian, in combination with Hilbert space techniques.

The last two chapters are essentially devoted to the spectral theory of bounded and unbounded self-adjoint operators, which is presented by using the Gelfand transform for Banach algebras. This spectral theory is illustrated with an introduction to the basic axioms of quantum mechanics, which motivated many studies in the Hilbert space theory.

Some very short historical comments have been included, mainly by means of footnotes. For a good overview of the evolution of functional analysis, J. Dieudonné's and A. F. Monna's books, [**10**] and [**31**], are two good references.

The limitation of space has forced us to leave out many other important topics that could, and probably should, have been included. Among them are the geometry of Banach spaces, a general theory of locally convex spaces and structure theory of Fréchet spaces, functional calculus of nonnormal operators, groups and semigroups of operators, invariant subspaces, index theory, von Neumann algebras, and scattering theory. Fortunately, many excellent texts dealing with these subjects are available and a few references have been selected for further study.

A small number of references have been gathered at the end of each chapter to focus the reader's attention on some appropriate items from a general bibliographical list of 44 items.

Almost 240 exercises are gathered at the end of the chapters and form an important part of the book. They are intended to help the reader to develop techniques and working knowledge of functional analysis. These exercises are highly nonuniform in difficulty. Some are very simple, to aid in better understanding of the concepts employed, whereas others are fairly challenging for the beginners. Hints and solutions are provided at the end of the book.

The prerequisites are very standard. Although it is assumed that the reader has some a priori knowledge of general topology, integral calculus with Lebesgue measure, and elementary aspects of normed or Hilbert spaces, a review of the basic aspects of these topics has been included in the first chapters.

I turn finally to the pleasant task of thanking those who helped me during the writing. Particular thanks are due to Javier Soria, who revised most of the manuscript and proposed important corrections and suggestions. I have also received valuable advice and criticism from María J. Carro and Joaquim Ortega-Cerdà. I have been very fortunate to have received their assistance.

Joan Cerdà
Universitat de Barcelona

Chapter 1

Introduction

The purpose of this introductory chapter is to fix some terminology that will be used throughout the book and to review the results from general topology and measure theory that will be needed later. It is intended as a reference chapter that initially may be skipped.

1.1. Topological spaces

Recall that a **metric** or **distance** on a nonempty set X is a function

$$d : X \times X \to [0, \infty)$$

with the following properties:

1. $d(x, y) = 0$ if and only if $x = y$,
2. $d(x, y) = d(y, x)$ for all $x, y \in X$, and
3. $d(x, y) \leq d(x, z) + d(z, y)$ for all $x, y, z \in X$ (triangle inequality).

The set X equipped with the distance d is called a **metric space**. If $x \in X$ and $r > 0$, the **open ball** of X with center x and radius r is the set $B_X(x, r) = \{y \in X;\ d(y, x) < r\}$, while $\bar{B}_X(x, r) = \{y \in X;\ d(y, x) \leq r\}$ denotes the corresponding **closed ball**.

Of course, a first example is the real n-dimensional Euclidean space, $\mathbf{R}^n$, with the **Euclidean distance** between two points $x, y \in \mathbf{R}^n$ defined to be

$$d(x, y) := |x - y| = \sqrt{\sum_{j=1}^{n} (x^j - y^j)^2}$$

where, for $x = (x^1, \dots, x^n)$,

$$|x| := \Big(\sum_{j=1}^{n} (x^j)^2\Big)^{1/2},$$

the **Euclidean norm** of x.

1.1.1. Topologies. Most continuity properties will be considered in the context of a metric, but we also need to consider the more general setting of topological spaces.

Recall that a nonempty set X is called a **topological space** if it is endowed with a collection $\mathcal{T}$ of sets having the following properties:

1. $\emptyset, X \in \mathcal{T}$,
2. $A, B \in \mathcal{T} \Rightarrow A \cap B \in \mathcal{T}$, and
3. $\bigcup_{A \in \mathcal{A}} A \in \mathcal{T}$ if $\mathcal{A} \subset \mathcal{T}$.

The elements of $\mathcal{T}$ are called the **open sets** of the space, and $\mathcal{T}$ is called the **topology**. The **closed sets** are the complements $G^c = X \setminus G$ of the open sets $G \in \mathcal{T}$.

A subset of X which contains an open set containing a point $x \in X$ is called a **neighborhood** of x in X. A collection $\mathcal{U}(x)$ of neighborhoods of a point x is a **neighborhood basis** of x if every neighborhood V of this point contains some $U(x) \in \mathcal{U}(x)$. Of course, the collection of all open sets that contain x is a neighborhood basis of this point.

The **interior** of $A \subset X$ is defined as the set $\operatorname{Int} A$ of all points x such that A contains some neighborhood $U(x)$ of x. It is the union of all open sets contained in A, and A is an open set if and only if $\operatorname{Int} A = A$.

The **closure** $\bar{A}$ of A is the set of all points $x \in X$ such that $U(x) \cap A \neq \emptyset$ for every neighborhood $U(x)$ of x. Obviously $\bar{A}$ is closed since, if $x \notin \bar{A}$, we can find an open neighborhood $U(x)$ contained in $\bar{A}^c$, so that this set is open. Moreover, $\bar{A}$ is contained in every closed set F that contains A, since F^c is an open set whose points do not belong to the closure of A.

A sequence $\{x_n\} \subset X$ **converges** to a point $x \in X$ if every neighborhood of x contains all but finitely many of the terms x_n. Then we write $\lim_{n\to\infty} x_n = x$ or $x_n \to x$ as $n \to \infty$.

All the topological spaces we are interested in will be **Hausdorff**,[1] which means that for two arbitrary distinct points $x, y \in X$ one can find disjoint neighborhoods of x and y. This implies that every point $\{x\}$ is a closed set,

[1]Named after the German mathematician Felix Hausdorff, one of the creators, in 1914, of a modern point set topology, and of measure theory. He worked at the Universities of Leipzig, Greifswald, and Bonn.

since any other point has a neighborhood that does not meet $\{x\}$, and $\{x\}^c$ is open.

Moreover, in a Hausdorff space, limits are unique. Indeed, if $x_n \to x$, then for any neighborhood $U(x)$ there exists N_x so that $x_n \in U(x)$ if $n \geq N$; for any other point y we can find disjoint neighborhoods $U(x)$ and $U(y)$ of x and y, and it is impossible that also $x_n \in U(y)$ if $n \geq N_y$.

Let us gather together some elementary facts concerning topological spaces:

(a) Suppose $(X_1, \mathcal{T}_1)$ and $(X_2, \mathcal{T}_2)$ are two topological spaces. A function $f : X_1 \to X_2$ is said to be **continuous at** $x \in X_1$ if, for every neighborhood $V(y)$ of $y = f(x) \in X_2$, there exists a neighborhood $U(x)$ of x such that $f(U(x)) \subset V(y)$. Obviously, we may always assume that $V(y) \in \mathcal{V}(y)$ and $U(x) \in \mathcal{U}(x)$ if $\mathcal{U}(x)$ and $\mathcal{V}(y)$ are neighborhood bases of x and y.

(b) If $f : X_1 \to X_2$ is continuous at every point x of X_1, f is said to be **continuous** on X_1. This happens if and only if, for every open set $G \subset X_2$, the inverse image $f^{-1}(G)$ is an open set of X_1, since $f(U(x)) \subset V(y) \subset G$ means that $U(x) \subset f^{-1}(G)$.

By taking complements, f is continuous if and only if the inverse images of closed sets are also closed. Moreover, in this case, $f(\bar{A}) \subset \overline{f(A)}$ for any subset A of X_1 since, if $U(x) \cap A \neq \emptyset$ when $U(x) \in \mathcal{U}(x)$, for every $U(f(x)) \in \mathcal{U}(f(x))$ we may choose $U(x)$ so that $f(U(x)) \subset U(f(x))$ and then $f(U(x)) \cap f(A) \subset U(f(x)) \cap f(A) \neq \emptyset$.

(c) Suppose two topologies $\mathcal{T}_1$ and $\mathcal{T}_2$ are defined on X. Then $\mathcal{T}_1$ is said to be **finer** than $\mathcal{T}_2$, or $\mathcal{T}_2$ is **coarser** than $\mathcal{T}_1$, if $\mathcal{T}_2 \subset \mathcal{T}_1$, which means that every $\mathcal{T}_2$-neighborhood is also a $\mathcal{T}_1$-neighborhood, or that the identity map $I : (X, \mathcal{T}_1) \to (X, \mathcal{T}_2)$ is continuous.

(d) If Y is a nonempty subset of the topological space X, then the topology $\mathcal{T}$ of X induces a topology on Y by taking the sets $G \cap Y$ $(G \in \mathcal{T})$ as the open sets in Y. With this new topology, we say that Y is a **topological subspace** of X. The closed sets of Y are the sets $F \cap Y$, with F closed in X.

(e) Many topologies encountered in this book can be defined by means of a distance. The **topology of a metric space**[2] X is the family of all subsets G with the property that every point $x \in G$ is the center of some open (or

---

[2]The name is due to F. Hausdorff (see footnote 1 in this chapter), but the concept of metric spaces was introduced in his dissertation by the French mathematician Maurice Fréchet (1906). See also footnote 2 in Chapter 3.

closed) ball contained in G. It is not hard to verify that the collection of these sets satisfies all the properties of a topology on X.

It is an easy exercise to check that open balls are open sets, closed balls are closed, $x_n \to x$ if and only if $d(x_n, x) \to 0$, and that, for a given point $x \in X$, the balls $B_X(x, 1/n)$ $(n \in \mathbf{N})$ form a countable neighborhood basis of x.

It follows from the triangle inequality that $B_X(x,r) \cap B_X(y,r) = \emptyset$ if $d(x,y) = 2r > 0$ and the topology of the metric space is Hausdorff.

Suppose A is a subset of the metric space X. Since $x \in \bar{A}$ if and only if $\bar{B}_X(x, 1/n) \cap A \neq \emptyset$ for every $n \in \mathbf{N}$, by taking $a_n \in \bar{B}_X(x, 1/n) \cap A$, we obtain that $x \in \bar{A}$ if and only if $x = \lim_n a_n$, i.e., $d(x, a_n) \to 0$, for some sequence $\{a_n\} \subset A$. That is, the closure is the "sequential closure".

Similarly, a function $f : X_1 \to X_2$ between two metric spaces is continuous at $x \in X_1$ if and only if $f(x) = \lim_n f(x_n)$ whenever $x = \lim_n x_n$ in X_1. Thus, f is continuous if it is "sequentially continuous" (see Exercise 1.9(a)).

But one should remember that knowledge of the converging sequences does not characterize what a topology is or when a function is continuous (cf. Exercise 1.9). A topological space is said to be **metrizable** when its topology can be defined by means of a distance.

1.1.2. Compact spaces. A Hausdorff topological space $(K, \mathcal{T})$ is said to be **compact** if, for every family $\{G_j\}_{j\in J}$ of open sets such that

$$K = \bigcup_{j\in J} G_j,$$

a finite subfamily $\{G_{j_1}, \ldots, G_{j_n}\}$ can be chosen so that

$$K = G_{j_1} \cup \cdots \cup G_{j_n}.$$

By considering complements, compactness is equivalent to the property that for a family of closed sets $F_j = G_j^c$ $(j \in J)$ such that every finite subfamily has a nonempty intersection (it is said that the family has the **finite intersection property**), $\bigcap_{j\in J} F_j$ is also nonempty.

A subset K of a Hausdorff topological space X is said to be compact if it is a compact subspace of X or, equivalently, if every cover of K by open subsets of X contains a finite subcover.

It is also a well-known fact that a metric space K is compact if and only if it is **sequentially compact**; this meaning that every sequence $\{x_n\} \subset K$ has a convergent subsequence.

In a metric space X it makes sense to consider a **Cauchy sequence** $\{x_k\}$, defined by the condition $d(x_p, x_q) \to 0$ as $p, q \to \infty$; that is, to every

$\varepsilon > 0$ there corresponds an integer N_ε such that $d(x_p, x_q) \le \varepsilon$ as soon as $p \ge N_\varepsilon$ and $q \ge N_\varepsilon$.

If $\{x_k\}$ is convergent in the metric space, so that $d(x_k, x) \to 0$ as $k \to \infty$, then $\{x_k\}$ is a Cauchy sequence, for $d(x_p, x_q) \le d(x_p, x) + d(x_q, x) \to 0$ as $p, q \to \infty$. The metric space is called **complete** if every Cauchy sequence in X converges to an element of X.

Every compact metric space K is complete, since the conditions $x_{n_k} \to x$ and $d(x_p, x_q) \to 0$, combined with the triangle property, imply that $x_n \to x$.

In a metric space, a set which is covered by a finite number of balls with an arbitrarily small radius is compact when the space is complete:

Theorem 1.1. *Suppose A is a subset of a complete metric space M. If for every $\varepsilon > 0$ a finite number of balls with radius ε cover A, then $\bar{A}$ is compact.*

Proof. Let us show that every sequence $\{a_n\} \subset A$ has a Cauchy subsequence.

Denote $\{a_{n,0}\} = \{a_n\}$. Since a finite number of balls with radius $1/2^{m+1}$ covers A, there is a ball $B(c, 1/2)$ which contains a subsequence $\{a_{n,1}\}$ of $\{a_n\}$. By induction, for every positive integer m, we obtain $\{a_{n,m+1}\} \subset B(c_{m+1}, 1/2^{m+1})$ which is a subsequence of $\{a_{n,m}\}$, since a finite number of balls with radius $1/2^{m+1}$ cover A.

The "diagonal subsequence" $\{a_{m,m}\}$ is then a Cauchy subsequence of $\{a_n\}$, since $a_{p,p} \in B(c_m, 1/2^m)$ if $p \ge m$, so that $d(a_{p,p}, a_{q,q}) \le 2/2^m$ if $p, q \ge m$.

Finally, if $\{x_n\}$ is any sequence in $\bar{X}$, by choosing $a_n \in A$ so that $d(a_n, x_n) < 1/n$ and a Cauchy subsequence $\{a_{m_n}\}$ of $\{a_n\}$, which converges to a point $x \in \bar{A}$, it is clear that also $x_{m_n} \to x$, since

$$d(x_{m_n}, x) \le d(x_{m_n}, a_{m_n}) + d(a_{m_n}, x) \to 0.$$

This shows that $\bar{A}$ is sequentially compact. □

In $\mathbf{R}^n$, a set is said to be bounded if it is contained in a ball and, by the **Heine-Borel theorem**, every closed and bounded set is compact. This is a typical fact of Euclidean spaces which is far from being true for a general metric space, even if it is complete.

The following properties are easily proved:

(a) In a Hausdorff topological space X, if a subset K is compact, then it is a closed subset of X.

(b) In a compact space, all closed subsets are compact.

(c) If $f : X_1 \to X_2$ is a continuous function between two Hausdorff topological spaces and K a compact subset of X_1, then $f(K)$ is a compact subset of X_2.

Property (a) is proved by assuming that there is some point $x \in \bar{K} \setminus K$. Then disjoint couples of open neighborhoods $U_y(x)$ of x and $V(y)$ of y for every $y \in K$ may be taken; by compactness, $K \subset \bigcup_{k=1}^N V(y_k)$ and $U := \bigcap_{k=1}^N U_{y_k}(x)$ would be a neighborhood of x disjoint with K, contrary to $x \in \bar{K}$.

To prove (b), complete any open covering of the closed subset with the complement of the subset, yielding an open covering of the whole space; then use the compactness of the space to select a finite covering.

In (c), the preimage by f of an open covering of $f(K)$ is an open covering of K, and a finite subcovering of this covering of K yields a corresponding finite subcovering for $f(K)$.

Suppose $(X_j, \mathcal{T}_j)$ is a family of topological spaces. The **product topology** $\mathcal{T}$ on the product set $X = \prod_{j \in J} X_j$ is defined as follows:

Let $\pi_j : X \to X_j$ be the projection on the jth component and, for every $x = \{x_j\}_{j \in J}$ in X, let $\mathcal{U}(x)$ denote the collection of the sets of the type

$$U(x) = \prod_{j \in J} U_j(x_j) = \bigcap_{j \in J} \pi_j^{-1}(U_j(x_j))$$

where $U_j(x_j)$ is an open neighborhood of x_j and $U_j(x_j) = X_j$ except for a finite number of indices $j \in J$. Then, by definition, $G \in \mathcal{T}$ if, for every $x \in G$, $x \in U(x) \subset G$ for some $U(x) \in \mathcal{U}(x)$. It is readily checked that $\mathcal{T}$ is a topology, which is Hausdorff if every $(X_j, \mathcal{T}_j)$ is a Hausdorff topological space.

Obviously the projections π_j are all continuous, and $\mathcal{T}$ is the "smallest" topology on X with this property, since for such a topology every set $\pi_j^{-1}(U_j(x_j))$ has to be a neighborhood of x.

Theorem 1.2 (Tychonoff[3]). *If K_j $(j \in J)$ is a family of compact spaces, then the product space $K = \prod_{j \in J} K_j$ is also compact.*

Proof. Let $\mathcal{F}$ be a family of closed sets of K with the finite intersection property and consider the collection Φ of all families of this type that contain $\mathcal{F}$, ordered by inclusion. By **Zorn's lemma**,[4] at least one of these families,

[3]Named after the Russian mathematician Andrey N. Tychonoff, who proved this theorem first in 1930 for powers of $[0, 1]$. He originally published in German, but the English transliteration Tichonov for his name is also commonly used. E. Čech proved the general case of the theorem in 1937.

[4]Zorn's lemma on partially ordered sets will be invoked several times in this book. It is equivalent to Zermelo's axiom of choice in set theory. A binary relation $\preceq$ on a set X is said to

$\mathcal{F}'$, is maximal, since, if $\{\mathcal{F}_\alpha\} \subset \Phi$ is totally ordered, then $\bigcup_\alpha \mathcal{F}_\alpha$ also has the finite intersection property and is an upper bound for $\{\mathcal{F}_\alpha\}$.

The closed sets $\overline{\pi_j(F)} \subset K_j$, where $F \in \mathcal{F}'$, also have the finite intersection property and every space K_j is compact. Hence we can find $x_j \in \bigcap_{F\in\mathcal{F}'} \overline{\pi_j(F)}$ and, if $U_j(x_j)$ is a closed neighborhood of $x_j \in K_j$, it follows that $\pi_j^{-1}(U_j(x_j)) \in \mathcal{F}'$ since $\mathcal{F}'$ is maximal. It follows that $\bigcap_{F\in\mathcal{F}'} \overline{\pi_j(F)} = \{x_j\}$.

Choosing $U_j(x_j) = K_j$ except for finitely many of the indices $j \in J$, we obtain

$$\prod_{j\in J} U_j(x_j) = \bigcap_{j\in J} \pi_j^{-1}(U_j(x_j)) \in \mathcal{F}'$$

and $\prod_{j\in J} U_j(x_j) \cap F \neq \emptyset$ for every $f \in \mathcal{F}'$. Then, if $F \in \mathcal{F}'$, every neighborhood of $x = \{x_j\}_{j\in J}$ intersects F, and $x \in F$ since F is closed. This proves that $\bigcap_{F\in\mathcal{F}} F \neq \emptyset$. □

1.1.3. Partitions of unity. A **locally compact space** is a Hausdorff topological space with the property that every point has a neighborhood basis of compact sets.

We suppose that X is a metric locally compact space, for instance any nonempty closed or open subset of $\mathbf{R}^n$, with the induced topology (see also Exercise 1.5). In this section, we use the letter G to denote an open subset of X, and we use K for a compact subset of X.

We will represent by $\mathcal{C}_c(X)$ the set of all real or complex continuous functions g on X whose **support**, $\operatorname{supp} g = \overline{\{g \neq 0\}}$, is a compact subset of X. We consider $\mathcal{C}_c(G) \subset \mathcal{C}_c(X)$ by defining $g(x) = 0$ when $x \notin G$, for every $g \in \mathcal{C}_c(G)$.

Note that if K is a compact subset of G, we can consider $K \subset \operatorname{Int} L \subset L$, where L is a second compact subset of G, since, if for every $x \in K$ we select a neighborhood $V(x)$ with compact closure $\overline{V(x)} \subset G$, we only need to take $L = \overline{V(x_1)} \cup \cdots \cup \overline{V(x_n)}$ if $K \subset V(x_1) \cup \cdots \cup V(x_n)$. Then

$$\varrho(x) := \frac{d(x, L^c)}{d(x, K) + d(x, L^c)}$$

defines a function $\varrho \in \mathcal{C}_c(G)$ such that $0 \leq \varrho \leq 1$ and $\varrho(x) = 1$ for all $x \in K$. We say that ϱ is a continuous **Urysohn function** for the couple $K \subset G$.

---

be a partial order if the following properties are satisfied: (a) $x \preceq x$, (b) if $x \preceq y$ and $y \preceq z$, then $x \preceq z$, and (c) if $x \preceq y$ and $y \preceq x$, then $x = y$. A subset Y of the partially ordered set X is said to be totally ordered if every pair $x, y \in Y$ satisfies either $x \preceq y$ or $y \preceq x$. According to Zorn's lemma, if every totally ordered subset Y of a nonempty partially ordered set X has an upper bound $z_Y \in X$ (this meaning that $y \preceq z_Y$ for every $y \in Y$), then X contains at least one maximal element (an element z such that $z \preceq x$ implies $z = x$).

With the notation $g \prec G$ we will mean that $g \in \mathcal{C}_c(G)$ and $0 \le g \le 1$, and $K \prec g$ will mean that $g \in \mathcal{C}_c(G)$ and $g = 1$ on a neighborhood of K and $0 \le g \le 1$. Thus, there is a Urysohn function g such that $K \prec g \prec G$.

Theorem 1.3. *Let $\{\Omega_1, \dots, \Omega_m\}$ be a finite family of open subsets of X that cover the compact set $K \subset X$. Then there exists a system of functions $\varphi_j \in \mathcal{C}_c(\Omega_j)$ which satisfies*

1. $0 \le \varphi_j \le 1$ *for every* $j = 1, \dots, m$ *and*
2. $\sum_{j=1}^m \varphi_j(x) = 1$ *for every* $x \in K$.

This system $\{\varphi_j\}_{j=1}^m$ is called a **partition of unity** *subordinate to the covering $\{\Omega_1, \dots, \Omega_m\}$ of K.*

Proof. Let us choose a system $K_j \subset \Omega_j$ $(1 \le j \le m)$ of compact sets that covers K, that can be constructed as follows: if

$$K \subset B_X(x_1, r_1) \cup \cdots \cup B_X(x_N, r_N)$$

with $\bar{B}_X(x_i, r_i) \subset \Omega_j$, just define

$$K_j := \bigcup \{\bar{B}_X(x_i, r_i) \cap K;\ \bar{B}_X(x_i, r_i) \subset \Omega_j\}.$$

For every j let ϱ_j be a Urysohn function for the couple $K_j \subset \Omega_j$. If we define

$$\varphi_1 := \varrho_1,\ \varphi_k := (1 - \varrho_1) \cdots (1 - \varrho_{k-1}) \varrho_k \qquad (1 < k \le m),$$

an induction argument shows that

$$\varphi_1 + \cdots + \varphi_k = 1 - (1 - \varrho_1)(1 - \varrho_2) \cdots (1 - \varrho_k)$$

if $1 \le k \le m$. If $x \in K$, $\varrho_j(x) = 1$ for some j. □

Remark 1.4. If G is open in $\mathbf{R}^n$, it is shown in Chapter 6 that the Urysohn functions ϱ_j can be chosen to be C^∞ (cf. Section 6.1). In this case, the functions φ_j are also smooth.

1.2. Measure and integration

The Riemann integral on $\mathbf{R}^n$, which may be historically grounded and useful for numerical computation and sufficient in many areas of mathematics, is far from being adequate for the requirements of functional analysis. Much more appropriate is the Lebesgue integral, based on computing the measure of level sets of functions.

We will summarize the Lebesgue construction for a general measure.

A **measurable space** is a nonempty set Ω where a distinguished collection Σ of subsets has been selected having the properties of a **σ-algebra**, meaning that the following axioms are satisfied:

1. $\Omega \in \Sigma$.
2. If $\{A_k\}_{k=1}^{\infty}$ is a sequence of sets in Σ, then $\bigcup_{k=1}^{\infty} A_k \in \Sigma$.
3. If $A \in \Sigma$, then its complement $A^c = \Omega \setminus A$ is also in Σ.

These properties also imply that $\emptyset = \Omega^c \in \Sigma$, that $\bigcap_{k=1}^{\infty} A_k \in \Sigma$ if $\{A_k\}_{k=1}^{\infty} \subset \Sigma$, and $A \setminus B = A \cap B^c \in \Sigma$ if $A, B \in \Sigma$. The sets in Σ are the **measurable sets** of the measurable space.

If Ω is any nonempty set, a trivial example of σ-algebra is the collection $\mathcal{P}(\Omega)$ of all subsets of Ω.

In our applications Ω can be assumed to be a locally compact metric space, or an open or closed subset of $\mathbf{R}^n$, and it can be assumed that the measurable sets are the **Borel sets** of Ω, the elements of the **Borel σ-algebra** $\mathcal{B}_\Omega$, which is the intersection of all the σ-algebras that contain all the open sets of Ω; that is, $\mathcal{B}_\Omega$ is the σ-algebra generated by the open sets of Ω.

A **measure** (sometimes also called a positive measure) μ on the measurable space (Ω, Σ) is a mapping

$$\mu : \Sigma \to [0, \infty]$$

such that $\mu(\emptyset) = 0$ and with the following σ-additivity property:

If $\{A_k\}_{k=1}^{\infty}$ is a sequence of disjoint sets in Σ, then

$$\mu\Big(\biguplus_{k=1}^{\infty} A_k\Big) = \sum_{k=1}^{\infty} \mu(A_k).$$

We use the symbol $\uplus$ to indicate a union of mutually disjoint sets.

A **measure space** is a measurable space (Ω, Σ) with a distinguished measure, $\mu : \Sigma \to [0, \infty]$.

The following properties are easy consequences of the definitions:

- Finite additivity, $\mu(A_1 \uplus \cdots \uplus A_m) = \mu(A_1) + \cdots + \mu(A_m)$, since the finite family can be completed with the addition of a sequence of empty sets to obtain a countable disjoint family.
- $\mu(A) \le \mu(B)$ if $A \subset B$, since $B = A \cup (B \setminus A)$ and $\mu(B) = \mu(A) + \mu(B \setminus A)$.
- If $A_m \uparrow A$ (i.e., A is the union of an increasing sequence of measurable sets, A_m), then $\mu(A) = \lim_m \mu(A_m)$, since $A = A_1 \uplus (A_2 \setminus A_1) \uplus (A_3 \setminus (A_2 \cup A_1)) \uplus \cdots$.
- If $A_m \downarrow A$ (i.e., A is the intersection of a decreasing sequence of measurable sets) and $\mu(A_1) < \infty$, then $\mu(A) = \lim_m \mu(A_m)$, since $\mu(A_1) - \mu(A) = \lim_m(\mu(A_1) - \mu(A_m))$.

All our measures will be **σ-finite**: $\Omega = \bigcup_{k=1}^{\infty} \Omega_k$ with $\mu(\Omega_k) < \infty$.

A simple example in any set Ω is the Dirac delta-measure, δ_p, located at an arbitrary point $p \in \Omega$, defined on any set $A \subset \Omega$ by

$$\delta_p(A) = \chi_A(p)$$

where χ_A stands for the characteristic function of the set A, such that $\chi_A(x) = 1$ if $x \in A$ and $\chi_A(x) = 0$ when $x \notin A$. Here the σ-algebra Σ is taken to be all subsets of Ω.

But the main example for us is the **Lebesgue measure**[5] $|A|$ of Borel sets of $\mathbf{R}^n$. The change of variables formula is supposed to be known and it shows that this measure is invariant by rigid displacements, that is, by translation, by rotation, and by symmetry. If A is an interval, $|A|$ is the volume of A (the length if $n = 1$, and the area if $n = 2$).

The Lebesgue measure is an example of **Borel measure**[6] on $\mathbf{R}^n$, that is, a measure μ on the Borel σ-algebra which is finite on compact sets. It is the only Borel measure on $\mathbf{R}^n$ which is invariant by translation, and such that the measure of the unite cube $[0,1]^n$ is 1.

We will see in Theorem 1.6 that every Borel measure on $\mathbf{R}^n$ is a **regular measure**, meaning that the following two properties hold:

(a) **Outer regularity**: For every Borel set B,

$$\mu(B) = \inf\{\mu(G);\ G \supset B, G \text{ an open subset of } \mathbf{R}^n\}.$$

(b) **Inner regularity**: For every Borel set B,

$$\mu(B) = \sup\{\mu(K);\ K \subset B, K \text{ a compact subset of } \mathbf{R}^n\}.$$

Given two σ-finite measure spaces, $(\Omega_1, \Sigma_1, \mu_1)$ and $(\Omega_2, \Sigma_2, \mu_2)$, the product σ-algebra $\Sigma = \Sigma_1 \otimes \Sigma_2$ is defined to be the smallest σ-algebra containing all the sets $A_1 \times A_2$ ($A_1 \in \Sigma_1$, $A_2 \in \Sigma_2$). It can be shown that there is a unique measure μ on Σ, the product measure, such that $\mu(A_1 \times A_2) = \mu(A_1)\mu(A_2)$.

In the special case of the Borel measures on $\mathbf{R}^n$, it easy to show that $\mathcal{B}_{\mathbf{R}^n} \otimes \mathcal{B}_{\mathbf{R}^m} = \mathcal{B}_{\mathbf{R}^{n+m}}$, and the product of the corresponding Lebesgue measures is the Lebesgue measure on the Borel sets in $\mathbf{R}^{n+m}$.

In measure theory only functions $f : \Omega \to \mathbf{R}$ such that every level set $\{f > r\} := \{\omega \in \Omega;\ f(\omega) > r\}$ ($r \in \mathbf{R}$) is measurable are admissible. They

[5] Described by the French mathematician Henry Lebesgue in his dissertation "Intégrale, longueur, aire", in 1902 at the University of Nancy. He then applied his integral to real analysis, with the study of Fourier series. In fact, this integral had been previously obtained by W. H. Young.

[6] Named after the French mathematician Émile Borel, one of the pioneers of measure theory and of modern probability theory.

are called **measurable functions** and, with the usual operations, they form a real vector space which is closed under pointwise limits, supremums and infimums for sequences of measurable functions. Simple functions,

$$s = \sum_{m=1}^{N} \lambda_m \chi_{A_m} \qquad (A_n \in \Sigma,\ \lambda_m \in \mathbf{R}),$$

where we may suppose that the measurable sets A_n are disjoint, are examples of measurable functions. In fact, every measurable function, f, is a pointwise limit of a sequence of simple functions s_n such that

$$|s_n(x)| \uparrow |f(x)| \text{ for every } x \in \Omega. \tag{1.1}$$

To obtain this sequence, consider $f = f^+ - f^-$ ($f^+(x) = \max(f(x), 0)$) and, if $f \geq 0$, define

$$s_n = \chi_{\{f \geq n\}} + \sum_{k=1}^{n2^n} \frac{k-1}{2^n} \chi_{\{\frac{k-1}{2^n} \leq f < \frac{k}{2^n}\}}. \tag{1.2}$$

If f is bounded, $s_n \to f$ uniformly on Ω as $n \to \infty$.

The **Lebesgue integral** $\int f\, d\mu$ (or $\int_\Omega f(x)\, dx$) is then defined as follows: If $f \geq 0$,

$$\int f\, d\mu := \sup \sum_{n=1}^{N} \alpha_j \mu(A_j) \in [0, \infty], \tag{1.3}$$

where the "sup" is extended over all simple functions

$$s = \sum_{j=1}^{N} \alpha_j \chi_{A_j} \qquad (N \in \mathbf{N},\ A_j = s^{-1}(\alpha_j) \in \Sigma)$$

such that $0 \leq s \leq f$.

This integral of nonnegative measurable functions is additive and positively homogeneous. Moreover $\int f\, d\mu = 0$ if and only if $\mu(\{f \neq 0\}) = 0$, that is, $f = 0$ almost everywhere (a.e.), and it satisfies the following fundamental property:

Monotone convergence theorem. If $0 \leq f_n(x) \uparrow f(x)$ $\forall x \in \Omega$ (or a.e.), then $\int f_n\, d\mu \uparrow \int f\, d\mu$.

If the convergence is not monotone, the following inequality still holds:

Fatou lemma.[7] If $f_n(x) \geq 0$ $\forall x \in \Omega$, then $\int \liminf f_n\, d\mu \leq \liminf \int f_n\, d\mu$.

[7]Obtained by the French mathematician Pierre Fatou (1906) in his dissertation, when working on the boundary problem of a harmonic function. Fatou also studied iterative processes, and in 1917 he presented a theory of iteration similar to the results of G. Julia which initiated the theory of complex dynamics.

If $f = f^+ - f^-$, the integral $\int f\,d\mu := \int f^+\,d\mu - \int f^-\,d\mu$ is defined if at least one of the integrals $\int f^\pm\,d\mu$ is finite, and we write $f \in \mathcal{L}^1(\mu)$ if both integrals are finite, that is, if $\int |f|\,d\mu = \int f^+\,d\mu + \int f^-\,d\mu < \infty$. In this case, f is said to be **integrable** or **absolutely integrable**.

Then, with the usual operations, $\mathcal{L}^1(\mu)$ is a real vector space. On this linear space, the integral is a **positive linear form**: $\int f\,d\mu \geq 0$ if $f \geq 0$. Hence, $\int f\,d\mu \leq \int g\,d\mu$ if $f \leq g$, and $\left|\int f\,d\mu\right| \leq \int |f|\,d\mu$.

Moreover, the following fundamental convergence result also holds:

Dominated convergence theorem. If $f_n(x) \to f(x)$ $\forall x \in \Omega$ (or a.e.) and $|f_n(x)| \leq g(x)$ $\forall x \in \Omega$ (or a.e.), where $g \in \mathcal{L}^1(\mu)$, then $\int f_n\,d\mu \to \int f\,d\mu$.

These convergence theorems, as well as the change of variable formula and the Fubini-Tonelli theorem on iterated integration on $\mathbf{R}^n$ or on a product measure space, will be freely used in this text.

For any measurable set A, we denote $\int_A f\,d\mu := \int \chi_A f\,d\mu$.

Lebesgue differentiation theorem. For the usual Lebesgue measure on $\mathbf{R}^n$, if $f \in \mathcal{L}^1(\mathbf{R}^n)$, then

$$\lim_{r\downarrow 0} \frac{1}{|B(x,r)|} \int_{B(x,r)} |f(y) - f(x)|\,dy = 0 \quad \text{a.e. on } \mathbf{R}^n.$$

When $n = 1$, $\lim_{r\downarrow 0} \frac{1}{2r} \int_{x-r}^{x+r} |f(y) - f(x)|\,dy = 0$ a.e. on $\mathbf{R}$.

A function F on an interval $[a,b] \subset \mathbf{R}$ such that

$$F(x) := \int_a^x f(t)\,dt + c \tag{1.4}$$

for some $f \in \mathcal{L}^1(\mathbf{R})$ and some constant c is called **absolutely continuous**. Obviously, it is continuous, and it follows from the Lebesgue differentiation theorem that $F'(x) = f(x)$ a.e. on $[a,b]$, since, assuming that $f = 0$ on $[a,b]^c$,

$$\left|\frac{F(x \pm r) - F(x)}{\pm r} - f(x)\right| \leq 2\frac{1}{2r} \int_{x-r}^{x+r} |f(t) - f(x)|\,dy \to 0 \text{ a.e. as } r \downarrow 0.$$

The integration by parts formula

$$\int_a^b F(x)G'(x)\,dx = (F(b)G(b) - F(a)G(a)) - \int_a^b F'(x)G(x)\,dx$$

holds if F and G are absolutely continuous.

To deal with a complex-valued function, just consider the decomposition into real and imaginary parts, $f = u + iv : \Omega \to \mathbf{C}$ (u and v real measurable functions) and define

$$\int f\,d\mu := \int u\,d\mu + i\int v\,d\mu$$

if $u, v \in \mathcal{L}^1(\mu)$ (that is, if $\int |f|\,d\mu < \infty$). This integral is a linear form on the class $\mathcal{L}^1(\mu)$ of all these complex integrable functions, which is a complex vector space, and also $\left|\int f\,d\mu\right| \le \int |f|\,d\mu$.

In measure theory, two functions are equivalent when they coincide a.e. If $\mathcal{N}(\mu) = \{f;\ f = 0 \text{ a.e.}\}$, we denote $L^1(\mu) := \mathcal{L}^1(\mu)/\mathcal{N}(\mu)$ and $\|f\|_1 := \int |f|\,d\mu$ does not depend on the representative of $f \in L^1(\mu)$.

If $1 \le p < \infty$, we also define

$$\|f\|_p := \left(\int |f|^p d\mu\right)^{1/p}$$

and

$$\|f\|_\infty := \min\{C \ge 0;\ |f(x)| \le C \text{ a.e.}\}.$$

Obviously, $\|f\|_p = 0$ if and only if $f = 0$ a.e., and $\|\lambda f\|_p = |\lambda|\|f\|_p$. We set $0 \cdot \infty := 0$.

The **Minkowski inequality**[8]

$$\|f + g\|_p \le \|f\|_p + \|g\|_p$$

is clear if $p = 1$ or $p = \infty$. In the remaining cases $1 < p < \infty$, it can be obtained from **Hölder's inequality**[9]

$$\int |fg|\,d\mu \le \|f\|_p\|g\|_{p'} \qquad \left(p' = \frac{p}{p-1},\ \text{ or } \frac{1}{p} + \frac{1}{p'} = 1\right),$$

which follows from the convexity of the exponential function, which allows us to set

$$ab = e^{\frac{1}{p}\log a^p + \frac{1}{p'}\log b^{p'}} \le \frac{1}{p}e^{\log a^p} + \frac{1}{p'}e^{\log b^{p'}} = \frac{1}{p}a^p + \frac{1}{p'}b^{p'}.$$

Then just take $a = |f(x)|/\|f\|_p$, $b = |g(x)|\|g\|_{p'}$ and integrate.

For $1 < p < \infty$, the Minkowski inequality is now obtained from

$$|f + g|^p \le |f||f + g|^{p-1} + |g||f + g|^{p-1},$$

since then an application of Hölder's inequality gives

$$\int |f + g|^p d\mu \le \|f\|_p \left(\int |f + g|^p d\mu\right)^{1/p'} + \|g\|_p \left(\int |f + g|^p d\mu\right)^{1/p'};$$

[8] Named after the Lithuanian-German mathematician Hermann Minkowski, who in 1896 used geometrical methods in number theory, in his "geometry of numbers". He used these geometrical methods to deal also with problems in mathematical physics. Minkowski taught at the Universities of Bonn, Göttingen, Königsberg, and Zürich.

[9] First found by the British mathematician Leonard James Rogers (1888) and independently rediscovered by Otto Hölder (1889).

that is,

$$\|f+g\|_p^p \le \|f\|_p\|f+g\|_p^{p/p'} + \|g\|_p\|f+g\|_p^{p/p'} = (\|f\|_p + \|g\|_p)\|f+g\|_p^{p/p'},$$

where $p - p/p' = 1$.

We will write

$$\langle f, g\rangle := \int fg\,d\mu$$

if the integral exists, and Hölder's inequality reads

$$|\langle f, g\rangle| \le \|f\|_p\|g\|_{p'}.$$

The collection of all real or complex measurable functions f such that $\|f\|_p < \infty$, with the usual operations, is a real or complex vector space, and the quotient space by $\mathcal{N}(\mu)$ is denoted $L^p(\mu)$. The value $\|f\|_p$ is the same for all the representatives f we may pick in an equivalence class, and $\|\cdot\|_p$ has on $L^p(\mu)$ all the typical properties of a norm.[10]

If $p = \infty$, note that a representative of f is bounded and then in (1.2) we obtain $s_n(x) \to f(x)$ uniformly.

When $p = 2$, note that $\|f\|_2 = \sqrt{(f, f)_2}$, where

$$(f, g)_2 := \int f\bar{g}\,d\mu \qquad \Big(\int fg\,d\mu \text{ in the real case}\Big)$$

is well-defined, and it is a scalar product that allows us to work with $L^2(\mu)$ as with a Euclidean space. This will be the basic example of Hilbert spaces.

1.2.1. Borel measures on a locally compact space X and positive linear forms on $\mathcal{C}_c(X)$. Let X be a locally compact metric space.

If μ is a Borel measure on X, then every $g \in \mathcal{C}_c(X)$ is μ-integrable, since $|g| \le C\chi_K$ for $K = \operatorname{supp} g$ and compact sets have finite measure. Note that, on $\mathcal{C}_c(X)$, the integral $\int g\,d\mu$ is linear and positive, that is, $\int g\,d\mu \ge 0$ if $g \ge 0$. These linear forms are called **Radon measures**, first obtained in 1913 by J. Radon[11] by considering Borel measures on a compact subset of $\mathbf{R}^n$.

[10] The name L^p for these spaces was coined by F. Riesz in honor of Lebesgue. See footnote 5 in this chapter.

[11] In France, N. Bourbaki chose as starting point of the integral the Radon measures on a locally compact space X, a point of view previously considered by W. H. Young in 1911 and by P. J. Daniel in 1918. It is worth noting that L. Schwartz constructed his distributions in the same spirit, by changing the test space $\mathcal{C}_c(X)$. If every open subset of X is a countable union of compact sets, then the Radon measures are in a bijective correspondence with the Borel measures through the Riesz-Markov representation Theorem 1.5.

Theorem 1.5 (Riesz-Markov representation theorem). *Let J be a positive linear form on $\mathcal{C}_c(X)$. Then there exists a uniquely determined Borel measure μ on X so that*

$$J(g) = \int g\, d\mu \qquad (g \in \mathcal{C}_c(X)) \tag{1.5}$$

and which satisfies the inner regularity property for open sets

$$\mu(G) = \sup\{\mu(K);\ K \subset G,\ \ K \text{ compact}\}.$$

This Borel measure is also outer regular: $\mu(B) = \inf\{\mu(G);\ G \supset B,\ G \text{ open}\}$.

Proof. We start by defining μ^* on open sets by

$$\mu^*(G) := \sup\{J(g);\ g \prec G\},$$

where we assume $\mu^*(\emptyset) = 0$.

This set function has the following properties:

(a) $\mu^*(G_1) \le \mu^*(G_2)$ if $G_1 \subset G_2$, since then $\mathcal{C}_c(G_1) \subset \mathcal{C}_c(G_2)$.

(b) $\mu^*(G_1 \cup G_2) \le \mu^*(G_1) + \mu^*(G_2)$, since, if K is the support of $g \prec (G_1 \cup G_2)$, for $j = 1, 2$ we can find $\varphi_j \in \mathcal{C}_c(G_j)$ such that $0 \le \varphi_j \le 1$ and $\sum_{j=1}^m \varphi_j(x) = 1$ for every $x \in K$, a partition of unity constructed as in Theorem 1.3; thus $g = g\varphi_1 + g\varphi_2$, $J(g) = J(g\varphi_1) + J(g\varphi_2) \le \mu^*(G_1) + \mu^*(G_2)$, and (b) follows.

(c) $\mu^*(\bigcup_{k=1}^\infty G_k) \le \sum_{k=1}^\infty \mu^*(G_k)$, since the support K of every $g \prec \bigcup_{k=1}^\infty G_k$ is contained in some finite union $\bigcup_{k=1}^N G_k$ so that, by (b),

$$J(g) \le \mu^*(\bigcup_{k=1}^N G_k) \le \sum_{k=1}^N \mu^*(G_k) \le \sum_{k=1}^\infty \mu^*(G_k),$$

and (c) follows.

Now we extend μ^* and for every set A we define

$$\mu^*(A) := \inf\{\mu^*(G);\ G \supset A\}.$$

This set function has the properties of an **outer measure**:

(a) $\mu^*(\emptyset) = 0$,

(b) $\mu^*(A) \le \mu^*(B)$ if $A \subset B$, so that it is an extension of μ^* previously defined on open sets, and

(c) $\mu^*(\bigcup_{k=1}^\infty A_k) \le \sum_{k=1}^\infty \mu^*(A_k)$ (σ-subadditivity).

To prove (c), take any $\varepsilon > 0$ and pick $A_k \subset G_k$ with $\mu^*(G_k) \le \mu^*(A_k) + \varepsilon/2^k$. Now, by (b) and from the σ-subadditivity on open sets,

$$\mu^*(\bigcup_{k=1}^\infty A_k) \le \mu^*(\bigcup_{k=1}^\infty G_k) \le \sum_{k=1}^\infty \mu^*(G_k) \le \sum_{k=1}^\infty \mu^*(A_k) + \varepsilon,$$

which yields (c) since $\varepsilon > 0$ was arbitrary.

Let us say that a set E is measurable if it satisfies the **Carathéodory condition**

$$\mu^*(A) \geq \mu^*(A \cap E) + \mu^*(A \setminus E)$$

and let μ be the restriction of the outer measure μ^* to measurable sets.[12]

It is a general fact in measure theory that the collection Σ of all measurable sets defined in this way is a σ-algebra and the restriction μ of μ^* to Σ is a measure. We need to show that any open set G is measurable, since then Σ will contain the Borel σ-algebra $\mathcal{B}_X$.

Thus, let us prove that G satisfies the Caratheodory condition when A is also an open set, U. Let $g \prec U \cap G$ so that $J(g) \geq \mu^*(U \cap G) - \varepsilon$. We have $G^c \subset (\operatorname{supp} g)^c$ and choose $h \prec U \setminus \operatorname{supp} g$ such that $J(h) \geq \mu^*(U \setminus \operatorname{supp} g) - \varepsilon$. Then

$$\mu^*(U) \geq J(g+h) \geq \mu^*(U\cap G)+\mu^*(U\setminus \operatorname{supp} g)-2\varepsilon \geq \mu^*(U\cap G)+\mu(U\setminus G)-2\varepsilon,$$

which yields $\mu^*(U) \geq \mu^*(U \cap G) + \mu^*(U \setminus G)$ since ε is arbitrary.

For any set A, if U is an open set with $A \subset U$, we have that

$$\mu^*(U) \geq \mu^*(U \cap G) + \mu^*(U \setminus G) \geq \mu^*(A \cap G) + \mu^*(A \setminus G)$$

and the Caratheodory condition for G follows by taking the infimum over these $U \supset A$.

With this construction we have built a Borel measure since, for any compact set K, we are going to prove that

$$\mu(K) = \inf\{J(g);\ K \prec g\}, \tag{1.6}$$

where the set on the right side is not empty.

Indeed, $\mu(K) = \inf_{G\supset K} \mu(G) \leq J(g)$ whenever $K \prec g$, since $K \subset \{g > 1 - \varepsilon\}$ if $0 < \varepsilon < 1$; for any $h \prec \{g > 1 - \varepsilon\}$ we have $h \leq g/(1 - \varepsilon)$ and $J(h) \leq J(g)/(1 - \varepsilon)$, by the positivity of J, which yields

$$\mu(K) \leq \mu(\{g > 1 - \varepsilon\}) = \sup_{h\prec\{g>1-\varepsilon\}} J(h) \leq J(g)/(1 - \varepsilon),$$

and then we let $\varepsilon \to 0$.

To prove (1.6), let $\varepsilon > 0$. Choose $G \supset K$ such that $\mu(K) \geq \mu(G) - \varepsilon$ and $K \prec g \prec G$. Then $\mu(K) \leq J(g) \leq \mu(G) \leq \mu(K) + \varepsilon$ and (1.6) follows.

By construction, μ satisfies the announced regularity properties.

We still need to prove the representation identity (1.5), where we can assume that $0 \leq g \leq 1$, since every $g \in \mathcal{C}_c(X)$ is a linear combination of such functions.

---

[12] In his book "Vorlesungen ̈ber reelle Funktionen" (1918), the Greek mathematician Constantin Carathéodory chose outer measures as the starting point for the construction of measure theory.

Let $K_0 = \operatorname{supp} g$. We will decompose g as a sum of N functions obtained by truncating g as follows. If $0 < j \le N$, let $K_j = \{g \ge j/N\}$ and define

$$g_j = \min\big(\max\{g - \frac{j-1}{N}, 0\}, \frac{1}{N}\big);$$

that is, $g_j(x) = 1/N$ if $x \in K_j$, $g_j(x) = g(x) - (j-1)/N$ if $x \in K_{j-1} \setminus K_j$, and $g_j(x) = 0$ otherwise. Then $g = \sum_{j=1}^N g_j$ and the estimates

$$\frac{1}{N}\mu(K_j) \le \int g_j \, d\mu \le \frac{1}{N}\mu(K_{j-1})$$

and

$$\frac{1}{N}\mu(K_j) \le J(g_j) \le \frac{1}{N}\mu(K_{j-1})$$

are readily checked. Then

$$\frac{1}{N}\sum_{j=1}^{N}\mu(K_j) \le \int g \, d\mu \le \frac{1}{N}\sum_{j=0}^{N-1}\mu(K_{j-1})$$

and

$$\frac{1}{N}\sum_{j=1}^{N}\mu(K_j) \le J(g) \le \frac{1}{N}\sum_{j=0}^{N-1}\mu(K_{j-1}),$$

so that

$$\Big|J(g) - \int g \, d\mu\Big| \le \frac{1}{N}(\mu(K_0) - \mu(K_N)) \le \frac{1}{N}\mu(K_0) \to 0$$

and $J(g) = \int g \, d\mu$.

To prove the uniqueness part, we only need to show that if μ is a Borel measure that satisfies the required regularity property, then it follows from the representation property $J(g) = \int g \, d\mu$ that $\mu(G) = \sup_{g \prec G} J(g)$.

Indeed, obviously $J(g) \le \int \chi_G \, d\mu = \mu(G)$ if $g \prec G$, and $\mu(G) \ge \sup_{g \prec G} J(g)$. Now, for every compact subset $K \subset G$ we choose $K \prec g \prec G$. Then $\mu(K) \le J(g) \le \mu(G)$ and, by the inner regularity property, $\mu(G) = \sup_{K \subset G} \mu(K)$. □

Theorem 1.6. *If X is a locally compact subset of $\mathbf{R}^n$ (so that every open set $G \subset X$ is the union of an increasing sequence of compact sets), then every Borel measure λ on X is regular:*

$$\lambda(B) = \inf_{G \supset B} \lambda(G) = \sup_{K \subset B} \lambda(K)$$

for every Borel set B.

Proof. We apply to $J_\lambda(g) := \int g \, d\lambda$ the Riesz-Markov theorem, so that

$$J_\lambda(g) = \int g \, d\mu \qquad (g \in \mathcal{C}_c(X)).$$

First we show that $\lambda(G) = \mu(G)$ for every open set G by considering $K_m \uparrow G$. Then we can choose $L_m \prec g_m \prec G$ with $L_1 = K_1$ and

$$L_m = \Big(\bigcup_{j=1}^{m} K_j\Big) \cup \Big(\bigcup_{j=1}^{m-1} \operatorname{supp} g_j\Big),$$

so that $g_m \uparrow \chi_G$ and $\lambda(G) = \mu(G)$ by monotone convergence.

We now study the outer regularity for any Borel set B.

Let $B = \biguplus_{j=1}^{\infty} B_j$ so that $\mu(B_j) < \infty$ and, by the regularity properties of μ, we can choose $G_j \supset B_j$ so that $\mu(G_j \backslash B_j) \leq \varepsilon/2^j$. Then $G = \bigcup_{j=1}^{\infty} G_j \supset B$ and $\mu(G \setminus B) \leq \varepsilon$.

Similarly, there exists an open set $U \supset B^c$ such that $\mu(U \setminus B^c) \leq \varepsilon$, and then the closed set $F = U^c$ satisfies $F \subset B$ and $\mu(B \setminus F) = \mu(U \setminus B^c) \leq \varepsilon$.

Therefore $\lambda(G \backslash F) = \mu(G \backslash F) \leq 2\varepsilon$, and it follows from $\lambda(G) \leq \lambda(B) + 2\varepsilon$ that λ is outer regular.

To show that λ is also inner regular, consider $K_m \uparrow F$, so that $\lambda(K_m) \to \lambda(F) \geq \lambda(B) - \varepsilon$. □

1.2.2. Complex measures. A **complex measure** on the measurable space (X, Σ) is a complex-valued set function $\mu : \Sigma \to \mathbf{C}$ which satisfies the σ-additivity condition

$$\mu\Big(\biguplus_{k=1}^{\infty} B_k\Big) = \sum_{k=1}^{\infty} \mu(B_k).$$

We will say that μ is a **real measure** if $\mu(B) \in \mathbf{R}$ for all $B \in \Sigma$.

Note that actually the convergence in $\mathbf{C}$ of the series $\sum_{k=1}^{\infty} \mu(B_k)$ is absolute, since the union of the sets B_k does not change with a permutation of the subscripts k.

The **total variation measure** of the complex measure μ is the set function defined on Σ by

$$|\mu|(B) := \sup\Big\{\sum_{k=1}^{\infty} |\mu(B_k)|;\ B = \biguplus_{k=1}^{\infty} B_k\Big\}.$$

In general, a complex measure is not a positive measure and, if it is a positive measure, it is finite.

Theorem 1.7. *The total variation $|\mu|$ of the complex measure μ is a finite measure that satisfies*

$$|\mu(B)| \leq |\mu|(B) \qquad (B \in \Sigma). \tag{1.7}$$

It is the smallest measure satisfying this property; that is, if λ is another measure such that $|\mu(B)| \leq \lambda(B)$ for every B, then $|\mu|(B) \leq \lambda(B)$ $\forall B \in \Sigma$.

Proof. The estimate (1.7) is obvious, since $B = B \cup \emptyset \cup \cdots \cup \emptyset \cup \cdots$. Moreover, if $|\mu(B)| \leq \lambda(B)$, then $\lambda(B) \geq |\mu|(B)$, since, if $B = \biguplus_{k=1}^{\infty} B_k$,

$$\lambda(B) = \sum_{k=1}^{\infty} \lambda(B_k) \geq \sum_{k=1}^{\infty} |\mu(B_k)|.$$

To show that $|\mu|$ is σ-additive, let $B = \biguplus_{k=1}^{\infty} B_k$ and consider any other partition $B = \biguplus_{j=1}^{\infty} A_j$, so that $A_j = \biguplus_{k=1}^{\infty}(A_j \cap B_k)$. Then

$$\sum_{j=1}^{\infty} |\mu(A_j)| \leq \sum_{j,k} |\mu(A_j \cap B_k)| \leq \sum_{k=1}^{\infty} |\mu|(B_k),$$

which implies $|\mu|(B) \leq \sum_{k=1}^{\infty} |\mu|(B_k)$.

To prove the opposite inequality, for $B = \biguplus_{k=1}^{\infty} B_k$, let $\delta_k < |\mu|(B_k)$ and $B_k = \biguplus_{j=1}^{\infty} B_{k,j}$ so that

$$\sum_{j=1}^{\infty} |\mu(B_{k,j})| > \delta_k.$$

Then

$$|\mu|(B) \geq \sum_{j,k} |\mu(B_{k,j})| \geq \sum_{k=1}^{\infty} \delta_k$$

and we obtain that $|\mu|(B) \geq \sum_{k=1}^{\infty} |\mu|(B_k)|$ by taking the supremum over all the possible δ_k.

Also $|\mu|(\emptyset) = 0$ and $|\mu|$ is a measure.

To show that $|\mu|(X) < \infty$, suppose that $|\mu|(B) = \infty$ for some $B \in \Sigma$. For every $t > 0$, we would find a partition $B = \biguplus_{k=1}^{\infty} B_k$ such that $\sum_{k=1}^{\infty} |\mu(B_k)| > t$ and then

$$\sum_{k=1}^{N} |\mu(B_k)| > t \tag{1.8}$$

for some N.

We **claim** that there is an absolute constant $c > 1$ such that

$$\sum_{k=1}^{N} |\mu(B_k)| \leq c \Big| \sum_{j \in J} \mu(B_j) \Big| \tag{1.9}$$

for some $J \subset \{1, \ldots, N\}$, so that, for $A = \bigcup_{j \in J} B_k \subset B$ we obtain

$$t < \sum_{k=1}^{N} |\mu(B_k)| \leq c|\mu(A)|,$$

and then $|\mu(A)| > t/c$. In (1.8) we choose $t \geq c$, so that $|\mu(A)| > 1$ and

$$|\mu(B \setminus A)| \geq |\mu(A)| - |\mu(B)| > \frac{t}{c} - |\mu(B)| = 1$$

if $t = c(1 + |\mu(B)|)$.

Now we have $B = A \uplus (B \setminus A)$ with $|\mu(A)|, |\mu(B \setminus A)| > 1$ and at least $|\mu|(A)$ or $|\mu|(B \setminus A)$ equal to ∞.

Suppose now that $|\mu|(X) = \infty$. Then we can successively split

$$X = A_1 \uplus B_1 = A_1 \uplus A_2 \uplus B_2 = \cdots = A_1 \uplus A_2 \uplus \cdots \uplus A_N \uplus B_N = \cdots$$

with $|\mu(A_j)| > 1$ and $|\mu|(B_N) = \infty$ for every j and N.

We should have in $\mathbf{C}$

$$\mu\Big(\biguplus_{j=1}^{\infty} A_j\Big) = \sum_{j=1}^{\infty} \mu(A_j),$$

but the series cannot converge, since $1 < |\mu(A_j)| \not\to 0$, which yields a contradiction. Therefore $|\mu|(X) < \infty$.

To prove the claim (1.9), let $z_k = \mu(B_k)$ $(1 \leq k \leq N)$ and $r = \sum_{k=1}^{N} |z_k|$. For at least one of the four quadrants of $\mathbf{C}$, Q, limited by $|y| = |x|$, we have $\sum_{z_k \in Q} |z_k| \geq r/4$. Denote $J = \{j;\ z_j \in Q\}$ and choose a rotation with angle ϑ so that $z_j' = e^{i\vartheta} z_j$ is in the quadrant $|y| \leq x$. Then

$$|\sum_{j \in J} z_j| = |\sum_{j \in J} z_j'| \geq \sum_{j \in J} \Re z_j' \geq \frac{1}{\sqrt{2}} \sum_{j \in J} |z_j'| \geq \frac{1}{4\sqrt{2}} r,$$

which proves (1.9) with $c = 4\sqrt{2}$. □

If μ is a real measure, then

$$\mu^+ := \frac{|\mu| + \mu}{2}, \quad \mu^- := \frac{|\mu| - \mu}{2}$$

are two (positive) finite measures such that

$$\mu = \mu^+ - \mu^-, \quad |\mu| = \mu^+ + \mu^-.$$

They are called the **positive and negative variations** of μ, respectively.

Every complex measure μ is a linear combination of four measures, since $\mu = \Re\mu + i\Im\mu$, where $\Re\mu$ and $\Im\mu$ are two real measures.

In Lemma 4.12 we will show that

$$\mu(B) = \int_B h\, d|\mu| \qquad (B \in \Sigma) \tag{1.10}$$

for a uniquely $|\mu|$-a.e. defined $|\mu|$-integrable function h such that $|h| = 1$, so we will be allowed to define $L^p(\mu) = L^p(|\mu|)$ and $\int f\, d\mu = \int fh\, d|\mu|$ for every $f \in L^1(\mu)$.

Every complex Borel measure on $\mathbf{R}$ is a linear combination of finite (positive) Borel measures, and such a measure, μ, is the Lebesgue-Stieltjes measure associated to the **distribution function**

$$F(t) := \mu\big((-\infty, t]\big),$$

which is increasing and right-continuous, so that $\mu\big((a,b]\big) = F(b) - F(a)$ and the Riemann-Stieltjes integrals of all Riemann-Stieltjes integrable functions on $[a,b]$ coincide with the Lebesgue integrals:

$$\int_a^b f(t)\,dF(t) = \int_{(a,b]} f\,d\mu.$$

Linear combinations of these integrals and measures are the corresponding integrals with complex distribution functions and complex measures that allow us to give the representation

$$\int_{\mathbf{R}} f(t)\,dF(t) = \int_{\mathbf{R}} f\,d\mu$$

for complex measures μ and, say, $f \in \mathcal{C}_c(\mathbf{R})$.

1.3. Exercises

Exercise 1.1. Prove that

$$d(r,s) = |\arctan s - \arctan r|$$

defines a distance on $\mathbf{R}$ whose topology is the usual one, but $\mathbf{R}$ is not complete with this distance. This shows that two distances which are topologically equivalent may not have the same Cauchy sequences.

Exercise 1.2. Prove that, in a metric space, every compact set is contained in a ball.

Exercise 1.3. Prove that, in a metric space M, the closure of a subset A is compact if and only if every sequence $\{a_k\} \subset A$ has a convergent subsequence in M.

Exercise 1.4. Prove that every point of a compact space K has a neighborhood basis of compact sets. That is, every compact space is locally compact.

Exercise 1.5. Prove that a nonempty subset X of $\mathbf{R}^n$ is a locally compact subspace if and only if it is the intersection of a closed and an open set, and that every open set of X is the union of an increasing sequence of compact sets.

Exercise 1.6. Let $I = [a, b]$ and $K = I^I = \prod_{t \in I} I$ endowed with the compact product topology. Note that $\{f(t)\}_{t \in I} \in K$ means that $f = \{f(t)\}_{t \in I}$ represents a function $f : I \to I$ and prove that

$$M := \{f = \{f(t)\}_{t \in I} \in K;\ \{x;\ f(x) \neq 0\} \text{ is at most countable}\}$$

with the topology induced by K is sequentially compact but not compact.

Exercise 1.7. Suppose $(X_j, \mathcal{T}_j)$ is a family of topological spaces and Y is another topological space. Show that for the product topology $\mathcal{T}$ on $X = \prod_{j \in J} X_j$, a mapping $f : Y \to X$ is continuous if and only if every $\pi_j \circ f : Y \to X_j$ is continuous.

Exercise 1.8. Assume that $\mathcal{T}$ and $\mathcal{T}'$ are two Hausdorff topologies on the same set K and $\mathcal{T}'$ is finer than $\mathcal{T}$. Prove that if $(K, \mathcal{T}')$ is compact, then $\mathcal{T} = \mathcal{T}'$.

Exercise 1.9. Suppose that $f : X \to Y$, where X and Y are two topological spaces, and $x_0 \in X$.

(a) If X is metrizable (or if x_0 has a countable neighborhood basis), prove that f is continuous at x_0 if and only if f is sequentially continuous at x_0.

(b) Let X be $\mathbf{R}$ endowed with the topology of all sets $G \subset \mathbf{R}$ such that G^c is countable, and let Y also be $\mathbf{R}$ but with the discrete topology (all subsets of $\mathbf{R}$ are open sets). Show that $Id : X \to Y$ is a sequentially continuous noncontinuous function.

Exercise 1.10. If f is a measurable function on a measurable space, show that

$$\operatorname{sgn} f(x) := \frac{f(x)}{|f(x)|}$$

(with $0/0 := 0$) defines another measurable function.

Exercise 1.11. If ν is the counting measure on a set X, so that $\nu(A) = n \in \mathbf{N}$ or $\nu(A) = \infty$ for any set $A \subset X$, prove that $f : X \to \mathbf{R}$ (or $\mathbf{C}$) is in $L^1(\nu)$ if and only if $N := \{f \neq 0\}$ is at most countable and $\sum_{k \in N} |f(k)| < \infty$. In this case, show that $\int f\, d\nu = \sum_{k \in N} f(k)$.

In this context one usually writes $\ell^1(X)$ or ℓ^1 for $L^1(\nu)$.

Exercise 1.12. Compute the limits of

$$\int_0^n \left(1 + \frac{x}{n}\right)^n e^x\, dx, \quad \int_0^n \left(1 - \frac{x}{n}\right)^n e^{-x}\, dx, \quad \text{and} \quad \int_0^n \left(1 - \frac{x}{2n}\right)^n e^x\, dx$$

as $n \to \infty$.

Exercise 1.13. Use the Fubini-Tonelli theorem to prove that the integral

$$I := \int_{(0,1)^2} \frac{1}{|x - y|^\alpha}\, dx\, dy$$

is finite if and only if $\alpha < 1$, and then show that $I = 2/(1-\alpha)(2-\alpha)$.

Exercise 1.14. If $F : [a,b] \to \mathbf{R}$ is absolutely continuous, prove that F satisfies the following property:

If $\varepsilon > 0$ is given, there is a $\delta > 0$ such that

$$\sum_k |F(b_k) - F(a_k)| \leq \varepsilon$$

for every finite sequence $\{(a_k, b_k)\}$ of nonoverlapping intervals contained in $[a,b]$ such that $\sum_k (b_k - a_k) \leq \delta$.

The converse is also true: If the above property holds, then F has a representation as in (1.4). See a proof in [**37**], [**39**], or [**6**].

Exercise 1.15. Let μ be a Borel measure on $\mathbf{R}$ and define

$$F(0) = 0, \quad F(t) = \mu((0,t]) \text{ if } t > 0, \quad F(t) = -\mu((t,0]) \text{ if } t < 0.$$

Then F is an increasing right continuous function and μ is the Lebesgue-Stieltges measure associated to F as a distribution function, that is,

$$\mu((a,b]) = F(b) - F(a).$$

Moreover, if $f \in \mathcal{C}[a,b]$, $\int f\,d\mu = \int_a^b f(t)\,dF(t)$, a Riemann-Stieltjes integral.

Exercise 1.16. If μ is a complex measure, prove that $\lim_k \mu(B_k) = \mu(B)$ if either $B_k \uparrow B$ or $B_k \downarrow B$.

Exercise 1.17. Show that, for any complex Borel measure μ,

$$|\mu|(B) = \sup\Big\{\Big|\int_B f\,d\mu\Big|;\ |f| \leq 1\Big\}.$$

Exercise 1.18. Suppose that $\{\lambda_k\} \in \ell^1(\mathbf{N})$ and that $\{a_k\}$ is a sequence in $\mathbf{R}^n$. Prove that there is a uniquely determined Borel measure μ on $\mathbf{R}^n$ such that

$$\int g\,d\mu = \sum_{k=1}^{\infty} \lambda_k g(a_k) \qquad (g \in \mathcal{C}_c(\mathbf{R}^n)).$$

Show that $\sum_{k=1}^{\infty} |\lambda_n| = \sup\{|\int g\,d\mu|;\ g \in \mathcal{C}_c(\mathbf{R}^n),\ |g| \leq 1\}$ and $|\mu|(F^c) = 0$ if $F = \{a_1, a_2, \dots\}$.

References for further reading:

J. Cerdà, *Análisis Real.*

P. R. Halmos, *Measure Theory.*

L. Kantorovitch and G. Akilov, *Analyse fonctionnelle.*

J. L. Kelley, *General Topology.*

A. N. Kolmogorov and S. V. Fomin, *Elements of the Theory of Functions and Functional Analysis.*

H. L. Royden, *Real Analysis.*

W. Rudin, *Real and Complex Analysis.*

Chapter 2

Normed spaces and operators

The objects in functional analysis are function spaces endowed with topologies that make the operations continuous as well as the operators between them. This chapter is devoted to the most basic facts concerning Banach spaces and bounded linear operators.

It can be useful for the reader to retain as a first model of function spaces the linear space $\mathcal{C}(L)$ of all real continuous functions on a compact set L in $\mathbf{R}^n$ with the uniform convergence, defined by the condition

$$\|f - f_n\|_L := \max_{t \in L} |f(t) - f_n(t)| \to 0.$$

Examples of operators on this space are the integral operators

$$Tf(x) = \int_L K(x, y) f(y)\, dy$$

where $K(x, y)$ is continuous on $L \times L$. Then $T : \mathcal{C}(L) \to \mathcal{C}(L)$ is linear and

$$\|Tf - Tf_n\|_L \leq \max_{x \in L} \int_L |K(x, y)||f(y) - f_n(y)|\, dy \leq M\|f - f_n\|_L,$$

so that T satisfies the continuity condition $\|Tf - Tf_n\|_L \to 0$ if $\|f - f_n\|_L \to 0$.

Note that if $L = [a, b]$, this space is infinite-dimensional, since it contains the linearly independent functions 1, x, x^2, etc. Two major differences with respect to the usual finite-dimensional Euclidean spaces are that a linear map between general Banach spaces is not necessarily continuous and that the closed balls are not compact.

We start this chapter with some basic definitions and, after the basic examples of L^p spaces and $\mathcal{C}(K)$, with the inclusion of the proof of the Weierstrass and the Stone-Weierstrass theorems, we consider the space of all bounded linear operators.

The use of Neumann series, which will appear again when studying the invertible elements in a Banach algebra and the spectrum of unbounded operators, in combination with Volterra integral operators gives an application for the solution of initial value problems for linear ordinary differential equations. The introduction of Green's function also allows us to solve boundary value problems by Fredholm integral operators. These applications are described in the last section.

The chapter includes a review of the most basic facts concerning orthogonality in a Hilbert space. Duality will be discussed in Chapter 4.

There is also a section on summability kernels that will be useful in later developments. They are applied here to show the density of the trigonometric polynomials in $L^p(0,1)$ and to prove the Riemann-Lebesgue lemma.

The section devoted to the Riesz-Thorin interpolation theorem of linear operators on L^p spaces is optional. It will be used only in Chapter 7 to include the nice proof of the L^p-continuity of the Hilbert transform due to M. Riesz.

2.1. Banach spaces

2.1.1. Topological vector spaces. In this book, a vector space will always be a linear space over the real field $\mathbf{R}$ or the complex field $\mathbf{C}$. The letter $\mathbf{K}$ will denote either of them, and $\mathbf{K}$ will be endowed with the usual topology defined by the distance $d(\lambda,\mu) = |\lambda - \mu|$, where $|\cdot|$ represents the absolute value. The collection of all the discs (intervals if $\mathbf{K} = \mathbf{R}$)

$$D(\lambda_0, \varepsilon) = \{\lambda \in \mathbf{K};\ |\lambda - \lambda_0| < \varepsilon\} = \lambda_0 + D(0,\varepsilon) \qquad (\varepsilon > 0)$$

is a neighborhood basis of the point $\lambda_0 \in \mathbf{K}$.

A **vector topology** $\mathcal{T}$ on a vector space E will be a Hausdorff topology such that the vector operations

$$(x,y) \in E \times E \mapsto x + y \in E, \qquad (\lambda, x) \in \mathbf{K} \times E \mapsto \lambda x \in E$$

are continuous when we endow $E \times E$ and $\mathbf{K} \times E$ with the corresponding product topologies. Then we say that E, or the couple $(E, \mathcal{T})$, is a **topological vector space**.

On a topological vector space E, every translation $\tau_{x_0} : x \in E \mapsto x + x_0 \in E$ is continuous, since it is obtained by fixing a variable in the sum: If $V(y)$ is a neighborhood of $y = x + x_0$, there is a neighborhood $U(x) \times U(x_0)$ of $(x, x_0) \in E \times E$ such that $U(x) + U(x_0) \subset V(y)$, and then $\tau_{x_0}(U(x)) \subset V(y)$.

Similarly, every multiplication $x \in E \mapsto \lambda_0 x \in E$ by a given scalar λ_0 is also continuous.

Since the inverse τ_{-x_0} of τ_{x_0} is continuous, U is a neighborhood of $0 \in E$ if and only if $U + x_0$ is a neighborhood of x_0. Thus, the topology of E is translation-invariant: $\mathcal{U}$ is a neighborhood basis of $0 \in E$ if and only if $\mathcal{U}(x_0) = \{U + x_0;\ U \in \mathcal{U}\}$ is a neighborhood basis of $x_0 \in E$. We will say that $\mathcal{U}$ is a **local basis** of E. An obvious example is the collection of all open sets that contain 0.

A **subspace** of a topological vector space $(E, \mathcal{T})$ is a vector subspace F with the topology that consists of all the sets $G \cap F$, $G \in \mathcal{T}$. Then F becomes a topological vector space.

Recall that $C \subset E$ is convex if and only if $0 < t < 1$ implies $tC + (1-t)C \subset C$.

Theorem 2.1. *Let E be a topological vector space. Then:*

(a) *The closure $\bar{F}$ of a vector subspace F of E is also a vector subspace.*

(b) *The closure $\bar{C}$ of a convex subset C of E is also convex.*

Proof. To prove (a), suppose $\lambda, \mu \in \mathbf{K}$. If $\lambda \neq 0$, multiplication by λ and by its inverse is continuous and we always have $\lambda\bar{F} = \overline{\lambda F}$. Then

$$\lambda\bar{F} + \mu\bar{F} = \overline{\lambda F} + \overline{\mu F} \subset \bar{F} + \bar{F} \subset \bar{F},$$

since, if $x, y \in \bar{F}$, for every neighborhood $x + y + U$ of $x + y$ (when U is in a local basis $\mathcal{U}$), the continuity of the sum allows us to take $V \in \mathcal{U}$ so that $V + V \subset U$. Hence, if $a \in (x + V) \cap F$ and $b \in (y + V) \cap F$, then $a + b \in x + y + V + V \subset x + y + U$ and also $x + y \in \bar{F}$.

To prove (b) consider $x, y \in \bar{C}$ and let $\alpha + \beta = 1$ ($\alpha, \beta \geq 0$). Using the same argument as in (a), $\alpha x + \beta y \in \bar{C}$. □

Theorem 2.2. *Suppose $\mathcal{U}$ and $\mathcal{V}$ are local bases of the topological vector spaces E and F. A linear mapping $T : E \to F$ is continuous if and only if, for every $V \in \mathcal{V}$, $T(U) \subset V$ for some $U \in \mathcal{U}$.*

Proof. $T(U) \subset V$ if and only if $T(x_0 + U) \subset T(x_0) + V$. □

2.1.2. Normed and Banach spaces. Normed spaces are the simplest and most useful topological vector spaces.

Recall that $\|\cdot\|_E : E \to [0, \infty)$ is a **norm** on the vector space E if it satisfies the following properties:

1. $\|x\|_E > 0$ if $x \neq 0$,
2. triangle inequality: $\|x + y\|_E \leq \|x\|_E + \|y\|_E$ if x, y in E, and
3. $\|\lambda x\|_E = |\lambda|\|x\|_E$ if $x \in E$ and $\lambda \in \mathbf{K}$.

A **normed space** is a vector space E endowed with a norm $\|\cdot\|_E$ defined on it, with the topology associated to the distance $d_E(x,y) := \|y-x\|_E$.

If E and F are two normed spaces, we endow $E\times F$ with the product norm

$$\|(x,y)\|_{E\times F} := \max(\|x\|_E, \|y\|_F)$$

and its distance

$$d((x_1,y_1),(x_2,y_2)) = \max(\|x_1-x_2\|_E, \|y_1-y_2\|_F),$$

which defines the product topology. A sequence $(x_n,y_n)\in E\times F$ ($n\in\mathbf{N}$) is convergent in the product space if and only if the component sequences $\{x_n\}$ and $\{y_n\}$ are convergent in the corresponding factor space.

Of course, these facts extend in the obvious way to finite products.

Theorem 2.3. *The topology of a normed space E is a vector topology, and the norm $\|\cdot\|_E$ is a continuous real function on E.*

Proof. The continuity of the vector operations follow from the inequalities

$$\begin{aligned}\|(x_1+x_2)-(y_1-y_2)\|_E &\le \|x_1-y_1\|_E + \|x_2-y_2\|_E\\ &\le 2\max(d_E(x_1,y_1), d_E(x_2,y_2))\end{aligned}$$

and

$$\|\lambda x_n - \mu x\|_E \le |\lambda-\mu|\|x_n\|_E + |\mu|\|x_n - x\|_E.$$

To show that the norm is also continuous, note that

$$\big|\|x\|_E - \|y\|_E\big| \le \|x-y\|_E \tag{2.1}$$

because $\|x\|_E \le \|x-y\|_E + \|y\|_E$ and $\|y\|_E \le \|x-y\|_E + \|x\|_E$. □

A topological vector space $(E,\mathcal{T}_E)$ is said to be **normable** if $\mathcal{T}_E$ is the topology defined by some norm on E.

A normed space E is called a **Banach space**[1] if it is complete; that is, whenever $\|x_p - x_q\|_E \to 0$, we can find $x\in E$ so that $\|x-x_k\|_E\to 0$. Completeness is a fundamental property of many normed spaces. Some basic theorems will apply only to complete spaces.

The calculus with numerical series is meaningful for vector-valued series in a Banach space E. If $\sum_{n=1}^\infty \|x_n\|_E < \infty$, then $\sum_{n=1}^\infty x_n$ is called absolutely convergent and, as in the case of numerical series, every absolutely

[1] The term was introduced by M. Fréchet to honor Stephan Banach's work around 1920, culminating in his 1932 book "Théorie des Opérations Linéaires" [**3**], the first monograph on the general theory of linear metric spaces.

convergent series is convergent in E since, if $s_N = \sum_{n=1}^N x_n$ and $p < q$,

$$\|s_q - s_p\|_E = \| \sum_{n=p+1}^{q} x_n\|_E \leq \sum_{n=p+1}^{\infty} \|x_n\|_E \to 0 \quad \text{when} \quad p \to \infty,$$

and the limit $x = \lim_{N\to\infty} s_N = \sum_{n=1}^\infty x_n$ exists in E.

We now present the most basic examples of Banach spaces. Many other spaces of analysis are obtained from them.

Example 2.4. The real or complex **Euclidean spaces $\mathbf{K}^n$** are the first examples of Banach spaces. They are finite-dimensional and their norm

$$|x| = \Big(\sum_{j=1}^{n} |x_j|^2\Big)^{1/2} \qquad (x = (x_1, \ldots, x_n)) \tag{2.2}$$

is associated to the usual Euclidean scalar product $x \cdot y = \sum_{k=1}^n x_k \bar{y}_k$ by the relation $|x| = \sqrt{x \cdot x}$.

Example 2.5. On the vector space $\mathbf{B}(X)$ of all $\mathbf{K}$-valued bounded functions on a given nonempty set X, we consider the **uniform norm**, or "sup" norm,

$$\|f\|_X = \sup_{x\in X} |f(x)|.$$

The convergence $\|f - f_n\|_X \to 0$ means that $f_n(x) \to f(x)$ uniformly on X.

A special case is the sequence space $\ell^\infty = \mathbf{B}(\mathbf{N})$ (or $\mathbf{B}(\mathbf{Z})$). In this context one usually writes

$$\|\{x_n\}\|_\infty = \sup_n |x_n|$$

for the sup norm.

These spaces are also complete.

Indeed, if $\{f_n\}_n$ is a Cauchy sequence in $\mathbf{B}(X)$, it is uniformly Cauchy and every $\{f_n(x)\}_n$ is a Cauchy sequence in $\mathbf{K}$, so that there exists $f(x) = \lim f_n(x)$. Then it follows from the uniform estimate

$$|f_p(x) - f_q(x)| \leq \varepsilon \qquad (p, q \geq n_0)$$

that

$$|f_p(x) - f(x)| \leq \varepsilon \qquad (p \geq n_0)$$

when $q \to \infty$. Thus, $\|f_p - f\|_X \leq \varepsilon$, $\|f\|_X \leq \varepsilon + \|f_p\|_X < \infty$, and $f_n \to f$ uniformly on X.

Example 2.6. A **subspace** F of a normed space E will be a vector subspace equipped with the restriction of the norm $\|\cdot\|_E$. If E is a Banach space, F is complete if and only if it is closed in E, since every Cauchy sequence of F is convergent in E and the limit is in F when F is closed.

Let X be a topological space. We denote by $\mathcal{C}_b(X)$ the subspace of $\mathbf{B}(X)$ of all $\mathbf{K}$-valued bounded and continuous functions f on X. As a closed subspace of the Banach space $\mathbf{B}(X)$, $\mathcal{C}_b(X)$ is also complete.

Note that every Cauchy sequence $\{g_n\}$ in $\mathcal{C}_b(X)$ is convergent in $\mathbf{B}(X)$ to a function f, which belongs to $\mathcal{C}_b(X)$ as a uniform limit of continuous functions.

Example 2.7. The product $E = E^1 \times \cdots \times E^n$ of a finite number of Banach spaces with the product norm

$$\|(x^1, \ldots, x^n)\|_E := \max_{1 \leq j \leq n} \|x^j\|_{E^j}$$

is complete, since in E we have coordinatewise convergence.

Let $m \in \mathbf{N}$ and suppose that Ω is a nonempty open set in $\mathbf{R}^n$. We call $\mathcal{C}^m(\bar{\Omega})$ the normed space of all $\mathcal{C}^m$ functions f on Ω such that every derivative $D^\alpha f$ of order $|\alpha| \leq m$ admits a continuous and bounded extension to $\bar{\Omega}$. With the norm

$$\|f\|_{\mathcal{C}^m} := \max_{|\alpha| \leq m} \|D^\alpha f\|_{\bar{\Omega}},$$

$\mathcal{C}^m(\bar{\Omega})$ is a Banach space.

The space $\mathcal{C}^m(\bar{\Omega})$ can be seen as a closed subspace of the finite product $E = \prod_{|\alpha| \leq m} \mathcal{C}_b(\bar{\Omega})$ of Banach spaces by means of the injection

$$f \in \mathcal{C}^m(\bar{\Omega}) \mapsto \{D^\alpha f\}_{|\alpha| \leq m} \in E.$$

Here, according to the simplifying notation introduced by H. Whitney, for every $\alpha = (\alpha_1, \ldots, \alpha_n) \in \mathbf{N}^n$, we denote

$$x^\alpha = x_1^{\alpha_1} \cdots x_n^{\alpha_n} \text{ if } x = (x_1, \ldots, x_n)$$

and

$$D^\alpha = \partial_1^{\alpha_1} \cdots \partial_n^{\alpha_n},$$

where $\partial_j = \partial / \partial x_1$ represents the partial derivative with respect to the jth coordinate, and $|\alpha| = \alpha_1 + \cdots + \alpha_n$ is the **order** of the differential operator D^α. With this notation, the n-dimensional **Leibniz formula** states that

$$D^\alpha(fg) = \sum_{\beta \leq \alpha} \binom{\alpha}{\beta} D^\beta f D^{\alpha - \beta} g$$

and it is easily proved by induction (cf. Exercise 2.1). Here $\beta \leq \alpha$ means $\beta_j \leq \alpha_j$ for $1 \leq j \leq n$ and $\alpha - \beta = (\alpha_1 - \beta_1, \ldots, \alpha_n - \beta_n)$.

Example 2.8. Let (Ω, Σ, μ) be a measure space and let $1 \le p \le \infty$. Consider $L^p(\mu)$ with the L^p-norm

$$\|f\|_p = \left(\int |f|^p d\mu\right)^{1/p},$$

when $1 \le p < \infty$ and $\|f\|_\infty := \min\{C \ge 0;\ |f(x)| \le C \text{ a.e.}\}$, as in Section 1.2. It will follow from the next theorem that this space is complete.

A similar example is $\ell^p = \ell^p(\mathbf{N})$ or $\ell^p(\mathbf{Z})$, the space of all numerical sequences $x = \{x_k\}$ ($k \in \mathbf{N}$ or $\mathbf{Z}$) such that

$$\|x\|_p = \Big(\sum_k |x_k|^p\Big)^{1/p} \qquad (\sup_k |x_k| \text{ if } p = \infty)$$

with the usual operations.[2]

Theorem 2.9. *Let $1 \le p < \infty$ and assume that all the functions f_k are measurable.*

(a) *Let $f_k(x) \to f(x)$ everywhere and $|f_k| \le g$ a.e. for some $g \in L^p(\mu)$. Then $f_k \to f$ in $L^p(\mu)$.*

(b) *If $\sum_{k=1}^\infty \|f_k\|_p < \infty$, then $\sum_{k=1}^\infty |f_k(x)| < \infty$ a.e., there exists a function $f \in L^p(\mu)$ such that $f(x) = \sum_{k=1}^\infty f_k(x)$ a.e., and $f = \sum_{k=1}^\infty f_k$ in $L^p(\mu)$.*

(c) *Every convergent sequence $f_k \to f$ in $L^p(\mu)$ has a subsequence which converges pointwise a.e. to f.*

Proof. (a) Since also $|f(x)| \le g(x)$, $|f_k - f|^p \le (2g)^p \in L^1(\mu)$ and $|f_k(x) - f(x)|^p \to 0$, and by dominated convergence, $\|f_k - f\|_p \to 0$.

(b) Let $M = \sum_{k=1}^\infty \|f_k\|_p$ and put $g_N(x) := \sum_{k=1}^N |f_k(x)| \uparrow g(x)$, so that $g_N^p \uparrow g^p$ and $\int g_N^p \, d\mu \le M^p$, by the triangle inequality. By monotone convergence, also $\int g^p \, d\mu \le M^p$ and $g(x) = \sum_{k=1}^N |f_k(x)| < \infty$ a.e. The sum $f(x) := \sum_{k=1}^\infty f_k(x)$ is defined a.e., or everywhere by picking equivalent representatives for the functions f_k. Now we apply (a) to the partial sums $\sum_{j=1}^k f_j$ to obtain $\sum_{j=1}^k f_j \to f$ in $L^p(\mu)$.

(c) Since $\|f_m - f_n\| \to 0$, we can select a subsequence $\{f_{k_n}\}$ such that

$$\|f_{k_{m+1}} - f_{k_m}\| \le \frac{1}{2^m}.$$

This subsequence is the sequence of partial sums of the series

$$f_{k_1} + (f_{k_2} - f_{k_1}) + (f_{k_3} - f_{k_2}) + \cdots + (f_{k_{m+1}} - f_{k_m}) + \cdots$$

---

[2] On an arbitrary set J endowed with the counting measure, one usually writes $\ell^p(J)$ instead of $L^p(J)$. A function f on J is in $\ell^p(J)$ ($1 \le p < \infty$) if and only if $\sum_{j \in J} |f(j)|^p < \infty$. If $\sum_{j \in J} |f(j)|^p < \infty$, then $N := \{j \in J;\ f(j) \ne 0\}$ is at most countable (cf. Exercise 1.11).
If $J = \{1, 2, \ldots, n\}$, $\ell^2(J)$ is the Euclidean space $\mathbf{K}^n$.

which is absolutely convergent on $L^p(\mu)$, since $\sum_{m=1}^{\infty} 1/2^m < \infty$. Now an application of (b) gives $f_{k_m} \to h$ in $L^p(\mu)$ and a.e. Obviously, $h = f$ a.e. □

Corollary 2.10. $L^p(\mu)$ is a Banach space.

Proof. Assume first $1 \leq p < \infty$ and let $\{f_k\}$ be a Cauchy sequence in $L^p(\mu)$. As in the preceding proof of (c), there exists a subsequence $\{f_{k_n}\}$ which is convergent to a function $h \in L^p(\mu)$. By the triangle inequality, we also obtain $f_k \to h$ in $L^p(\mu)$.

In $L^\infty(\mu)$, if $\{f_k\}$ is a Cauchy sequence, the sets $B_k := \{x;\, |f_k(x)| > \|f_k\|_\infty\}$ and $B_{p,q} := \{x;\, |f_p(x) - f_q(x)| > \|f_p - f_q\|_\infty\}$ have measure 0, and also $\mu(B) = 0$ if B is the union of all of them. Then we have $\lim_k f_k(x) = f(x)$ uniformly on B^c and $\lim_k f_k = f$ in $L^\infty(\mu)$. □

Remark 2.11. In a Banach space, every absolutely convergent series is convergent, and the converse is also true: If every absolutely convergent series of a normed space E is convergent, then the space is complete. To show this fact, just follow the proof of (c) in Theorem 2.9 for a Cauchy sequence $\{f_k\} \subset E$; the terms f_{k_n} are the partial sums of the absolutely convergent series

$$f_{k_1} + (f_{k_2} - f_{k_1}) + (f_{k_3} - f_{k_2}) + \cdots + (f_{k_{m+1}} - f_{k_m}) + \cdots,$$

which is convergent.

Corollary 2.12. Let (Ω, Σ, μ) be a measure space and $1 \leq p < \infty$. Then every $f \in L^p(\mu)$ is the limit in $L^p(\mu)$ of a sequence $\{s_n\} \subset L^p(\mu)$ of simple functions.

Proof. Just consider s_n such that $|s_n(x)| \uparrow |f(x)|$ for every $x \in \Omega$ as in (1.1) and apply Theorem 2.9(a). □

Corollary 2.13. Suppose μ is a Borel measure on a locally compact metric space X and let $1 \leq p < \infty$. Then every $f \in L^p(\mu)$ is the limit in $L^p(\mu)$ of a sequence $\{g_k\} \subset \mathcal{C}_c(X)$, meaning that every g_k has a continuous with compact support and that $\lim_{k\to\infty} \|f - g_k\|_p = 0$.

Proof. Since simple functions are dense in $L^p(\mu)$, we only need to approximate every Borel set B with finite measure by functions in $\mathcal{C}_c(X)$. By regularity, we can find $K \subset B \subset G$ such that $\mu(G \setminus K) \leq \varepsilon^p$ and choose $K \prec g \prec G$. Then

$$\int |\chi_B - g|^p \, d\mu \leq \int_{G\setminus K} 1 \, d\mu \leq \varepsilon^p.$$

□

A very important property of $L^p(\mathbf{R}^n)$ space is the continuity of translations:

Theorem 2.14. *Assume that* $f \in L^p(\mathbf{R}^n)$ $(1 \le p < \infty)$ *and denote* $(\tau_h f)(x) = f(x-h)$. *Then the* L^p*-valued function* $h \in \mathbf{R}^n \mapsto \tau_h f \in L^p(\mathbf{R}^n)$ *is continuous; that is,*

$$\lim_{h\to h_0} \|\tau_h f - \tau_{h_0} f\|_p = 0.$$

Proof. Since $\|\tau_h f - \tau_{h_0} f\|_p = \|\tau_{h-h_0} f - f\|_p$, we can assume $h_0 = 0$.

Start first with $g \in \mathcal{C}_c(\mathbf{R}^n)$ supported by K and let $K(1) = K + \bar{B}(0,1)$. If $\varepsilon > 0$, by the uniform continuity of g, we can find $\delta > 0$ so that

$$\|\tau_h g - g\|_\infty \le \varepsilon/|K(1)|^{1/p} \text{ if } |h| \le \delta.$$

Then it follows that $\|\tau_h g - g\|_p \le \varepsilon$ if $|h| \le \delta$.

With Corollary 2.13 in hand, we choose $g \in \mathcal{C}_c(\mathbf{R}^n)$ so that $\|f-g\|_p \le \varepsilon$, and then

$$\|\tau_h f - f\|_p \le \|\tau_h(f-g)\|_p + \|\tau_h g - g\|_p + \|f-g\|_p \le 3\varepsilon.$$

□

Every normed space E has a **completion**, defined as a Banach space $\tilde{E}$ with an isometric linear embedding $J_{\tilde{E}} : E \hookrightarrow \tilde{E}$ with dense image $J_{\tilde{E}}(E)$ in $\tilde{E}$. A completion can be obtained either in the same way as $\mathbf{R}$ is defined as the completion of $\mathbf{Q}$ or using duality, as in Theorem 4.25.

The completion is unique in the sense that, if $J_F : E \hookrightarrow F$ is a second one, there exists a unique isomorphism $\Phi : F \to \tilde{E}$ such that $\Phi \circ J_F : E \to \tilde{E}$ is the embedding $J_{\tilde{E}} : E \hookrightarrow \tilde{E}$ and Φ is an isometry: For every Cauchy sequence $\{x_n\} \subset E$, the images $\{J_{\tilde{E}} x_n\} \subset \tilde{E}$ and $\{J_F x_n\} \subset F$ are also Cauchy and convergent in $\tilde{E}$ and F, respectively, and $\Phi(\lim_n J_{\tilde{E}} x_n) = \lim_n J_F x_n$ is the only possible definition of Φ. It is clear that Φ is an isometric isomorphism.

It is customary to identify E with the subspace $J_{\tilde{E}}(E)$ of the completion $\tilde{E}$.

2.1.3. The space $\mathcal{C}(K)$ and the Stone-Weierstrass theorem.

By $\mathcal{C}(K)$ we represent either the real or the complex Banach space of all real-valued or complex-valued continuous functions on a compact topological space K, endowed with the uniform norm. When confusion is possible, we write $\mathcal{C}(K;\mathbf{R})$ or $\mathcal{C}(K;\mathbf{C})$, respectively.

Let A be a subalgebra of $\mathcal{C}(K)$; that is, A is a vector subspace of $\mathcal{C}(K)$ and the product of two functions in A belongs to A.

The closure of A is also a subalgebra of $\mathcal{C}(K)$, since it is a vector subspace (cf. Theorem 2.1) and $f_n \to f$ and $g_n \to g$ uniformly on K imply $f_n g_n \to fg$. To prove this fact, just consider

$$\begin{aligned}\|fg - f_n g_n\|_K &\leq \|fg - fg_n\|_K + \|fg_n - f_n g_n\|_K \\ &\leq \|f\|_K \|g - g_n\|_K + \|g_n\|_K \|f - f_n\|_K \to 0.\end{aligned}$$

We say that A **separates points of** K if, given $x \neq y$ in K, there is a function $g \in A$ such that $g(x) \neq g(y)$, and we say that A **does not vanish at any point of** K if, given $x \in K$, there is a function $g \in A$ such that $g(x) \neq 0$.

Lemma 2.15. *If a subalgebra A of $\mathcal{C}(K)$ separates points and does not vanish at any point of K, then the following interpolation property holds:*

For any two different points $x, y \in K$ and two given numbers α and β, there is a function $f \in A$ such that $f(x) = \alpha$ and $f(y) = \beta$.

Proof. Let $g, h_x, h_y \in A$ such that $g(x) \neq g(y)$, $h_x(x) \neq 0$, and $h_y(y) \neq 0$. Then

$$g_1 := gh_y - g(x)h_y, \quad g_2 := gh_x - g(y)h_x$$

belong to A and we define

$$f = \frac{\alpha}{g_2(x)} g_2 + \frac{\beta}{g_1(y)} g_1.$$

This function also belongs to A and satisfies the required interpolation property. □

If K is a compact subset of $\mathbf{R}^n$, then the set $\mathcal{P}(K)$ of all real polynomial functions on K is a subalgebra of $\mathcal{C}(K; \mathbf{R})$ that separates points since, if $a, b \in K$ are two different points, at least $a_j \neq b_j$ for one coordinate j and then the corresponding monomial x_j has different values on a and b. Moreover $1 \in \mathcal{P}(K)$ and this subalgebra does not vanish at any point of K.

Theorem 2.19 will show that these facts will imply that $\mathcal{P}(K)$ is dense in $\mathcal{C}(K; \mathbf{R})$, but let us first consider the special case $K = [a, b]$ of one variable and prove the classical **Weierstrass theorem.**[3]

[3] First proved in 1885 using the summability kernel W_t of Exercise 2.27 by one of the fathers of modern analysis, the German mathematician Karl Weierstrass, who taught at Gewerbeinstitut in Berlin.

We are going to present the constructive proof based on the **Bernstein polynomials**[4] $B_n(f)$ associated to a continuous function f on $[0,1]$:

$$B_nf(x) := \sum_{k=0}^{n} \binom{n}{k} f\Big(\frac{k}{n}\Big) x^k (1-x)^{n-k}.$$

For every n, the linear operator B_n on $\mathcal{C}[0,1]$ is positive; that is, $B_nf \geq 0$ if $f \geq 0$, so that $B_nf \geq B_ng$ if $f \geq g$ and $|B_nf| \leq B_n|f|$.

Theorem 2.16. *Every continuous function f on $[a,b]$ is the uniform limit on $[a,b]$ of a sequence of polynomials.*

Proof. The linear change of variables $x = (b-a)t + a$ allows us to assume $[a,b] = [0,1]$, and we will prove that $f \in \mathcal{C}[0,1]$ can be uniformly approximated on $[0,1]$ by the Bernstein polynomials B_nf.

It is clear that $B_n 1 = 1$ and we are going to show that, for $I(x) = x$ and $I^2(x) = x^2$,

$$B_nI = I, \quad B_nI^2 = \frac{n-1}{n} I^2 + \frac{1}{n} I,$$

and then $B_nI^2 \to I^2$ uniformly on $[0,1]$, since $\sup_{0 \leq x \leq 1} |B_nI^2(x) - I^2(x)| \leq 1/n$.

Indeed, differentiating

$$(x+y)^n = \sum_{k=0}^{n} \binom{n}{k} x^k y^{n-k} \tag{2.3}$$

with respect to x and multiplying by x, we obtain

$$nx(x+y)^{n-1} = \sum_{k=1}^{n} k \binom{n}{k} x^k y^{n-k},$$

which for $y = 1-x$ reads $nx = \sum_{k=1}^{n} k\binom{n}{k} x^k (1-x)^{n-k} = nB_nx$.

Also, differentiating (2.3) once again with respect to x and now multipying by x^2,

$$n(n-1)x^2(x+y)^{n-2} = \sum_{k=2}^{n} k(k-1) \binom{n}{k} x^k y^{n-k}.$$

[4]The Russian mathematician Sergei N. Bernstein (1880 – 1968), who solved Hilbert's nineteenth problem on the analytic solution of elliptic differential equations, found his constructive proof of the Weierstrass theorem using probabilistic methods in 1912: He considered a random variable X with a binomial distribution with parameters n and x, so that the expected value $E(X/n)$ is x, and he combined the use of the weak law of large numbers, which made it possible to show that $\lim_n P(|f(X/n) - f(x)| > \varepsilon) = 0$ for every continuous function f on $[0,1]$, with the remark that $E(f(X/n))$ is precisely the polynomial $B_nf(x)$.

Hence, for $y = 1 - x$,

$$\begin{aligned} n(n-1)x^2 &= \sum_{k=2}^{n} k(k-1)\binom{n}{k} x^k(1-x)^{n-k} \\ &= n^2 \sum_{k=2}^{n} \left(\frac{k}{n}\right)^2 \binom{n}{k} x^k (1-x)^{n-k} - n \sum_{k=2}^{n} \frac{k}{n}\binom{n}{k} x^k(1-x)^{n-k}, \end{aligned}$$

or $(n-1)x^2 = nB_n x^2 - B_n x$, which is equivalent to the identity announced for $B_n x^2$.

To prove the theorem, we may assume that $|f| \le 1$ on $[0,1]$ and, since f is uniformly continuous, for every $\varepsilon > 0$ we can choose $\delta > 0$ so that $|f(x) - f(y)| \le \varepsilon$ if $|x - y| \le \delta$, and then

$$|f(x) - f(y)| \le \varepsilon + \frac{2}{\delta^2}(x-y)^2$$

also when $|x - y| > \delta$.

We look at y as a parameter and x as the variable. Then, from the properties of B_n,

$$\begin{aligned} |B_n f - f(y)| &= |B_n(f - f(y)1)| \le B_n(|f - f(y)|) \\ &\le B_n\big(\varepsilon + \frac{2}{\delta^2}(x-y)^2\big) \le \varepsilon + \frac{2}{\delta^2}(B_n I^2 - 2yI + y^2). \end{aligned}$$

Finally, if we evaluate these functions at y,

$$|(B_n f)(y) - f(y)| \le \varepsilon + \frac{2}{\delta^2}((B_n I^2)y - y^2) \to \varepsilon$$

uniformly on $y \in [0,1]$ as $n \to \infty$. Hence $\|B_n f - f\|_{[0,1]} \le 2\varepsilon$ as $n \ge n_0$, for some n_0. □

Corollary 2.17. $\mathcal{C}[a,b]$ is separable; that is, it contains a countable dense set.

Proof. Every $g \in \mathcal{C}[a,b]$ is the uniform limit on $[a,b]$ of a sequence of polynomials, and every polynomial $P(x) = \sum_{k=1}^{N} a_k x^k$ is the uniform limit of a sequence $P_m(x) = \sum_{k=1}^{N} q_{k,m} x^k$ of polynomials with rational coefficients. Just take $\mathbf{Q} \ni q_{k,m} \to a_k$ in $\mathbf{R}$ ($q_{k,m} \in \mathbf{Q} + i\mathbf{Q}$ in the complex case). Then the collection of these polynomials with rational coefficients is dense and it is countable. □

Exercise 2.9 extends Corollary 2.17 to $\mathcal{C}(K)$ if K is a compact metric space.

A subset of a normed space is called **total** if its linear span is dense. With the argument of Corollary 2.17, if a normed space contains a countable total set Z, it is separable, since the set of all linear combinations of elements in Z with rational coefficients is dense.

Obviously the product of two separable normed spaces is also separable, and it is readily seen that a subspace F of a separable space E is also separable.

Indeed, if Z is a countable dense subset of E, the collection of all the balls $B_E(z, 1/m)$ with $z \in Z$ and $m \in \mathbf{N}$ is countable and covers E. By choosing a point from every nonempty set $B_E(z, 1/m) \cap F$, we obtain a countable subset A of F which is dense in F since, for every $y \in F$ and every $m \in \mathbf{N}$, there exists some $z_m \in Z$ such that $\|z_m - y\|_E \leq 1/2m$ and some $a_m \in A$ such that $\|z_m - a_m\|_E \leq 1/2m$; then $\|y - a_m\|_E \leq 1/m$.

Remark 2.18. The fact of being separable indicates that a normed space E is somehow "not too large". It cannot contain an uncountable set $\{x_\alpha\}_{\alpha \in A}$ such that $\|x_\alpha - x_b\|_E \geq \delta$ if $\alpha \neq \beta$, for some $\delta > 0$, since if $\{y_n\}_{n \in \mathbf{N}}$ is dense in E, for every $\alpha \in A$, $\|x_\alpha - y_{n(\alpha)}\|_E < \delta/2$ for some $n(\alpha) \in \mathbf{N}$, and the mapping $\alpha \in A \mapsto n(\alpha) \in \mathbf{N}$ is injective.

Next we prove the extension of the Weierstrass Theorem 2.16, due in 1937 to M. H. Stone, which includes the proof of the density of the set of all polynomials in $\mathcal{C}(K)$ when K is a compact subset of $\mathbf{R}^n$.

Theorem 2.19 (Stone-Weierstrass). *Let K be a compact space. If a subalgebra A of $\mathcal{C}(K; \mathbf{R})$ separates points and does not vanish at any point of K, then $\bar{A} = \mathcal{C}(K; \mathbf{R})$.*

Proof. Let $f \in \mathcal{C}(K; \mathbf{R})$ and consider any positive number $\varepsilon > 0$. We will prove that $\|f - g\|_K < \varepsilon$ for some $g \in \bar{A}$ in four steps.

(1) If $f \in A$, then $|f| \in \bar{A}$.

Let $a < 0$ and $b > 0$ be such that $f(K) \subset [a, b]$ and $v(x) = |x|$ on $[a, b]$. According to the Weierstrass theorem, we can find $Q_n \in \mathcal{P}[a, b]$ so that

$$\lim_n \|Q_n - v\|_{[a,b]} \to 0.$$

Since $Q_n(0) \to v(0) = 0$, the polynomials $P_n = Q_n - Q_n(0)$ are such that $P_n(0) = 0$ and $\|P_n - v\|_{[a,b]} \to 0$ as $n \to \infty$. Hence $P_n(x) = \sum_{k=1}^{N(n)} a_k x^k$, so that $P_n(f) = a_1 f + a_2 f^2 + \cdots + a_n f^n \in A$ and

$$\|P_n(f) - |f|\|_K \leq \|P_n - v\|_{[a,b]} \to 0.$$

(2) If $f, g \in A$, then $\sup\{f, g\}, \inf\{f, g\} \in \bar{A}$, since according to (1)

$$\sup\{f, g\}(x) := \max\{f(x), g(x)\} = \frac{f + g + |f - g|}{2} \in \bar{A}$$

and

$$\inf\{f,g\}(x) := \min\{f(x),g(x)\} = \frac{f+g-|f-g|}{2} \in \bar{A}.$$

(3) If $x \in K$, we can find $g_x \in \bar{A}$ so that $g_x(x) = f(x)$ and $g_x < f+\varepsilon$ on K.

According to Lemma 2.15, for every $y \in K$ we can choose $f_y \in A$ so that

$$f_y(x) = f(x) \quad \text{and} \quad f_y(y) = f(y).$$

By continuity, $f_y < f+\varepsilon$ on a neighborhood $U(y)$ of y and, since K is compact,

$$K = U(y_1) \cup \cdots \cup U(y_N).$$

An application of (2) gives

$$g_x := \inf\{f_{y_1}, \ldots, f_{y_N}\} \in \bar{A},$$

and $g_x(x) = \min\{f(x), \ldots, f(x)\} = f(x)$. For every $z \in K$, $g_x(z) \le f_{y_j}(z) < f(z)+\varepsilon$ if $z \in U(y_j)$. Thus, $g_x < f+\varepsilon$.

(4) Finally, if we choose g_x as in (3) for every $x \in K$, then $g_x > f-\varepsilon$ on a neighborhood $V(x)$ of x and

$$K = V(x_1) \cup \cdots \cup V(x_k).$$

It follows as in (3) that

$$g := \sup\{g_{x_1}, \ldots, g_{x_k}\} \in \bar{A}$$

satisfies $\|g-f\|_K < \varepsilon$, since $g_{x_1}, \ldots, g_{x_k} < f+\varepsilon$ and every $z \in K$ belongs to some $V(x_j)$, and $g(z) > g_{x_j}(z) > f(z)+\varepsilon$. □

There is also a **complex form of the Stone-Weierstrass theorem**:

Corollary 2.20. Let A be subalgebra of the Banach space $\mathcal{C}(K;\mathbf{C})$ of all complex-valued continuous functions on a compact space K that separates points and does not vanish at any point of K. If A is self-conjugate, that is $\bar{f} \in A$ if $f \in A$, then $\bar{A} = \mathcal{C}(K;\mathbf{C})$.

Proof. Since A is self-conjugate, if $f \in A$ and $u = \Re f$, then also $u = \Im(if) = (f+\bar{f})/2 \in A$ and

$$A_0 := \{\Re f;\ f \in A\} = \{\Im f;\ f \in A\}.$$

is a subalgebra of $\mathcal{C}(K;\mathbf{R})$, since $u = \Re f$ and $v = \Re g \in A_0$ imply $uv = \Re(fg+f\bar{g})/2 \in A_0$.

Moreover A_0 separates points and does not vanish at any point, since $x, y \in K$ and $f(x) \ne f(y)$ for some $f \in A$ implies $\Re f(x) \ne \Re f(y)$ or $\Im f(x) \ne \Im f(y)$, and $\Re f(x) \ne 0$ or $\Im f(x) \ne 0$ if $f(x) \ne 0$. Hence A_0 is dense in $\mathcal{C}(K;\mathbf{R})$ and, if $f \in \mathcal{C}(K;\mathbf{C})$, we can find two sequences $\{u_n\}$ and $\{v_n\}$

in A_0 that approximate $\Re f$ and $\Im f$, so $f_n = u_n + iv_n \in A_0 + iA_1 \subset A$ and $f_n \to f$ uniformly on K. □

2.2. Linear operators

2.2.1. Bounded linear operators.

Theorem 2.21. *A linear mapping $T : E \to F$ between two normed spaces is continuous if and only if*

$$\|Tx\|_F \le C\|x\|_E \tag{2.4}$$

for some finite constant $C \ge 0$.

Proof. We know from Theorem 2.2 that T is continuous if and only if, for every $\varepsilon > 0$, we can choose $\delta > 0$ so that

$$T(\bar{B}_E(0,\delta)) \subset \bar{B}_F(0,\varepsilon).$$

Hence, $\|Tx\|_F \le \varepsilon$ if $\|x\|_E \le \delta$, or $\|Tx\|_F \le C\|x\|_E$ for any $x \in E$, with $C = \varepsilon/\delta$. □

If a set $A \subset E$ is contained in a ball $\bar{B}_E(0,R)$, we say that A is **bounded**.

A continuous linear mapping T between two normed spaces E and F is also called a **bounded linear operator**, since condition (2.4) means that $\|Tx\|_F \le C$ when $x \in \bar{B}_E(0,1)$ and $T\big(\bar{B}_E(0,1)\big)$ is bounded in F. That is, T is bounded on the unit ball of E.

By denoting $\|T\| = \sup_{\|x\|_E \le 1} \|Tx\|_F$, the smallest constant C in (2.4), T is continuous if and only if $\|T\|$ is finite.

Example 2.22 (Fredholm operator[5]). If $K : [a,b] \times [c,d] \to \mathbf{C}$ is a continuous function and $T_K f(x) = \int_c^d K(x,y)f(y)\,dy$, then

$$T_K : \mathcal{C}[c,d] \to \mathcal{C}[a,b]$$

is a bounded linear operator, since $|T_K f(x)| \le (d-c)\|K\|_{[c,d]\times[a,b]}\|f\|_{[c,d]}$, so that $\|T_K f\|_{[a,b]} \le C\|f\|_{[c,d]}$ with $C = (d-c)\|K\|_{[c,d]\times[a,b]}$.

---

[5]The Swedish mathematician Erik Ivar Fredholm, in 1900, created the first theory on linear equations in infinite-dimensional spaces, establishing the modern theory of integral equations

$$f(x) + \int_0^1 K(x,y)f(y)\,dy = g(y) \qquad (0 \le x \le 1)$$

as the limiting case of linear systems

$$f(x_i) + \sum_{j=1}^{n} K(x_i,y_j)f(y) = g(y_j) \qquad (1 \le i \le n),$$

and found his "alternative", which we will meet in Theorem 4.33. He based his fundamental paper published in 1903 on the determinant named after him which is associated to the Fredholm operators. His method was immediately followed by the work of Hilbert, Schmidt, Poincaré, F. Riesz, and many others.

Example 2.23 (Volterra operator[6]). Similarly, if K is a continuous function on the triangle $\Delta := \{(x,y) \in [a,b] \times [a,b]; a \leq y \leq x \leq b\}$ and

$$T_K f(x) := \int_a^x K(x,y) f(y)\, dy,$$

then

$$T_K : \mathcal{C}[a,b] \to \mathcal{C}[a,b]$$

is a bounded linear operator, and $\|T_K\| \leq (b-a)\|K\|_\Delta$.

Let $T : E \to F$ be a bijective linear mapping between two normed spaces. We say that T is an **isomorphism of normed spaces** if and only if T and T^{-1} are continuous. By (2.4), the continuity of T and T^{-1} means that we can find two constants $\alpha, \beta > 0$ that satisfy

$$\alpha\|x\|_E \leq \|Tx\|_F \leq \beta\|x\|_E.$$

Two norms $\|\cdot\|_1$ and $\|\cdot\|_2$ on a vector space E are said to be equivalent if they define the same topology, that is, if the identity $I : (E, \|\cdot\|_1) \to (E, \|\cdot\|_2)$ is an isomorphism, so that

$$\alpha\|x\|_1 \leq \|x\|_2 \leq \beta\|x\|_1. \tag{2.5}$$

Example 2.24. On $\mathcal{C}[a,b]$, the usual norm $\|f\|_{[a,b]}$ and $\|f\|_1 = \int_a^b |f(t)|\, dt$ satisfy

$$\|f\|_1 \leq (b-a)\|f\|_{[a,b]},$$

and we say that $\|\cdot\|_{[a,b]}$ is **finer** than $\|\cdot\|_1$, since $I : (E, \|\cdot\|_{[a,b]}) \to (E, \|\cdot\|_1)$ is continuous and the topology of $\|\cdot\|_{[a,b]}$ is finer than the one of $\|\cdot\|_1$.

It is readily checked that, if $|\cdot|$ is the Euclidean norm (2.2), $\|x\|_\infty := \max_{j=1}^n |x_j|$, and $\|x\|_1 := \sum_{j=1}^n |x_j|$, then

$$\|x\|_\infty \leq |x| \leq \sqrt{n}\|x\|_\infty, \quad \|x\|_\infty \leq \|x\|_1 \leq n\|x\|_\infty.$$

Next we will prove that, in fact, the norms on $\mathbf{K}^n$ are all equivalent.

Theorem 2.25. *If E is a normed space of finite dimension n, then every linear bijection $T : \mathbf{K}^n \to E$ is an isomorphism.*[7]

Thus, E is complete and, on E, two norms are always equivalent.

---

[6] In 1896 the Italian mathematician and physicist Vito Volterra published papers on what is now called "an integral equation of Volterra type". His main contributions in the area of integral and integro-differential equations is contained in his 1930 book "Theory of Functionals and of Integral and Integro-Differential Equations".

[7] This result holds for every topological vector space. The case $n = 1$ is included in Exercise 4.5. The proof is by induction on n (cf. Berberian [**4**, (23.1)]).

Proof. On $\mathbf{K}^n$ we consider the Euclidean norm $|\cdot|$ and the norm $\|x\| := \|Tx\|_E$. If $x = \sum_{j=1}^n x^j u_j$, where $\{u_1, \dots, u_n\}$ is the canonical basis in $\mathbf{K}^n$, then

$$\|x\| \leq \sum_{j=1}^n |x^j| \|u_j\| \leq C|x| \qquad (C = \sum_{j=1}^n \|u_j\|),$$

since $|x^j| \leq |x|$, so that $\|Tx\|_E \leq C|x|$.

To prove the reverse estimate, we will use the fact that the unit Euclidean sphere $S = \{x;\ |x| = 1\}$ is compact and that the function $f(x) := \|x\|$ is continuous on S, since $|f(x) - f(y)| \leq \|x - y\| \leq C|x - y|$. This function has a minimum value $c = \min f = \|x_0\| > 0$ for some $x_0 \in S$, and then $\|x/|x|\| \geq c$, so that $|x| \leq c^{-1}\|x\|$. □

Every compact subset K of a normed space E is closed, and it is bounded, since $K \subset \bigcup_{m=1}^{\infty} B_E(0, m)$, so that $K \subset \bigcup_{m=1}^{N} B_E(0, m) = B_E(0, N)$ for some $N > 0$.

The converse is also true when E is a finite-dimensional space, since K is then homeomorphic to a closed and bounded set of a Euclidean space $\mathbf{K}^n$, which is compact by the Heine-Borel theorem.

An important fact is that this property characterizes finite-dimensional normed spaces; that is, if the unit ball B_E is compact, then $\dim E < \infty$. The proof will be based on the existence of "**nearly orthogonal elements**" to any closed subspace of a normed space.

Lemma 2.26 (F. Riesz). *Suppose E is a normed space and M a closed subspace, $M \neq E$, and let $0 < \varepsilon < 1$. Then there exists $u \in E$ such that $\|u\|_E = 1$ and $d(u, M) \geq 1 - \varepsilon$.*

Proof. Let $v \in E \setminus M$, $d = d(v, M) > 0$ (M is closed), and choose $m_0 \in M$ so that $\|v - m_0\|_E \leq d/(1 - \varepsilon)$. The element $u = (v - m_0)/\|v - m_0\|_E$ satisfies the required conditions, since, if $m \in M$,

$$\|u - m\|_E = \|v - (m_0 + \|v - m_0\|_E m)\|_E / \|v - m_0\|_E \geq 1 - \varepsilon.$$

□

Remark 2.27. If we have an increasing sequence of closed subspaces M_n of a normed space, then there exists a sequence $\{u_n\}$ such that $u_n \in M_n$, $\|u_n\|_E = 1$, and $d(u_{n+1}, M_n) \geq 1/2$, so that $\{u_n\}$ has no Cauchy subsequence, since $\|u_p - u_q\|_E \geq 1/2$ if $p \neq q$.

A similar remark holds for a decreasing sequence of closed subspaces.

Theorem 2.28. *If the unit sphere $S_E = \{x \in E;\ \|x\|_E = 1\}$ of a normed space E is compact, then E is of finite dimension.*

Proof. Assume that E has a sequence $\{x_n\}$ of linearly independent elements. We can apply Remark 2.27 to the subspaces $M_n = [x_1, \ldots, x_n]$, which are closed since they are complete by Theorem 2.25. Then $\{u_n\} \subset S_E$ has no convergent subsequence. □

2.2.2. The space of bounded linear operators. With the usual vector operations, the set $\mathcal{L}(E;F)$ of all bounded linear operators between two normed spaces E and F is a vector space, and it becomes a normed space with the **operator norm**

$$\|T\| := \sup_{\|x\|_E \le 1} \|Tx\|_F,$$

since all the properties of a norm are satisfied:

1. If $\|T\| = 0$, then $\|Tx\|_F \le \|T\|\|x\|_E = 0$ for all $x \in E$ and $T = 0$,
2. $\|\lambda T\| = \sup_{\|x\|_E \le 1} |\lambda|\|Tx\|_F = |\lambda| \sup_{\|x\|_E \le 1} \|Tx\|_F = |\lambda|\|T\|$, and
3. $\|(T_1 + T_2)x\|_F \le (\|T_1\| + \|T_2\|)\|x\|_E$ and then $\|T_1 + T_2\| \le \|T_1\| + \|T_2\|$.

Moreover, the norm of the product ST of two bounded linear operators $T : E \to F$ and $S : F \to G$ is submultiplicative,

$$\|ST\| \le \|S\|\|T\|, \tag{2.6}$$

since

$$\|STx\|_G \le \|S\|\|Tx\|_F \le \|S\|\|T\|\|x\|_E.$$

Note that

$$\|T\| = \sup_{\|x\|_E < 1} \|Tx\|_F = \sup_{\|x\|_E = 1} \|Tx\|_F,$$

since, if $\|x\|_E \le 1$,

$$\|Tx\|_F = \lim_{\varepsilon \downarrow 0} (1-\varepsilon)\|Tx\|_F = \lim_{\varepsilon \downarrow 0} \|T((1-\varepsilon)x)\|_F \le \sup_{\|x\|_E < 1} \|Tx\|_F$$

and then $\|T\| = \sup_{\|x\|_E < 1} \|Tx\|_F$. Also, if $0 < \|x\|_E < 1$,

$$\|Tx\|_F = \|x\|_E \|T(x/\|x\|_E)\|_F \le \sup_{\|x\|_E = 1} \|Tx\|_F,$$

and $\|T\| = \sup_{\|x\|_E = 1} \|Tx\|_F$.

Theorem 2.29. *If F is a Banach space, then $\mathcal{L}(E;F)$ is also complete.*

Proof. Let $\{T_n\}$ be a Cauchy sequence in $\mathcal{L}(E;F)$. For every $x \in E$, $\{T_n x\}$ is a Cauchy sequence in the complete space F, since $\|T_p x - T_q x\|_F \le \|T_p - T_q\|\|x\|_E$. We define $Tx := \lim T_n x$ and, by the continuity of the vector operations, $T : E \to F$ is linear. Moreover, for every $\varepsilon > 0$ we can find $N > 0$ so that

$$\|T_p x - T_q x\|_F \le \|T_p - T_q\|\|x\|_E \le \varepsilon \qquad \forall p, q \ge N, \ \forall \|x\|_E \le 1$$

and, by letting $q \to \infty$, $\|T_p x - Tx\|_F \leq \varepsilon$. Therefore $T \in \mathcal{L}(E; F)$ and $\|T_p - T\| \to 0$ if $q \to \infty$. □

In the special case $F = \mathbf{K}$, the Banach space of all bounded linear forms $E' := \mathcal{L}(E; \mathbf{K})$ with the norm

$$\|u\| = \sup_{\|x\|_E \leq 1} |u(x)|$$

is called the **dual space** of E.

We will use the notation $\mathcal{L}(E)$ for $\mathcal{L}(E; E)$. Every element of $\mathcal{L}(E)$ is called a bounded linear operator on E.

2.2.3. Neumann series. Let us consider the problem of solving the equation

$$Tu - \lambda u = v \qquad (T \in \mathcal{L}(E),\ v \in E),$$

where E is any Banach space and $v \in E$ and $0 \neq \lambda \in \mathbf{K}$ are given.

We may obtain the inverse of $T - \lambda I = -\lambda(1 - T/\lambda)$ using the **Neumann series**[8]

$$-\frac{1}{\lambda} \sum_{n=0}^{\infty} \frac{1}{\lambda^n} T^n,$$

similar to a numerical geometric series. If $\|T\|/|\lambda| < 1$,

$$\sum_{n=0}^{\infty} \|(1/\lambda^n) T^n\| \leq \sum_{n=0}^{\infty} \|(1/\lambda) T\|^n = \sum_{n=0}^{\infty} (\|T\|/|\lambda|)^n < \infty,$$

and the Neumann series is convergent. It is shown that

$$(T - \lambda I)^{-1} = -\frac{1}{\lambda} \sum_{n=0}^{\infty} \frac{1}{\lambda^n} T^n \qquad (\|T\| < |\lambda|), \tag{2.7}$$

by checking that, if S is the sum of the series, $S(T - \lambda I) = (T - \lambda I)S = I$. For instance, if $S_N := -\sum_{n=0}^{N} \lambda^{-n-1} T^n$, a partial sum of the series, then $S(T - \lambda I) = \lim_N S_N(T - \lambda I)$ by (2.6), and

$$S_N(T - \lambda I) = -\sum_{n=0}^{N} \frac{T^{n+1}}{\lambda^{n+1}} + \sum_{n=0}^{N} \frac{T^n}{\lambda^n} = I - \frac{T^{N+1}}{\lambda^{N+1}} \to I$$

when $N \uparrow \infty$, since we have $\|T^n/\lambda^n\| \leq (\|T\|/|\lambda|)^n \to 0$.

An interesting application refers to the example of the Volterra operators defined in Example 2.23. If T_K and T_H are defined by the integral kernels

[8] The German mathematician Carl G. Neumann, who worked on the Dirichlet principle, used this series in 1877 in the context of potential theory.

$K, H \in \mathcal{C}(\Delta)$, then $T_K T_H$ is also a Volterra operator T_L, defined by the composition L of K with H,

$$L(x,z) = \int_a^x K(x,y)H(y,z)\,dy$$

since, assuming that $K(x,y) = H(x,y) = 0$ if $y > x$,

$$\begin{aligned} T_K T_H f(x) &= \int_a^b K(x,y) T_H f(y)\,dy = \int_a^b K(x,y) \int_a^b H(y,z) f(z)\,dz\,dy \\ &= \int_a^b \Big(\int_a^b K(x,y)H(y,z)\,dy \Big) f(z)\,dz \end{aligned}$$

and, if $a \leq x < z \leq b$, in the last integral we have $K_1(x,y) = 0$ or $K_2(y,z) = 0$ and then $K(x,z) = 0$. It is worth observing how this composition of kernels of integral operators corresponds to the product of matrices of linear mappings in linear algebra.

The continuity of

$$L(x,z) = \int_a^x K(x,y)H(y,z)\,dy$$

defined on Δ follows from the continuity of the restriction of K and H to this triangle.

We can compose T_K n times with itself to obtain

$$T_K^n f(x) = \int_a^x K_n(x,y) f(y)\,dy,$$

where $K_1 = K$ and

$$K_n(x,y) = \int_a^b K(x,z)K_{n-1}(z,y)\,dz = \int_y^x K(x,z)K_{n-1}(z,y)\,dz$$

when $y \leq x$ and $n \geq 2$. If $|K(x,y)| \leq M$, by induction,

$$|K_n(x,y)| \leq \frac{M^n}{(n-1)!}|x-y|^{n-1} \tag{2.8}$$

since $K_{n+1}(x,y) = \int_y^x K(x,z)K_n(z,y)\,dz$ and, if $y \leq x$,

$$|K_{n+1}(x,y)| \leq \int_y^x M \frac{M^n}{(n-1)!}(x-y)^{n-1}\,dz = \frac{M^{n+1}}{n!}(x-y)^n.$$

Theorem 2.30. *Let $T_K : \mathcal{C}[a,b] \longrightarrow \mathcal{C}[a,b]$ be the Volterra operator defined by the kernel $K \in \mathcal{C}(\Delta)$. Then, for any $\lambda \neq 0$,*

$$(T_K - \lambda I)^{-1} = -\frac{1}{\lambda} \sum_{n=0}^{\infty} \frac{T_K^n}{\lambda^n}, \tag{2.9}$$

and the series is absolutely convergent in $\mathcal{L}(\mathcal{C}[a,b])$.

Proof. From (2.8),

$$\|T_K^n\| \le (b-a)\|K_n\| \le \frac{M^n(b-a)^n}{(n-1)!}$$

and

$$\|(1/\lambda)^n T_K^n\| \le \frac{M^n(b-a)^n}{(n-1)!}\frac{1}{|\lambda|^n},$$

which is the general term of a convergent numerical series.

The identity (2.9) is obtained as (2.7). □

In Theorem 2.48 and Exercise 2.20, the reader will find interesting applications of Theorem 2.30 to find the unique solution for the Cauchy problem of a linear differential equation.

2.3. Hilbert spaces

Let us review some basic facts concerning Hilbert spaces.

2.3.1. Scalar products. A **scalar product** or **inner product** in a real or complex vector space H is a $\mathbf{K}$-valued function on $H \times H$,

$$(x, y) \in H \times H \mapsto (x, y)_H \in \mathbf{K},$$

having the following properties:

(1) It is a **sesquilinear form**, meaning that, for every $x \in H$, $(\cdot, x)_H$ is a linear form on H and $(x, \cdot)_H$ is **skewlinear**, that is,

$$(x, y_1 + y_2)_H = (x, y_1)_H + (x, y_2)_H, \qquad (x, \lambda y)_H = \bar{\lambda}(x, y)_H.$$

In the real case, $\mathbf{K} = \mathbf{R}$, $(\cdot, \cdot)_H$ is a **bilinear form**, since $\bar{\lambda} = \lambda$.

(2) $(y, x)_H = \overline{(x, y)_H}$, so that, if $\mathbf{K} = \mathbf{R}$, $(\cdot, \cdot)_H$ is symmetric.

(3) $(x, x)_H > 0$ if $x \neq 0$.

By (1), $(x, 0)_H = (0, y)_H = 0$.

Given this scalar product, the associated norm on H is

$$\|x\|_H := (x, x)_H^{1/2}. \tag{2.10}$$

Obviously $\|\lambda x\|_H = |\lambda|\|x\|_H$ and $\|x\|_H > 0$ if $x \neq 0$. The subadditivity follows from the fundamental **Schwarz inequality**[9]

$$|(x, y)_H| \le \|x\|_H \|y\|_H \tag{2.11}$$

[9] Also called the Cauchy-Bunyakovsky-Schwarz inequality; first published by the French mathematician Augustin Louis Cauchy for sums (1821), and for integrals stated by the Ukrainian mathematician Viktor Bunyakovsky (1859) and rediscovered by Hermann A. Schwarz (1888), who worked on function theory, differential geometry, and the calculus of variations in Halle, Göttingen, and Berlin.

by considering $\|x+y\|_H^2 = (x+y, x+y)_H$ and the properties of the inner product to obtain

$$\begin{aligned} \|x+y\|_H^2 &= \|x\|_H^2 + \|y\|_H^2 + 2\Re(x,y)_H \le \|x\|_H^2 + \|y\|_H^2 + 2|(x,y)_H| \\ &\le \|x\|_H^2 + \|y\|_H^2 + 2\|x\|_H\|y\|_H, \end{aligned}$$

that is, $\|x+y\|_H^2 \le (\|x\|_H + \|y\|_H)^2$.

To prove the Schwarz inequality, which is obvious if $y = 0$, in

$$0 \le \|x+\lambda y\|_H^2 = \|x\|_H^2 + |\lambda|^2\|y\|_H^2 + \bar{\lambda}(x,y)_H + \lambda(y,x)_H$$

we only need to choose $\lambda = -(x,y)_H/\|y\|_H^2$ and then multiply by $\|y\|_H^2$.

It follows from the Schwarz inequality that the inner product is continuous at any point (a,b) of $H \times H$ since

$$\begin{aligned} |(x_n, y_n)_H - (a,b)_H| &= |(x_n - a, y_n)_H + (a, y_n - b)_H| \\ &\le \|x_n - a\|_H\|y_n\|_H + \|a\|_H\|y_n - b\|_H \to 0 \end{aligned}$$

if $(x_n, y_n) \to (a,b)$ in $H \times H$.

A **Hilbert space**[10] is a Banach space, H, whose norm is induced by an inner product as in (2.10).

Remark 2.31. The completion $\tilde{H}$ of the normed space H with a norm defined by a scalar product as in (2.10) is a Hilbert space.

If $\tilde{x}, \tilde{y} \in \tilde{H}$ with $x_n \to x$ and $y_n \to y$ $(x_n, y_n \in H)$, we obtain a scalar product on $\tilde{H}$ by defining

$$(\tilde{x}, \tilde{y})_{\tilde{H}} := \lim_{n\to\infty} (x_n, y_n)_H,$$

since $\|x_n\|_H, \|y_m\|_H \le C$ and, by the Schwarz inequality,

$$|(x_n, y_n)_H - (x_m, y_m)_H| \le \|x_n - x_m\|_H\|y_m\|_H + \|x_m\|_H\|y_n - y_m\|_H \to 0$$

as $m, n \to \infty$.

This definition does not depend on the sequences $x_n \to x$ and $y_n \to y$ since, if also $x'_n \to x$ and $y'_n \to y$, then $\{(x_n, y_n)_H\}$ and $\{(x'_n, y'_n)_H\}$ are subsequences of a similar one obtained by mixing both of them.

Note that

$$\|\tilde{x}\|_{\tilde{H}}^2 = \lim_n \|x_n\|_H^2 = \lim_n (x_n, x_n)_H = (\tilde{x}, \tilde{x})_{\tilde{H}}.$$

[10]The name was coined in 1926 by Hilbert's student J. von Neumann, who included the condition of separability in the definition, when working on the mathematical foundation of quantum mechanics. Hilbert used the name of infinite-dimensional Euclidean space when dealing with integral equations, around 1909. It was another student of Hilbert, Erhard Schmidt, beginning his 1905 dissertation in Göttingen, who completed the theory of Hilbert spaces for ℓ^2 by introducing the language of Euclidean geometry.

Example 2.32. The **Euclidean space $\mathbf{K}^n$**, with the norm

$$|(x_1, \dots, x_n)| = \sqrt{|x_1|^2 + \cdots + |x_n|^2}$$

induced by the **Euclidean inner product** $x \cdot y = \sum_{k=1}^{n} x_j \bar{y}_j$, is the simplest Hilbert space.

Example 2.33. Another fundamental example is $L^2(\mu)$, with

$$\|f\|_2 = \Big(\int |f|^2 \, d\mu\Big)^{1/2}$$

induced by the scalar product $(f, g)_2 = \int f\bar{g}\, d\mu$.

Example 2.34. The space ℓ^2 of all sequences $x = \{x_n\}_{n\in\mathbf{N}}$ (or $\{x_n\}_{n\in\mathbf{Z}}$) that satisfy

$$\|x\|_2^2 = \sum_n |x_n|^2 < \infty$$

is the Hilbert space whose norm is induced by $(x, y)_2 = \sum_n x_n \bar{y}_n$. It is the L^2 space on $\mathbf{N}$ (or $\mathbf{Z}$) with the counting measure.

2.3.2. Orthogonal projections.

The elementary properties and terminology of Euclidean spaces extend to any Hilbert space H:

It is said that $a, b \in H$ are orthogonal if $(a, b)_H = 0$, and the **orthogonal space** of a subset A of H is defined as

$$A^\perp = \{z \in H;\ (z, a)_H = 0\, \forall a \in A\}.$$

It is a closed subspace of H, since, by the Schwarz inequality, every $(\cdot, a)_H$ is a continuous linear form on H and $A^\perp = \bigcap_{a\in A} \operatorname{Ker}(\cdot, a)_H$, an intersection of closed subspaces.

For a finite number of points $x_1, \dots, x_n$ that are pairwise orthogonal, the relation

$$\|x_1 + \cdots + x_n\|_H^2 = \|x_1\|_H^2 + \cdots + \|x_n\|_H^2$$

is the **Pythagorean theorem**, and a useful formula is the **parallelogram identity**

$$2\|a\|_H^2 + 2\|b\|_H^2 = \|a + b\|_H^2 + \|a - b\|_H^2.$$

They follow immediately from the definition (2.10) of the norm. The parallelogram identity will be useful to prove the existence and uniqueness of an **optimal projection** on a closed convex set for every point in a Hilbert space:

Theorem 2.35 (Projection theorem). (a) *Suppose C is a nonempty closed and convex subset of the Hilbert space H and x any point in H. Then there is a unique point $P_C(x)$ in C that satisfies $\|x - P_C(x)\|_H = d(x, C)$, where $d(x, C) = \inf_{y\in C} \|y - x\|_H$.*

A point $y \in C$ is this optimal projection $P_C(x)$ of x if and only if

$$\Re(c - y, x - y) \le 0 \quad \forall c \in C. \tag{2.12}$$

(b) *If F is a closed subspace of the Hilbert space H, then $H = F \oplus F^\perp$ (direct sum), and $x = y + z$ with $y \in F$ and $z \in F^\perp$ if and only if $y = P_F(x)$ and $z = P_{F^\perp}(x)$.*

Moreover P_F is a bounded linear operator $P_F : H \to H$ with norm 1 (if $F \ne \{0\}$), $\operatorname{Ker} P_F = F^\perp$, $\operatorname{Im} P_F = F$, $(P_F(x_1), x_2)_H = (x_1, P_F(x_2))_H$, and $P_F^2 = P_F$.

Proof. (a) Let $d = d(x, C)$ and choose $y_n \in C$ so that $d_n = \|x - y_n\| \to d$. Since C is convex, $(y_p + y_q)/2 \in C$ and, by an application of the parallelogram identity to $a = (x - y_p)/2$ and $b = (x - y_q)/2$,

$$\frac{1}{2}(d_p^2 + d_q^2) = \|x - \frac{1}{2}(y_p + y_q)\|_H^2 + \frac{1}{4}\|y_p - y_q\|_H^2 \ge d^2 + \frac{1}{4}\|y_p - y_q\|_H^2.$$

By letting $p, q \to \infty$, $d^2 \ge d^2 + \lim_{p,q\to\infty} \frac{1}{4}\|y_p - y_q\|_H^2$ and $\|y_p - y_q\|_H^2 \to 0$. Since C is complete, $y_n \to y \in C$ and $\|x - y\|_H = \lim_n \|x - y_n\|_H = d$.

The uniqueness of the minimizer y follows from the fact that if z is also a minimizer, the foregoing argument shows that $\{y, z, y, z, y, z, \ldots\}$ converges and $y = z$.

To prove (2.12), suppose that $c \in C$ and $0 \le t \le 1$. Then $(1-t)y + tc \in C$ by the convexity of C, and

$$\|x - y\|_H^2 \le \|x - [(1-t)y - tc]\|_H^2 = \|x - y - t(c - y)\|_H^2,$$

where

$$\|x - y - t(c - y)\|_H^2 = \|x - y\|_H^2 - 2t\Re(c - y, x - y)_H + t^2\|c - y\|_H^2. \tag{2.13}$$

Hence $2\Re(c - y, x - y)_H \le t\|c - y\|_H^2$ and it follows that $\Re(c - y, x - y)_H \le 0$ by letting $t \to 0$.

Conversely, if $\Re(c - y, x - y)_H \le 0$ and in (2.13) we put $t = 1$,

$$\|x - y\|_H^2 - \|c - y\|_H^2 = 2\Re(c - y, x - y)_H - \|y - c\|_H^2 \le 0.$$

(b) Since $F \cap F^\perp = \{0\}$, it is sufficient to decompose every $x \in H$ into $x = y + z$ with $y \in F$ and $z \in F^\perp$. We choose $y = P_F(x)$ and $z = x - y$, so that we need to prove that $z \in F^\perp$. Indeed, by (a) we have $\Re(c - y, z)_H \le 0$ for any $u = c - y \in F$ and also $\Re(\lambda u, z)_H \le 0$, so that $(u, z)_H = 0$ for every $u = c - y \in F$ and $z \in F^\perp$.

Let $y_j = P_F(x_j)$ and $z_j = x_j - y_j$ $(j = 1, 2)$. Since $z_j \in F^\perp$,

$$(P_F(x_1), x_2)_H = (y_1, y_2 + z_2)_H = (y_1 + z_1, y_2)_H = (x_1, P(x_2))_H.$$

The linearity of P_F follows very easily from this identity, and it is also clear that $P_F^2(x) = P_F(y) = y = P_F(x)$, $\operatorname{Ker} P_F = F^\perp$, and $\operatorname{Im} P_F = F$.

From $\|y+z\|_H^2 = \|y\|_H^2 + \|z\|_H^2$ we obtain $\|P_F(x)\|_H^2 \leq \|x\|_H^2$, so that $\|P_F\| \leq 1$, since $\|P_F(y)\|_H = \|y\|_H$ if $y \in F$. □

If F is a closed vector subspace of H, we call $F^\perp$ the **orthogonal complement** of F, and the optimal projection P_F is called the **orthogonal projection** on F.

The projection theorem contains the fact that $F^\perp \neq \{0\}$ if $F \neq H$. This will be used to prove Theorem 4.1, the Riesz representation theorem for the dual space of H.

Theorem 2.36. *Let A be a subset of H. Then the closed linear span $\overline{[A]}$ of A coincides with $A^{\perp\perp}$, so that A is total in H if and only if $A^\perp = \{0\}$.*

Thus, a vector subspace F of H is closed if and only if $F^{\perp\perp} = F$.

Proof. It is clear that $A^\perp = [A]^\perp$ and, by continuity $[A]^\perp = \overline{[A]}^\perp$; thus, if $F = \overline{[A]}$, we need to prove that $F^{\perp\perp} = F$.

We have $F \subset F^{\perp\perp}$ and, if $x \notin F$, it follows from Theorem 2.35 that we can choose $z \in F^\perp$ so that $(x, z)_H \neq 0$; just take $z = P_{F^\perp}(x)$. This shows that also $x \notin F^{\perp\perp}$.

Note that A is total when $F = E$ and, by Theorem 2.35, this happens if and only if $A^\perp = F^\perp = \{0\}$. □

2.3.3. Orthonormal bases. A subset S of the Hilbert space H is called an **orthonormal system** if the elements of S are mutually orthogonal and they are all of norm 1.

Suppose $F = [e_1, \ldots, e_n]$, where $\{e_1, \ldots, e_n\}$ is a finite orthonormal set in H. Then

$$P_F(x) = \sum_{k=1}^{n} (x, e_k)_H e_k$$

since $z = x - \sum_{k=1}^{n} (x, e_k)_H e_k \in F^\perp$. It follows from $\|P_F(x)\|_H^2 \leq \|x\|_H^2$ that

$$\sum_{k=1}^{n} |(x, e_k)_H|^2 \leq \|x\|_H^2.$$

This estimate, which is known as **Bessel's inequality**, is valid for any orthonormal system $S = \{e_j\}_{j\in J}$; just take the supremum over all the finite sums[11]:

$$\sum_{j\in J} |(x, e_j)_H|^2 \le \|x\|_H^2. \tag{2.14}$$

The numbers $\widehat{x}(j) := (x, e_j)_H$ are called the **Fourier coefficients** of x with respect to S.

An **orthonormal basis** of H is a maximal orthonormal system, which is also said to be complete. That is, the orthonormal system $S = \{e_j\}_{j\in J}$ is an orthonormal basis if and only if

$$\widehat{x}(j) := (x, e_j)_H = 0 \quad \forall j \in J \Rightarrow x = 0,$$

which means that $S^\perp = 0$ and S is total.

By an application of Zorn's lemma, it can be proved that every orthonormal system can be extended to a maximal one.

In our examples, Hilbert spaces will be separable, so that an orthonormal system $S = \{e_j\}_{j\in J}$ is finite or countable, since $\|e_i - e_j\|_H = \sqrt{2}$ if $i \neq j$ (cf. Remark 2.18).

We will only consider the separable case and write $\ell^2 = \ell^2(J)$ if $J = \{1, 2, \ldots, N\}$, $\mathbf{N}$, or $\mathbf{Z}$, but the results extend easily to any Hilbert space.

By Bessel's inequality $x \in H \mapsto \widehat{x} = \{\widehat{x}(j)\} \in \ell^2$ is a linear transform such that $\widehat{x} = \{\widehat{x}(j)\} \in \ell^2$ and $\|\widehat{x}\|_2 \le \|x\|_H$. When S is complete, this mapping is clearly injective, since in this case $\widehat{x}(j) = 0$ for all $j \in J$ implies $x = 0$ by definition. Moreover, every $x \in H$ is recovered from its Fourier coefficients by adding the **Fourier series** $\sum_{j\in J} \widehat{x}(j) e_j$:

Theorem 2.37 (Fischer-Riesz[12]). *Suppose $S = \{e_j\}_{j\in J}$ is an orthonormal system of H. Then the following statements are equivalent:*

(a) *S is an orthonormal basis of H.*

(b) *$x = \sum_{j\in J} \widehat{x}(j) e_j$ in H for every $x \in H$.*

(c) *$\|x\|_H^2 = \|\widehat{x}\|_2$ for every $x \in H$ or, equivalently, $(x, y)_H = (\widehat{x}, \widehat{y})_2$ for all $x, y \in H$* (**Parseval's relation**).

Proof. Suppose $J = \mathbf{N}$ and let $c = \{c_j\} \in \ell^2$.

---

[11] According to footnote 2 in this chapter, if $f(j) = (x, e_j)_H$, then

$$\|f\|_2^2 := \sum_{j\in J} |f(j)|^2 = \sup\Big\{ \sum_{k\in F} |f(k)|^2;\ F \subset J,\ F \text{ finite}\Big\}$$

is the integral of $|f|^2$ relative to the counting measure on J, and $f \in \ell^2(J)$ if $\|f\|_2 < \infty$.

[12] Found independently in 1907 by the Hungarian mathematician Frigyes Riesz and the Austrian mathematician Ernst Fischer.

We claim that $z := \sum_{j=1}^{\infty} c_j e_j$ exists and $\widehat{z} = c$. Indeed, if $S_N = \sum_{j=1}^{N} c_j e_j$ and $p < q$,

$$\|S_q - S_p\|_H^2 = \Big\| \sum_{p<j\le q} c_j e_j \Big\|_H^2 = \sum_{p<j\le q} |c_j|^2$$

and we obtain the convergence of the series $\sum_{j=1}^{\infty} c_j e_j$ to some z in H. Moreover

$$\widehat{z}(k) = (\lim_N S_N, e_k)_H = \lim_N (S_N, e_k)_H = c_k,$$

since $(S_N, e_k)_H = c_k$ if $N \ge k$.

Now (b) follows from (a): By Bessel's inequality $\widehat{x} = \{\widehat{x}(j)\} \in \ell^2$ and, if $z = \sum_{j=1}^{\infty} \widehat{x}(j) e_j$, then $\widehat{z} = \widehat{x}$ and $\widehat{z - x}(j)$ for all j, so that $x = y$ by (a).

From (b), $\|x\|_H^2 = \lim_N \|S_N\|_H^2 = \lim_N \sum_{j=1}^{N} |\widehat{x}(j)|^2 = \|\widehat{x}\|_2$. Then also $(x, y)_H = (\widehat{x}, \widehat{y})_2$ by the **polarization identity**

$$(x, y)_H = \frac{1}{4}(\|x + y\|_H^2 - \|x - y\|_H^2) \tag{2.15}$$

if $\mathbf{K} = \mathbf{R}$, and

$$(x, y)_H = \frac{1}{4}(\|x + y\|_H^2 - \|x - y\|_H^2 + i\|x + iy\|_H^2 - i\|x - iy\|_H^2) \tag{2.16}$$

in the complex case. They are both checked by expanding the squared norms as scalar products.

Finally, if $\widehat{x} = 0$, from (c) we obtain $x = 0$ and (a) follows. □

Remark 2.38. We have proved that, when $\{e_j\}_{j\in J}$ is an orthonormal basis, the linear map $x \in H \mapsto \widehat{x} \in \ell^2$ is a bijective isometry, and that (b) defines its inverse.

The identity (b) is the expansion of x in a Fourier series. The classical best-known example is the following:

Example 2.39. In the Hilbert space $L^2(\mathbf{T}) = L^2(0, 2\pi)$, where for convenience we define the scalar product as

$$(f, g)_2 := \frac{1}{2\pi} \int_0^{2\pi} f(t)\overline{g(t)}\, dt = \int_{\mathbf{T}} f\bar{g},$$

the trigonometric system $e_k(t) := e^{ikt}$ $(k \in \mathbf{Z})$ is orthonormal. It is well known that it is complete and a proof of this fact, known as the **uniqueness theorem for Fourier coefficients**, follows from (2.27), where a constructive proof of the density of the trigonometric polynomials is given. Another proof based on the Stone-Weierstrass theorem is contained in Exercise 2.10.

The corresponding Parseval relation[13] is

$$\frac{1}{2\pi}\int_0^{2\pi} |f(t)|^2\,dt = \sum_{k=-\infty}^{\infty} \left|\frac{1}{2\pi}\int_0^{2\pi} f(t)e^{-ikt}\,dt\right|^2.$$

2.4. Convolutions and summability kernels

We are going to consider examples of linear operators T between complex L^p spaces on σ-finite measure spaces X and Y. The reader may assume that X and Y are two Borel subsets of $\mathbf{R}^n$ and $\mathbf{R}^m$, respectively, with the corresponding Lebesgue measures.

Assume that the domain $D(T)$ of T contains all complex integrable simple functions on X and that T takes values that are measurable functions on Y. If there exists a constant $M > 0$ such that

$$\|Tf\|_q \le M\|f\|_p \qquad (f \in D(T) \cap L^p(X)),$$

we say that T is of **type** (p,q) **with constant** M. As usual, we assume $1 \le p, q \le \infty$.

If T is of type (p,q) and $D(T) \cap L^p(X)$ is a dense vector subspace of $L^p(X)$, we keep the same notation T to represent the uniquely determined continuous extension $T : L^p(X) \to L^q(Y)$ of this operator.

2.4.1. Integral operators. Let $K(x,y)$ be an **integral kernel**, a complex-valued measurable function on $X \times Y$. The associated integral operator T_K is defined by

$$T_K f(x) := \int_Y K(x,y) f(y)\,dy.$$

Theorem 2.40. (a) *Under the condition*

$$\int_Y |K(x,y)|\,dy \le C < \infty \ \text{ a.e. on} X, \tag{2.17}$$

T_K is well-defined on L^∞ and it is of type (∞,∞) with constant C.

(b) *If*

$$\int_X |K(x,y)|\,dx \le C \ \text{ a.e. on} Y, \tag{2.18}$$

then T_K is well-defined on $L^1(X)$ and it is of type $(1,1)$ with constant C.

(c) *If both requirements (2.17) and (2.18) are satisfied, then T_K is a bounded linear operator on $T_K : L^p(X) \to L^p(Y)$ for every $p \in [1,\infty]$, and $\|T_K\| \le C$.*

---

[13] In 1806 Marc-Antoine Parseval published an identity for series as a self-evident fact, which he later applied to the Fourier series.

Proof. The first result follows from

$$|Tf(x)| \leq \int_Y |K(x,y)||f(y)|\,dy \leq \|f\|_\infty \int_Y |K(x,y)|\,dy$$

and from assumption (2.17).

Similarly, $\|Tf\|_1 \leq \int_X \int_Y |K(x,y)||f(y)|\,dy\,dx \leq C\|f\|_1$ in the second case.

A direct proof of the remaining case (c) is left as an exercise (Exercise 2.29). It will also be a trivial corollary of the Riesz-Thorin Theorem 2.45; see Exercise 2.30. □

Convolution operators are special instances of integral operators.

Recall that the **convolution** $f * g$ of two functions f and g on $\mathbf{R}^n$ is defined by

$$(f * g)(x) := \int_{\mathbf{R}^n} f(x-y)g(y)\,dy.$$

One has to be careful to make sure that this integral is meaningful a.e. and that it defines a measurable function. In this case we say that f and g are convolvable.

The convolution operator $f*$ is the integral operator associated to the integral kernel $K(x,y) = f(x-y)$, which is measurable on $\mathbf{R}^{2n}$ if f is measurable on $\mathbf{R}^n$. Indeed, if $F(x,y) = f(y)$, then $K = F \circ T$ is measurable on $\mathbf{R}^{2n}$ and $T(x,y) = (x+y, x-y)$ is a homeomorphism of $\mathbf{R}^{2n}$.

The following properties for convolvable functions are readily obtained from the definition:

(a) $f * g = g * f$ a.e.

(b) $\{f * g \neq 0\} \subset \{f \neq 0\} + \{g \neq 0\}$, so that, if $\operatorname{supp} f$ is compact, then

(2.19) $$\operatorname{supp} f * g \subset \operatorname{supp} f + \operatorname{supp} g.$$

(c) $\partial_k(f * g) = f * \partial_k g$ if $f \in L^1(\mathbf{R}^n)$ and $g \in C^1(\mathbf{R}^n)$ is bounded with a bounded partial derivative $\partial_k g$.

To prove (b), note that if $x \notin \{f \neq 0\} + \{g \neq 0\}$, then

$$(f * g)(x) := \int_{\{g \neq 0\}} f(x-y)g(y)\,dy$$

and for every $y \in \{g \neq 0\}$ we obtain $f(x-y) = 0$, so that $(f * g)(x) = 0$. If $\operatorname{supp} f$ is compact, then $\operatorname{supp} f + \operatorname{supp} g$ is closed, since from $x = \lim_n a_n + b_n$ with $a_n \in \operatorname{supp} f$ and $b_n \in \operatorname{supp} g$, we obtain $a_{n_k} \to a \in \operatorname{supp} f$ and $x - a = \lim_k(a_{n_k} + b_{n_k} - a_{n_k}) = b \in \operatorname{supp} g$.

As a corollary of Theorem 2.40 with $K(x,y) = g(x-y)$ and $C = \|g\|_1$, we obtain **Young's inequality**

(2.20) $\|f * g\|_p \leq \|f\|_p \|g\|_1 \qquad (1 \leq p \leq \infty,\ f \in L^p(\mathbf{R}^n),\ g \in L^1(\mathbf{R}^n)).$

By Hölder's inequality, we also have

(2.21) $\|f * g\|_\infty \leq \|f\|_p \|g\|_{p'} \qquad (1 \leq p \leq \infty,\ f \in L^p(\mathbf{R}^n),\ g \in L^{p'}(\mathbf{R}^n)).$

2.4.2. Summability kernels on $\mathbf{R}^n$. A **summability kernel** on $\mathbf{R}^n$ is a family $\{K_\lambda\}_{\lambda\in\Lambda}$ of integrable functions which satisfy

(1) $\Lambda \subset (0,\infty)$ and $0 \in \bar{\Lambda}$,

(2) $\int_{\mathbf{R}^n} K_\lambda(x)\,dx = 1$,

(3) $\sup_\lambda \|K_\lambda\|_1 < \infty$,

(4) $\lim_{\lambda\to 0} \int_{|x|\geq R} |K_\lambda(x)|\,dx = 0$ for all $R > 0$.

Of course, for positive summability kernels assumption (3) is redundant.

The following result justifies our also saying that a summability kernel is an **approximation of the identity**:

Theorem 2.41. *Let $\{K_\lambda\}_{\lambda\in\Lambda}$ be a summability kernel on $\mathbf{R}^n$.*

(a) *If f is a continuous function on $\mathbf{R}^n$ and $\lim_{|x|\to\infty} f(x) = 0$, then $\lim_{\lambda\to 0} K_\lambda * f = f$ uniformly on $\mathbf{R}^n$.*

(b) *If $f \in L^p(\mathbf{R}^n)$ for some $1 \leq p < \infty$, then $\lim_{\lambda\to 0} \|K_\lambda * f - f\|_p = 0$.*

Proof. (a) For $M > 0$, let

$$I(M) := \int_{|y|\leq M} |f(x-y) - f(y)||K_\lambda(y)|\,dy$$

and

$$J(M) := \int_{|y|\geq M} |f(x-y) - f(y)||K_\lambda(y)|\,dy.$$

Then, using property (2) of a summability kernel,

$$|(f * K_\lambda)(x) - f(x)| = \Big|\int_{\mathbf{R}^n} [f(x-y) - f(y)]K_\lambda(y)\,dy\Big| \leq I(M) + J(M).$$

For any $\varepsilon > 0$, since f is uniformly continuous, we can find M such that, by (3),

$$I(M) \leq \sup_{|y|\leq M,\, x\in\mathbf{R}^n} |f(x-y) - f(y)|\|K_\lambda\|_1 \leq \frac{\varepsilon}{2}.$$

Then, by property (4), we can select $\delta > 0$ so that, if $|\lambda| \leq \delta$,

$$J(M) \leq 2\|f\|_\infty \int_{|y|\geq M} |K_\lambda(y)|dy \leq \frac{\varepsilon}{2}$$

and it follows that $|f * K_\lambda(x) - f(x)| \leq \varepsilon$ for all $x \in \mathbf{R}^n$.

(b) Using (2) as before, we obtain

$$\|K_\lambda * f - f\|_p^p \le \int \Big(\int |(\tau_y f - f)(x)||K_\lambda(y)|\,dy \Big)^p\,dx$$

and, by writing $|K_\lambda| = |K_\lambda|^{1/p}|K_\lambda|^{1/p'}$ if $1 < p < \infty$, an application of Hölder's inequality and property (3) gives

$$\|K_\lambda * f - f\|_p^p \le C \int_{\mathbf{R}^n} \|\tau_y f - f\|_p^p |K_\lambda(y)|\,dy$$

with $C = \sup_\lambda \|K_\lambda\|_1^{p/p'}$. We continue as in (a) and we write

$$\int_{\mathbf{R}^n} \|\tau_y f - f\|_p^p |K_\lambda(y)|\,dy \le I(M) + J(M),$$

with M small so that

$$I(M) = \int_{|y|\le M} \|\tau_y f - f\|_p^p |K_\lambda(y)|\,dy \le \varepsilon^p/2,$$

since $\|\tau_y f - f\|_p^p \to 0$ if $y \to 0$, by Theorem 2.14. The proof ends by observing that

$$J(M) \le 2\|f\|_p^p \int_{|y|\ge M} |K_\lambda(y)|\,dy \le \varepsilon^p/2$$

for λ large enough. □

It is readily checked that a summability kernel on $\mathbf{R}^n$ is obtained from a single positive integrable function K such that $\int_{\mathbf{R}^n} K(x)\,dx = 1$ by defining

$$K_t(x) := \frac{1}{t^n} K\Big(\frac{x}{t}\Big) \qquad (t > 0). \tag{2.22}$$

Example 2.42. The **Poisson kernel** on $\mathbf{R}$ is the summability kernel

$$P_t(x) = \frac{1}{\pi}\frac{x}{t^2 + x^2} \qquad (t > 0),$$

obtained from the function

$$P(x) = \frac{1}{\pi}\frac{x}{1 + x^2}.$$

2.4.3. Periodic summability kernels.

Summability kernels can also be considered on a finite interval $I \subset \mathbf{R}$, say $I = [a, a+T)$. To define a convolution, we extend every function $f : I \to \mathbf{C}$ to the whole line $\mathbf{R}$ by periodicity, and we associate to every T-periodic function $f : \mathbf{R} \to \mathbf{C}$ a function $F : \mathbf{T} \to \mathbf{C}$ on $\mathbf{T} = \{z \in \mathbf{C};\ |z| = 1\}$ by the relation

$$f(t) = F(e^{2\pi i t/T}).$$

This is a bijective correspondence, and $F \in \mathcal{C}(\mathbf{T})$ if and only if $f \in \mathcal{C}_T(\mathbf{R})$, this notation meaning that f is continuous on $\mathbf{R}$ and T-periodic.

We will also write $L^p(\mathbf{T})$ to represent $L^p(a, a+T)$ or $L^p_T(\mathbf{R})$, the linear space of all T-periodic functions which are in L^p when restricted to an interval $(a, a+T)$. In the case $1 \leq p < \infty$, since $\mathcal{C}_c(a, a+T)$ is dense in $L^p(a, a+T)$, $\mathcal{C}(\mathbf{T})$ is also dense in $L^p(\mathbf{T})$.

As an example, note that $F(z) = z^k$ means that $f(t) = e^{2\pi ikt/T}$ $(k \in \mathbf{Z})$ and $P(z) = \sum_{k=-N}^{k=N} c_k z^k$ represents a trigonometric polynomial. We will denote $e_k(t) = e^{2\pi it/T}$, so that $e_k = z^k$ when we identify f and F, and $e_{-k} = \bar{z}^k$.

It will be convenient to use the notation

$$\int_{\mathbf{T}} f(t)\,dt := \frac{1}{T}\int_a^{a+T} f(t)\,dt$$

and to define the norm of $L^p(\mathbf{T})$ as $\|f\|_p = (\int_{\mathbf{T}} |f(t)|^p\,dt)^{1/p}$ if $1 \leq p < \infty$. Then $\|e_k\|_p = 1$ and $\|f\|_1 \leq \|f\|_p$.

The convolution on $\mathbf{T}$ is defined by

$$(f * g)(t) = \int_T f(s)g(t-s)\,ds$$

when $f, g \in L^1(\mathbf{T})$.

A **summability kernel** or **approximation of the identity** on $\mathbf{T}$ is a family of functions $\{K_\lambda\}_{\lambda\in\Lambda}$ in $L^1(\mathbf{T})$ satisfying the following:

(1) Λ is an unbounded subset of $(0, \infty)$ (for convenience we will let $\lambda \to \infty$).

(2) $\int_{\mathbf{T}} K_\lambda(x)\,dx = 1$.

(3) $\sup_\lambda \|K_\lambda\|_1 < \infty$.

(4) $\lim_{\lambda\to\infty} \int_\delta^{T-\delta} |K_\lambda(x)|\,dx = 0$ for all $0 < \delta < \pi$.

The proof of the following result is exactly the same as that of Theorem 2.41:

Theorem 2.43. *Let $\{K_\lambda\}_{\lambda\in\Lambda}$ be a summability kernel on $\mathbf{T}$.*

(a) *If $f \in \mathcal{C}(\mathbf{T})$, then $\lim_{\lambda\to\infty} K_\lambda * f = f$ uniformly on $\mathbf{T}$.*

(b) *If $f \in L^p(\mathbf{T})$ for some $1 \leq p < \infty$, then $\lim_{\lambda\to\infty} \|K_\lambda * f - f\|_p = 0$.*

If $f \in L^1(\mathbf{T})$,

$$f \sim \sum_{k=-\infty}^{\infty} c_k(f)e^{2\pi ikt/T}$$

means that $c_k(f) = \int_T f(t)e_{-k}(t)\,dt$ are the Fourier coefficients of f and that $\sum_{k=-\infty}^{\infty} c_k(f)e^{2k\pi it/T}$ is the classical Fourier series of f.

Our next aim is to show how summability kernels appear when studying the convergence of these Fourier series.

To study the possible convergence of the Fourier sums

$$\begin{aligned} S_N(f,x) &= \sum_{k=-N}^{N} \int_{\mathbf{T}} f(t)e^{-2\pi ikt/T} dt\, e^{2\pi ikx/T} \\ &= \int_{\mathbf{T}} f(t)\Big(\sum_{k=-N}^{N} e^{2\pi ik(x-t)/T}\Big)\, dt, \end{aligned}$$

by the change of variable $y = 2\pi x/T$ we can and will assume that $T = 2\pi$.

We will denote by

$$D_N(t) := \sum_{k=-N}^{N} e^{kit} = 1 + 2\sum_{n=1}^{N} \cos(nt) \tag{2.23}$$

a sequence of trigonometric polynomials which is called the **Dirichlet kernel**, to write

$$S_N(f) = f * D_N.$$

By adding the geometric sequence in (2.23), we also obtain, if $0 < |t| \le \pi$,

$$\begin{aligned} D_N(t) &= \frac{e^{i(N+1)t} - e^{-iNt}}{e^{it} - 1} = \frac{\sin[(N+\frac{1}{2})t]}{\sin\frac{t}{2}} \\ &= \frac{\sin[(N+1/2)t]}{\sin(t/2)}. \end{aligned}$$

Note that $\int_{\mathbf{T}} D_N(t)\, dt = 1$, since $\int_{\mathbf{T}} e_k(t)\, dt = 0$ if $k \neq 0$, and $D_N(-t) = D_N(t)$. For every $\delta > 0$, $\{D_N\}$ is uniformly bounded on $\delta \le |t| \le \pi$:

$$|D_N(t)| \le \frac{1}{\sin(\delta/2)} \qquad (\delta \le t \le \pi). \tag{2.24}$$

Property (3) of summability kernels fails for the Dirichlet kernel, which is not an approximation of the identity, but we obtain a summability kernel, the **Fejér kernel**, by making the averages

$$F_N := \frac{1}{N}\sum_{n=0}^{N-1} D_n. \tag{2.25}$$

Indeed,

$$F_N(t) = \frac{1}{N}\frac{\sin^2(Nt2)}{\sin^2(t/2)}$$

will follow from the identity

$$2\sin\alpha\sin\beta = \cos(\alpha-\beta) - \cos(\alpha+\beta) \tag{2.26}$$

and the properties

(a) $F_N \geq 0$,

(b) $F_N(-t) = F_N(t)$,

(c) $\frac{1}{T}\int_{-T/2}^{T/2} F_N(t)\,dt = 1$, and

(d) $\lim_{N\to\infty} \max_{\delta\leq|t|\leq T/2} F_N(t) = 0$ for every $0 < \delta < T/2$

are easily checked.

For instance, we multiply both sides of

$$NF_N(x) = \sum_{n=0}^{N-1} \frac{\sin[(n+1/2)x]}{\sin(x/2)}$$

by $2\sin^2(x/2)$ to obtain

$$2N\sin^2(x/2)F_N(x) = \sum_{n=0}^{N-1} 2\sin[(n+1/2)x]\sin(x/2),$$

and an application of (2.26) with $\alpha = (n+1/2)x$ and $\beta = x/2$ gives

$$2N\sin^2(x/2)F_N(x) = \sum_{n=0}^{N-1} [\cos(nx) - \cos((n+1)x)] = 1 - \cos(Nx).$$

Then, again by (2.26), but now with $\alpha = \beta = Nx/2$,

$$2\sin^2(Nx/2) = 1 - \cos(Nx),$$

and from both identities we obtain

$$N\sin^2(x/2)F_N(x) = \sin^2(Nx/2).$$

Now (a) and (b) follow immediately, and (c) also:

$$\frac{1}{T}\int_{-L}^{L} F_N(t)\,dt = \frac{1}{N}\sum_{n=0}^{N-1}\frac{1}{T}\int_{-L}^{L} D_n(t)\,dt = \frac{1}{N}\sum_{n=0}^{N-1} 1 = 1.$$

To check (d), note that, if $0 < \delta \leq t \leq \pi$,

$$0 \leq F_N(t) \leq \frac{1}{N}\frac{1}{\sin^2(t/2)} \leq \frac{1}{N}\frac{1}{\sin^2(\delta/2)},$$

since $\sin^2(t/2)$ is increasing on $(0,\pi]$.

The **Cesàro sums** of $f \in L^1(\mathbf{T})$ are the trigonometric polynomials

$$\sigma_N(f,x) := \frac{1}{N}\sum_{n=0}^{N-1} S_n(f,x) = f * F_N(x)$$

and according to Theorem 2.43, if $f \in L^p(\mathbf{T})$ $(1 \leq p < \infty)$, then

$$\lim_{N\to\infty} \|f - \sigma_N(f)\|_p = 0. \tag{2.27}$$

This shows that the trigonometric system $\{e_k\}_{k\in\mathbf{Z}}$ is total in $L^p(\mathbf{T})$ $(1 \leq p < \infty)$ and, as an application, we give a proof of the **Riemann-Lebesgue lemma**:

For every $f \in L^1(\mathbf{T})$, $\lim_{|k|\to\infty} c_k(f) = 0$, since this is obviously true for the trigonometric polynomials $\sigma_N(f)$ and we have $\|c(f - \sigma_N(f))\|_\infty \leq \varepsilon$ if $\|f - \sigma_N(f)\|_1 \leq \varepsilon$, so that $|c_k(f)| = |c_k(f - \sigma_N(f))| \leq \varepsilon$ for every $|k| > N$.

If $f, g \in L^1(\mathbf{T})$ have the same Fourier coefficients, then $f = g$. This fact, known as the **uniqueness theorem for Fourier coefficients**, also follows from (2.27), since $f = \lim_N \sigma_N(f) = \lim_N \sigma_N(g) = g$.

The closed subspace of ℓ^∞ which contains all the sequences $\{c_k\}$ with limit zero will be denoted c_0. Then $f \in L^1(\mathbf{T}) \mapsto \widehat{c}_0$ is injective and continuous.

2.5. The Riesz-Thorin interpolation theorem

Sometimes it is easy to show the continuity of a certain operator T when acting between two couples of Banach spaces, say $T : L^{p_0} \to L^{q_0}$ and $T : L^{p_1} \to L^{q_1}$. The interpolation theorems show that then T is also bounded between certain intermediate couples L^p and L^q.

We are going to prove Thorin's extension of the classical M. Riesz interpolation result, known as the convexity theorem, by combining techniques of real analysis with the maximum modulus property of analytic functions.[14]

The following result will be used in the proof of the interpolation theorem:

Theorem 2.44 (Three lines theorem). *Let f be a bounded analytic function in the unit strip $\mathbf{S} = \{z \in \mathbf{C};\ 0 < \Re z < 1\}$ that extends continuously to $\bar{\mathbf{S}} = \{z \in \mathbf{C};\ 0 \leq \Re z \leq 1\}$, and denote*

$$M(\vartheta) := \sup_{\Re z = \vartheta} |f(z)|.$$

Then

$$M(\vartheta) \leq M(0)^{1-\vartheta} M(1)^{\vartheta} \qquad (0 < \vartheta < 1).$$

Proof. To prove that $|f(\vartheta + iy)| \leq M(0)^{1-\vartheta} M(1)^{\vartheta}$, we can assume that neither $M(0)$ nor $M(1)$ is zero, since we can consider $M(0)+\varepsilon$ and $M(1)+\varepsilon$ and then $\varepsilon \downarrow 0$.

[14]The first interpolation theorem was proved in 1911 by the Belarusian mathematician Issai Schur, who worked in Germany for most of his life, for operators between ℓ^p spaces of type $(1,1)$ and (∞,∞) in terms of bilinear forms. The convexity theorem was proved in 1927 by Marcel Riesz, the younger brother of Frigyes Riesz who worked in Sweden for most of his life. It was extended in 1938 by Riesz's student Olov V. Thorin with a very ingenious proof, considered by Littlewood "the most impudent idea in Analysis". This theorem refers to couples of L^p spaces but, with the ideas contained in Thorin's proof, in the 1960s the Argentinian-American Alberto Calderón developed an abstract complex interpolation method for general couples of Banach spaces.

Let $F(z) = f(z)/(M_0^{1-z}M_1^z)$, which is similar to f but with $M(0) = M(1) = 1$ and $|F| \le K$ on $0 \le \Re z \le 1$. We only need to show that $|F| \le 1$.

For every $\varepsilon > 0$, let us consider $F_\varepsilon(z) = F(z)/(1+\varepsilon z)$ on the rectangle R_ε which is the fragment of $\mathbf{S}$ lying between the lines $y = \pm iK/\varepsilon$. On the boundary of $\mathbf{S}$,

$$|F_\varepsilon(z)| \le \frac{|F(z)|}{1+\varepsilon x} \le \frac{1}{1+\varepsilon x} \le 1 \qquad (x = \Re z)$$

and, when $|y| = K/\varepsilon$, also

$$|F_\varepsilon(z)| \le \frac{|F(z)|}{\varepsilon|y|} \le \frac{K}{\varepsilon|y|} = 1.$$

By the maximum modulus theorem, $|F_\varepsilon| \le 1$ on R_ε, and the estimate

$$|F_\varepsilon(z)| \le 1$$

is also valid everywhere on $\mathbf{S}$. Then we conclude that $|F(z)| \le 1 + \varepsilon|z|$ on $\mathbf{S}$, and $|F(z)| \le 1$ by letting $\varepsilon \to 0$. $\square$

Let us apply this result to prove the interpolation theorem for operators between L^p spaces on X and Y, two measurable spaces endowed with the σ-finite measures μ and ν.

We denote by $S(X)$ the class of all simple integrable complex functions on X and by $M(Y)$ the class of all complex measurable functions on Y, and we recall that a linear operator $T : D(X) \to M(Y)$ ($D(X) \subset M(X)$) is said to be of type (p, q) with constant M if $\|Tf\|_q \le M\|f\|_p$ for every $f \in D(X)$.

We will use the fact that

$$\|f\|_q = \sup_{\|g\|_{q'} \le 1} |\langle f, g\rangle| = \sup_{\|g\|_{q'} = 1} |\langle f, g\rangle|, \text{[15]} \tag{2.28}$$

where $\langle f, g\rangle = \int_Y fg\,d\nu$. The proof is obtained from Hölder's inequality as follows:

Obviously $\sup_{\|g\|_{q'} \le 1} |\langle f, g\rangle| \le \|f\|_q$. To prove the converse estimate, normalization allows us to suppose that $\|f\|_q = 1$, and we write $f = |f|s$ with $|s| = 1$. When $1 \le q < \infty$, define $g_0 = |f|^{q-1}\bar{s}$; then $\|g_0\|_{q'} = 1$ and $\langle f, g_0\rangle = 1$.

If $q = \infty$, suppose that $M := \sup_{\|g\|_1 = 1} |\langle f, g\rangle| > \|f\|_\infty$, so that, for some $m > 0$, we can choose $A \subset \{|f| > M + 1/m\}$ such that $0 < \nu(A) < \infty$. Then $g_0 := \nu(A)^{-1}\bar{s}\chi_A \in L^1(\nu)$ would satisfy $|\langle f, g_0\rangle| > \|f\|_\infty$, which contradicts Hölder's inequality.

[15]This is the description of the L^p norm by duality. See also (4.6).

Theorem 2.45 (Riesz-Thorin). *Let $T : S(X) \to M(Y)$ be a linear operator of types (p_0, q_0) and (p_1, q_1) with constants M_0 and M_1, respectively. Then, if $0 < \vartheta < 1$, T is also of type $(p_\vartheta, q_\vartheta)$ with constant $M(\vartheta)$, where*

$$\frac{1}{p_\vartheta} = \frac{1-\vartheta}{p_0} + \frac{\vartheta}{p_1}, \quad \frac{1}{q_\vartheta} = \frac{1-\vartheta}{q_0} + \frac{\vartheta}{q_1},$$

and $M(\vartheta) \le M_0^{1-\vartheta} M_1^{\vartheta}$.

Proof. (a) Let $p = p_\vartheta$ and $q = q_\vartheta$, and consider first the case $p_0 = p_1 = p$. Note that we only need to show that, if $q_0 \le q \le q_1$, then

$$\|g\|_q \le \|g\|_{q_0}^{1-\vartheta} \|g\|_{q_1}^{\vartheta}, \tag{2.29}$$

since then, for $g = Tf$, we obtain

$$\|Tf\|_q \le \|g\|_{q_0}^{1-\vartheta} \|g\|_{q_1}^{\vartheta} \le M_0^{1-\vartheta} M_1^{\vartheta} \|f\|_{q_0}^{1-\vartheta} \|f\|_{q_1}^{\vartheta}.$$

To prove (2.29) when $1 \le q_1 < \infty$, we use Hölder's inequality with the exponents $r = q_0/(1-\vartheta)q$ and $r' = q_1/\vartheta q$ to obtain

$$\|g\|_q^q = \int |g|^{(1-\vartheta)q} |g|^{\vartheta q} \le \Big(\int |g|^{q_0}\Big)^{(1-\vartheta)q/q_0} \Big(\int |g|^{q_1}\Big)^{\vartheta q/q_1},$$

that is, $\|g\|_q^q \le \|g\|_{q_0}^{(1-\vartheta)q} \|g\|_{q_1}^{\vartheta q}$.

If $p_0 = p_1 = p$ and $q_0 < q \le q_1 = \infty$, then $\vartheta = (q - q_0)/q$ and $1 - \vartheta = q_0/q$. Hence,

$$\|g\|_q^q = \int |g|^q \le \|g\|_\infty^{q-q_0} \int |g|^{q_0} = \|g\|_\infty^{q\vartheta} \|g\|_{q_0}^{q(1-\vartheta)}.$$

(b) Now assume that $p_0 \ne p_1$, and then $1 < p < \infty$. We denote

$$\alpha(z) = \frac{1-z}{p_0} + \frac{z}{p_1}, \quad \beta(z) = \frac{1-z}{q_0} + \frac{z}{q_1},$$

so that $\alpha(\vartheta) = 1/p$ and $\beta(\vartheta) = 1/q$.

Since $\|Ts\|_q = \sup_{\|g\|_{q'}=1} |\int_Y (Ts) g \, d\nu|$, we need to prove that

$$\Big|\int_Y (Ts) g \, d\nu\Big| \le M_0^{1-\vartheta} M_1^{\vartheta} \tag{2.30}$$

for all simple functions s and g satisfying $\|s\|_p = 1$ and $\|g\|_{q'} = 1$.

Suppose first that q' is finite. Then

$$s = \sum_{n=1}^{N} a_n \chi_{A_n}, \quad g = \sum_{k=1}^{K} b_k \chi_{B_k}$$

where the A_n (and the B_k) are disjoint sets of finite measure and $a_n \neq 0 \neq b_k$. Moreover $\|s\|_p = 1$ and $\|g\|_{q'} = 1$ give us that

$$\sum_{n=1}^{N} |a_n|^p \mu(A_n) = 1, \qquad \sum_{k=1}^{K} |b_k|^p \mu(B_k) = 1.$$

Write $s(x) = |s(x)|\sigma(x)$, so that $s(x) = |a_n| \exp(i \arg a_n)$ if $x \in A_n$, and also $g(y) = |g(y)|\gamma(y)$. Then for every $z \in \mathbf{C}$ we define the simple functions

$$s_z = |s|^{\alpha(z)/\alpha(\vartheta)}\sigma, \quad g_z = |g|^{(1-\beta(z))/(1-\beta(\vartheta))}\sigma$$

and

$$F(z) = \int_Y (Ts_z)g_z \, d\nu = \sum_{n,k} |a_n|^{\alpha(z)/\alpha(\vartheta)} |b_k|^{(1-\beta(z))/(1-\beta(\vartheta))} C_{n,k}$$

with $C_{n,k} = \int_Y (T\chi_{A_n})\chi_{B_k} \exp(i \arg a_n + i \arg b_k)$, which is an entire function as a linear combination of exponentials.

The real parts of $\alpha(z)$ and $\beta(z)$ are bounded on $\mathbf{S}$ and then F is also bounded. The announced estimate will be obtained from Theorem 2.44 if we show that

$$|F(iy)| \leq M_0, \quad |F(1+iy)| \leq M_1, \tag{2.31}$$

since $F(\vartheta) = \int_Y (Ts)g \, d\nu$ and then (2.30) will follow.

From the definition of $F(z)$, T being of type (p_0, q_0) with constant M_0, Hölder's inequality gives

$$|F(iy)| \leq \|Ts_{iy}\|_{q_0} \|g_{iy}\|_{q_0'} \leq M_0 \|s_{iy}\|_{p_0} \|g_{iy}\|_{q_0'},$$

where

$$\|s_{iy}\|_{p_0}^{p_0} = \sum_{n=1}^{N} \left| |a_n|^{\alpha(iy)/\alpha(\vartheta)} \right|^{p_0} \mu(A_n) = \sum_{n=1}^{N} |a_n|^p \mu(A_n) = 1,$$

since $\Re\alpha(iy) = 1/p_0$ and $\alpha(\vartheta) = 1/p$. Similarly, $\|g\|_{q_0'}^{q_0'} = 1$ and we arrive at the first estimate $|F(iy)| \leq M_0$ in (2.31).

The same argument using the fact that T is of type (p_1, q_1) with constant M_1 yields the second estimate $|F(1+iy)| \leq M_1$. This completes the proof for q' finite.

In the case $q' = \infty$, take $g_z = g$ for all z. □

Corollary 2.46. Let T be a linear operator on $D(T) = L^{p_0}(X) + L^{p_1}(X)$ of types (p_0, q_0) and (p_1, q_1) with constants M_0 and M_1. Then, with the same notation as in Theorem 2.45, $T : L^{p_\vartheta}(X) \to L^{q_\vartheta}(Y)$ is of type $(p_\vartheta, q_\vartheta)$ with constant $M(\vartheta) \leq M_0^{1-\vartheta} M_1^{\vartheta}$, for every $0 < \vartheta < 1$.

Proof. We can assume that $p_0 < p = p_\vartheta < p_1$, since, when $p_0 = p_1$, part (a) in the proof of Theorem 2.45 applies.

Let $f \in L^p$ and consider a sequence $\{s_k\}$ of simple functions such that $|s_k| \le |f|$ and $s_k \to f$. If $E = \{|f| \ge 1\}$, we define $g = f\chi_E$, $s_k^1 = s_k\chi_E$, $h = f - g$, and $s_k^2 = s_k - s_k^1$. By dominated convergence, $s_k \to f$, $s_k^1 \to g$, and $s_k^2 \to h$ in L^p.

By taking subsequences if necessary, $Ts_k^1 \to Tg$ and $Ts_k^2 \to Th$ in L^p and a.e., and also $Ts_k \to Tf$ a.e.

If $q < \infty$, by the Fatou lemma and Theorem 2.45,

$$\|Tf\|_q \le \liminf_k \|Ts_k\|_q \le M(\vartheta)\|s_k\|_p = M(\vartheta)\|f\|_p.$$

If $q = \infty$, then $q_0 = q_1 = \infty$, and also $\|Tf\|_\infty \le \liminf_k \|Ts_k\|_\infty$. □

As an application we will prove an extension of the Young inequalities (2.20) and (2.21):

Theorem 2.47. *Let* $1 \le p, q, r \le \infty$ *and let*

$$\frac{1}{p} + \frac{1}{q} = \frac{1}{r} + 1.$$

If $f \in L^p(\mathbf{R}^n)$ *and* $g \in L^q(\mathbf{R}^n)$, *then* f *and* g *are convolvable,* $f*g \in L^r(\mathbf{R}^n)$, *and*

$$\|f * g\|_r \le \|f\|_p\|g\|_q.$$

Proof. Assume that $p \le q$, so that also $p \le r$. By (2.20), $f*$ is of type $(1, p)$ with constant $\|f\|_p$, and by (2.21) it is of type (p', ∞) with the same constant. It follows from the Riesz-Thorin Theorem 2.45 that it is also of type (p, q) with constant $\|f\|_p$ by choosing $\vartheta = p/r$, since then $1/q = (1-\vartheta)/p' + \vartheta/1$ and $1/r = 1/\vartheta$. □

2.6. Applications to linear differential equations

The aim of this section is to present some applications of the preceding methods to solve initial value problems and boundary value problems for a second order linear equation with continuous coefficients,

$$u'' + a_1(x)u' + a_0(x)u = c(x),$$

on a bounded interval $[a, b]$. Functions are assumed to be real-valued.

This equation can be written in what is called **self-adjoint form**,

$$(pu')' - qu = f \qquad (0 < p \in \mathcal{C}^1[a,b];\ q, f \in \mathcal{C}[a,b]), \tag{2.32}$$

since, for $p(x) = \exp(\int a_1)$ such that $p' = pa_1$, our equation is equivalent to $pu'' + pa_1u' + pa_0u = pc$, that is, $(pu')' - p'u' + pa_1u' + pa_0u = pc$, so that we only need to consider $q = -pa_0$ and $f = pc$.

We write $Lu = (pu')' - qu$ for short, and by a solution of $Lu = f$ we mean a function $u \in \mathcal{C}^2[a,b]$ such that $Lu(x) = f(x)$ for every $x \in [a,b]$.

Note that the operator $L : \mathcal{C}^2[a,b] \to \mathcal{C}[a,b]$ satisfies the **Lagrange identity**

$$uLv - vLu = [p(uv' - u'v)]' = (pW)', \tag{2.33}$$

where $W = uv' - u'v$ is the **Wronskian determinant**.[16]

2.6.1. An initial value problem. Next we consider the **Cauchy problem**

$$Lu = f, \quad u(a) = \alpha, \quad u'(a) = \beta \tag{2.34}$$

where $\alpha, \beta \in \mathbf{R}$ are given and L is as above.

Integration and the assumption $u'(a) = \beta$ show that to solve (2.34) involves finding the solutions of

$$p(y)u'(y) - p(a)\beta - \int_a^y q(t)u(t)\,dt = \int_a^y f(t)\,dt \qquad (u \in \mathcal{C}^1[a,b],\ u(a) = \alpha).$$

After dividing by p, it turns out that integration and the condition $u(a) = \alpha$ show that this is equivalent to

$$u(x) = \int_a^x \frac{1}{p(s)} \int_a^s q(y)u(y)\,dyds - g(x) \qquad (u \in \mathcal{C}[a,b]),$$

where

$$g(x) = -\alpha - p(a)\beta \int_a^x \frac{ds}{p(s)} - \int_a^x \frac{1}{p(s)} \int_a^s f(y)\,dyds. \tag{2.35}$$

Note that, if $a < x \leq b$, we are dealing with the integral of a continuous function on the triangle

$$\Delta := \{(s,y);\ a \leq y \leq s \leq x\}$$

and, according to Fubini's theorem,

$$\int_\Delta v(s)w(y)\,dsdy = \int_a^x v(s) \int_a^t w(y)\,dyds = \int_a^x w(y) \int_y^x v(s)\,dsdy,$$

so that we are reduced to solving the integral equation

$$u(x) = \int_a^x q(y)u(y) \int_y^x \frac{ds}{p(s)}\,dy - g(x) \qquad (u \in \mathcal{C}[a,b]).$$

[16]If $\operatorname{supp} u \subset (a,b)$, then $\int_a^b (pW)' = 0$ and $\int_a^b u(x)(Lv)(x)\,dx = \int_a^b (Lu)(x)v(x)\,dx$, a self-adjointness property of L.

Let us consider

$$K(x,y) := q(y)\int_y^x \frac{1}{p(t)}\,dt \qquad (a \le y \le x \le b),$$

which is a continuous function on the triangle defined by $a \le y \le x \le b$, and suppose that $g \in \mathcal{C}[a,b]$ is as in (2.35). With Theorem 2.30 we have proved the following result:

Theorem 2.48. *The function $u \in \mathcal{C}^2[a,b]$ is a solution of the Cauchy problem (2.34) if and only if $u \in \mathcal{C}[a,b]$ and it satisfies the Volterra integral equation*

$$T_K u - u = g,$$

whose unique solution is

$$u = (T_K - I)^{-1} g = -\sum_{n=0}^{\infty} T_K^n g.$$

As shown in Exercise 2.20, a similar result holds for linear differential equations of higher order with continuous coefficients.

Let us recall how, from the existence and uniqueness of solutions for the Cauchy problem (2.34), it can be proved that $\operatorname{Ker} L$, the set of all solutions for the homogeneous equation, is a two-dimensional vector space:

Lemma 2.49. *Two solutions y_1 and y_2 of the equation $Lu = 0$ are linearly dependent if and only if their Wronskian, $W = y_1 y_2' - y_1' y_2$, vanishes at one point. In this case, $W(x) = 0$ for all $x \in [a,b]$.*

Proof. Indeed, if $W(\xi) = 0$ at a point $\xi \in [a,b]$, then the system

$$y_1(\xi)c_1 + y_2(\xi)c_2 = 0, \quad y_1'(\xi)c_1 + y_2'(\xi)c_2 = 0$$

has a solution $(c_1, c_2) \neq (0,0)$ and the function $v := c_1 y_1 + c_2 y_2$ is a solution of the boundary problems $Lv = 0$, $v(\xi) = v'(\xi) = 0$ on $[a,\xi]$ and on $[\xi, b]$. By the uniqueness of solutions for these Cauchy problems, it follows that $v = 0$ on $[a,b]$ and y_1, y_2 are linearly dependent.

Conversely, if $c_1 y_1 + c_2 y_2 = 0$ with $(c_1, c_2) \neq (0,0)$, then also $c_1 y_1' + c_2 y_2' = 0$, and $W(x) = 0$ for every $x \in [a,b]$ as the determinant of the homogeneous linear system

$$y_1(x)c_1 + y_2(x)c_2 = 0, \quad y_1'(x)c_1 + y_2'(x)c_2 = 0$$

with a nonzero solution. □

Theorem 2.50. *Let y_1 and y_2 be the solutions of*

$$Ly_1 = 0, \quad y_1(a) = 1, \quad y_1'(a) = 0$$

and

$$Ly_2 = 0, \quad y_2(a) = 0, \quad y_2'(a) = 1.$$

Then $\{y_1, y_2\}$ *is a basis of* $\operatorname{Ker} L$.

Proof. Since $W(a) = 1$, according to Lemma 2.49, the functions y_1 and y_2 are linearly independent.

To show that any other solution u of $Lu = 0$ is a linear combination of y_1 and y_2, note that $v := u(a)y_1 + u'(a)y_2$ is the solution of

$$Lv = 0, \quad v(a) = u(a), \quad v'(a) = u'(a),$$

and then $u = v = u(a)y_1 + u'(a)y_2$. □

2.6.2. A boundary value problem. We shall now restrict our attention to the homogeneous **boundary problem**

$$Lu = g, \quad B_1(u) = 0, \quad B_2(u) = 0, \tag{2.36}$$

where

$$Lu \equiv (pu')' - qu \qquad (0 < p \in \mathcal{C}^1[a,b];\ q \in \mathcal{C}[a,b])$$

as in (2.32) and

$$B_1(u) := A_1 u(a) + A_2 u'(a) = 0 \qquad (|A_1| + |A_2| \neq 0)$$

and

$$B_2(u) := B_1 u(b) + B_2 u'(b) = 0 \qquad (|B_1| + |B_2| \neq 0)$$

are two **separated boundary conditions** involving the two endpoints.

Note that

$$\mathcal{D} := \{u \in \mathcal{C}^2[a,b];\ B_1(u) = 0,\ B_2(u) = 0\}$$

is a closed subspace of $\mathcal{C}^2[a,b]$, endowed with the norm

$$\|u\| = \max(\|u\|_{[a,b]}, \|u'\|_{[a,b]}, \|u''\|_{[a,b]}),$$

and the restriction of L to $\mathcal{D}$, $L_{\mathcal{D}} : \mathcal{D} \to \mathcal{C}[a,b]$, is a bounded linear operator.

Our boundary problem will be solved for every $g \in \mathcal{C}[a,b]$ if we can construct an inverse for $L_{\mathcal{D}}$. To this end we need to **suppose that** L **is one-to-one on** $\mathcal{D}$; that is, we assume that $u = 0$ is the unique solution of $Lu = 0$ in $\mathcal{C}^2[a,b]$ such that $B_1(u) = B_2(u) = 0$.

Lemma 2.51. *There exist two linearly independent functions* $y_1, y_2 \in \mathcal{C}^2[a,b]$ *which are solutions of* $Lu = 0$ *and satisfy* $B_1(y_1) = 0$ *and* $B_2(y_2) = 0$.

Proof. Let y_1 and y_2 be nonzero solutions of

$$Ly_1 = 0, \qquad B_1(y_1) = 0$$

and

$$Ly_2 = 0, \qquad B_2(y_2) = 0.$$

By our assumptions, there are no nonzero solutions of $L_{\mathcal{D}}u = 0$, so that $B_1(y_2) \neq 0$ and $B_2(y_1) \neq 0$, and y_1, y_2 are linearly independent.

By the existence theorem for the Cauchy problem (Theorem 2.48) we can always find these functions y_1 and y_2. □

It is worth observing that, if $y_1, y_2 \in C^2[a,b]$ are two linearly independent solutions of the homogeneous equation $Lu = 0$, then pW is a nonzero constant, since $(pW)' = 0$ by (2.33) and $W \neq 0$ by the linear independence of y_1 and y_2.

Hence, if y_1 and y_2 are as in Lemma 2.51, then $pW = C$ is a constant. Our aim is to show that the boundary value problem (2.36) is solved by the Fredholm operator defined by the kernel

$$G(x,\xi) := \begin{cases} \frac{1}{C} y_1(\xi) y_2(x) & \text{if } a \le x \le \xi \le b, \\ \frac{1}{C} y_1(x) y_2(\xi) & \text{if } a \le \xi \le x \le b, \end{cases} \tag{2.37}$$

which is called the **Green's function**[17] of the differential operator L for the boundary conditions $B_1(u) = 0$, $B_2(u) = 0$. Note that G is a real-valued continuous function on $[a,b] \times [a,b]$ and $G(x,\xi) = G(\xi,x)$.

Theorem 2.52. *Under the assumption of L being one-to-one on $\mathcal{D}$, for every $\xi \in (a,b)$ the function $G(\cdot,\xi)$ is uniquely determined by the following conditions:*

(a) *$G(\cdot,\xi) \in C^2([a,\xi) \cup (\xi,b])$, $LG(\cdot,\xi) = 0$ on $[a,\xi) \cup (\xi,b]$, and $G(\cdot,\xi)$ satisfies the boundary conditions $B_1(G(\cdot,\xi)) = 0$, $B_2(G(\cdot,\xi)) = 0$.*

(b) *$G(\cdot,\xi) \in C[a,b]$.*

(c) *The right side and left side derivatives of $G(\cdot,\xi)$ exist at $x = \xi$ and*
$$\partial_x G(\xi+,\xi) - \partial_x G(\xi-,\xi) = \frac{1}{p(\xi)}.$$

Proof. That G satisfies (a)–(c) follows easily from the definition. Note that
$$C\partial_x G(\xi+,\xi) = y_1(\xi) y_2'(\xi), \qquad C\partial_x G(\xi-,\xi) = y_1'(\xi) y_2(\xi)$$
and then $C(\partial_x G(\xi+,\xi) - \partial_x G(\xi-,\xi)) = W$, which is equivalent to (c).

To show that G is uniquely determined by (a)-(c), we choose $\{y_1, y_2\}$ as in Lemma 2.51. By conditions (a) and since also $B_1(y_1) = B_2(y_2) = 0$, it follows that $a_2 B_1(y_2) = 0$ and $b_1 B_2(y_1) = 0$ with $B_1(y_2) \neq 0$ and $B_2(y_1) \neq 0$. Hence
$$G(\cdot,\xi) := \begin{cases} a_1(\xi) y_1 & \text{on } [a,\xi], \\ b_2(\xi) y_2 & \text{on } [\xi,b]. \end{cases}$$

[17]Named after the self-taught mathematician and physicist George Green who, in 1828, in "An Essay on the Application of Mathematical Analysis to the Theories of Electricity and Magnetism" introduced several important concepts, such as a theorem similar to Green's theorem, the idea of potential functions as used in physics, and the concept of what we call Green's functions.

Now, from our assumptions (b) and (c),

$$\begin{aligned} a_1(\xi)y_1(\xi) &= b_2(\xi)y_2(\xi), \\ b_2(\xi)y_2'(\xi) &- a_1(\xi)y_1'(\xi) = \frac{1}{p(\xi)}, \end{aligned}$$

a linear system with $W(\xi) \neq 0$ that determines $a_1(\xi)$ and $b_2(\xi)$. □

Theorem 2.53. *Under the assumption of $L_{\mathcal{D}}$ being injective, the Green's function G is defined by (2.37) and the Fredholm operator*

$$\begin{aligned} Tf(x) &:= \int_a^b G(x,\xi)f(\xi)\,d\xi \\ &= \frac{1}{C}\Big(y_1(x)\int_a^x y_2(\xi)f(\xi)\,d\xi + y_2(x)\int_a^x y_1(\xi)f(\xi)\,d\xi\Big) \end{aligned}$$

is the inverse operator $T : \mathcal{C}[a,b] \to \mathcal{D} \subset \mathcal{C}^2[a,b]$ of $L_{\mathcal{D}}$. That is, $u \in \mathcal{C}^2[a,b]$ is the unique solution of the boundary value problem

$$Lu = g, \quad B_1(u) = 0, \quad B_2(u) = 0$$

if and only if

$$u(x) = \int_a^b G(x,\xi)g(\xi)\,d\xi.$$

Proof. Suppose u is a solution of the boundary value problem and apply Green's formula

$$\int_x^y (uLv - vLu) = \Big[p(uv' - vu')\Big]_x^y,$$

obtained from the Lagrange identity (2.33) by integration, to the solution u and to $v = G(\cdot,\xi)$ on the intervals $[a, \xi - \varepsilon]$ and $[\xi + \varepsilon, b]$. Allowing $\varepsilon \downarrow 0$, we obtain

$$\int_a^\xi \Big(uLG(\cdot,\xi) - G(\cdot,\xi)Lu\Big) = \Big[p(uG(\cdot,\xi)' - G(\cdot,\xi)u')\Big]_a^{\xi-}$$

and

$$\int_\xi^b \Big(uLG(\cdot,\xi) - G(\cdot,\xi)Lu\Big) = \Big[p(uG(\cdot,\xi)' - G(\cdot,\xi)u')\Big]_{\xi+}^b.$$

Since $Lu = g$ and $LG(\cdot,\xi) = 0$, the sum of the left sides is $-\int_a^b G(\cdot,\xi)g$.

According to the properties of G, for the sum of the right sides we obtain

$$-pu\frac{1}{p} + \Big[p(uG(\cdot,\xi)' - G(\cdot,\xi)u')\Big]_a^b = -u$$

and $u(x) = \int_a^b G(x,\xi)g(\xi)\,d\xi$ holds.

Conversely, according to the definition of G, it follows by differentiation from

$$u(x) = \frac{1}{C}y_1(x) \int_a^x y_2(\xi)g(\xi)\,d\xi + \frac{1}{C}y_2(x) \int_a^x y_1(\xi)g(\xi)\,d\xi$$

that

$$\begin{aligned} u(x) &= \int_a^{x-} \partial_x G(x,\xi)g(\xi)\,d\xi + \int_{x+}^b \partial_x G(x,\xi)g(\xi)\,d\xi \\ &= \int_a^b \partial_x G(x,\xi)g(\xi)\,d\xi. \end{aligned}$$

With a similar computation, $(pu')'$ can be evaluated and, by using the properties of G, one gets $Lu(x) = \int_a^b L(G(\cdot,\xi))(x)\cdot g(\xi)\,d\xi + g(x) = g(x)$. □

2.7. Exercises

Exercise 2.1 (Leibniz formula). For one variable,

$$(fg)^{(m)} = \sum_{k=0}^{m} \binom{m}{k} f^{(k)} g^{(m-k)}.$$

In the case of n variables,

$$D^\alpha(fg) = \sum_{\beta\leq\alpha} \binom{\alpha}{\beta} (D^\beta f) D^{\alpha-\beta} g.$$

Exercise 2.2. Prove that if a vector subspace F of a topological vector space E has an interior point, then $F = E$.

Exercise 2.3. Equip a real or complex vector space E with the discrete topology, that is, the topology whose open sets are all the subsets of E. Is E a topological vector space?

Exercise 2.4. Prove that the interior of a convex subset K in a topological vector space E is also convex.

Exercise 2.5. Let c_{00} be the normed space of all finitely nonzero real-valued sequences with the "sup" norm $\|(x_1,\ldots,x_N,\ldots)\|_\infty = \sup_n |x_n|$. Find an unbounded linear form $u : c_{00} \to \mathbf{R}$ and an unbounded linear operator $T : c_{00} \to c_{00}$.

Exercise 2.6. Show that the Banach space of all bounded continuous functions on $\mathbf{R}^n$ with the "sup" norm can be considered a closed subspace of $L^\infty(\mathbf{R}^n)$, and prove that this space is not separable.

Exercise 2.7. Let $1 \leq p < \infty$. Prove that $L^p(\mathbf{R}^n)$ is separable.

Exercise 2.8. Prove that the normed space $\mathcal{C}_0(\mathbf{R}^n)$ of all continuous functions f on $\mathbf{R}^n$ such that $\lim_{|x|\to\infty} f(x) = 0$, with the usual operations and the "sup" norm, is a completion of the vector subspace $\mathcal{C}_c(\mathbf{R}^n)$ of all continuous functions f on $\mathbf{R}^n$ with a compact support endowed with the "sup" norm and that it is separable.

Exercise 2.9. Let K be a compact metric space.

(a) Prove that K is separable.

(b) Prove that $\mathcal{C}(K)$ is also separable by showing that, if $\{x_n\}$ is a dense sequence in K and $\varphi_{m,n}(x) = \max(1/m - d(x, x_n), 0)$, the countable set $\{\varphi_{m,n};\ m, n \geq 1\}$ is total in $\mathcal{C}(K)$.

Exercise 2.10. Prove that the trigonometric system $\{e_k\}_{k\in\mathbf{Z}}$ is total in $L^2(\mathbf{T})$ by showing first that $\mathcal{C}(\mathbf{T})$ is dense in $L^2(\mathbf{T})$ and that the polynomials $\sum_{k=-N}^{N} c_k z^k$ ($N \in \mathbf{N}$, $c_k \in \mathbf{C}$) are dense in $\mathcal{C}(\mathbf{T})$.

Exercise 2.11. Prove that the vector space of all C^1 functions on $[a, b]$ with the usual operations and the norm $\|f\| := \max(\|f\|_{[a,b]}, \|f'\|_{[a,b]})$ is a separable Banach space.

Exercise 2.12. For the norms $\|x\|_p := \big(\sum_{k=1}^n |x_k|^p\big)^{1/p}$ $(1 \leq p < \infty)$ and $\|x\|_\infty := \max_{k=1}^n |x_n|$ on $\mathbf{R}^n$, find the best constants α and β such that $\alpha\|x\|_\infty \leq \|x\|_p \leq \beta\|x\|_\infty$.

Exercise 2.13. If $\mathcal{C} = \mathcal{C}([0, 1]; \mathbf{R})$, compute the norm of $u \in \mathcal{C}'$ defined as

$$u(f) := \int_0^1 f(t)g(t)\,dt$$

with $g := \sum_{n=1}^{\infty} (-1)^n \chi_{(1/(n+1),1/n)}$ and show that $u(B_{\mathcal{C}}) = (-1, 1)$, so the norm of u is not attained by $|u(f)|$ on the closed unit ball $B_{\mathcal{C}}$ of $\mathcal{C}[0, 1]$.

Exercise 2.14. If $T_K : \mathcal{C}[c, d] \to \mathcal{C}[a, b]$ is a Fredholm operator (see 2.22) and $K \geq 0$, then $\|T_K\| = \sup_{a\leq x\leq b} \int_c^d K(x, y)\,dy$. Similarly, if T_K is a Volterra operator and $K \geq 0$, then $\|T_K\| = \sup_{a\leq x\leq b} \int_a^x K(x, y)\,dy$.

Exercise 2.15. Consider $K \in \mathcal{C}([0, 1]^2)$ and $T : L^1(0, 1) \to L^1(0, 1)$ such that

$$Tf(x) := \int_0^1 K(x, y)f(y)\,dy.$$

Prove that $\|T\| = \max_{0\leq y\leq 1} \int_0^1 |K(x, y)|\,dx$.

Exercise 2.16. In Theorem 2.40, assume that K is nonnegative. Then prove that the norm of $T_K : L^\infty \to L^\infty$ is precisely C if $\int_Y K(x, y)\,dx = C$ a.e. Similarly, if $C = \int_X K(x, y)\,dx$ a.e., show that the norm of $T_K : L^1 \to L^1$ is C.

Exercise 2.17. On $[-\pi,\pi]\times[-\pi,\pi]$ we define the integral kernel

$$K(x,y) := \sum_{n=1}^{N} a_n \cos(nx)\sin(ny)$$

and, for a given $v_0 \in \mathcal{C}[-\pi,\pi]$, consider the integral equation on $\mathcal{C}[-\pi,\pi]$

$$u(x) - \int_{-\pi}^{\pi} K(x,y)u(x)\,dx = v_0(x). \tag{2.38}$$

Prove that the Neumann series gives $(I-T)^{-1} = I+T$ for the Fredholm operator $T = T_K$. Then show that

$$u(x) := v_0(x) + \int_{-\pi}^{\pi} K(x,y)v_0(y)\,dy$$

is the unique solution of (2.38). If v_0 is an even function, then $u = v_0$.

Exercise 2.18. Find the Volterra integral equations that solve the following Cauchy problems on $[0,1]$:

(a) $u'' + u = 0, \quad u(0) = 0, \quad u'(0) = 1.$

(b) $u'' + u = \cos x, \quad u(0) = 0, \quad u'(0) = 1.$

(c) $u'' + a_1 u' + a_0 u = 0, \quad u(0) = \alpha, \quad u'(0) = \beta \ (\alpha,\beta \in \mathbf{R}).$

(d) $u'' + xu' + u = 0, \quad u(0) = 1, \quad u'(0) = 0.$

Exercise 2.19. Prove that

$$Tf(x) = \int_a^x dx_{n-1} \int_a^{x_{n-1}} dx_{n-2} \cdots \int_a^{x_1} f(t)\,dt$$

defines a Volterra operator on $\mathcal{C}[a,b]$ whose integral kernel is

$$K(x,t) = \frac{1}{(n-1)!}(x-t)^{n-1} \qquad (a \le t \le x \le b).$$

Exercise 2.20. On $[a,b]$, consider the Cauchy problem

$$u^{(n)} + a_1 u^{(n-1)} + \cdots + a_n u = f, \quad u^{(j)}(a) = c_j \quad (0 \le j \le n-1),$$

with $a_1 \ldots a_n, f \in \mathcal{C}[a,b]$ and $c_0 \ldots c_{n-1} \in \mathbf{R}$.

(a) By denoting $v = u^{(n)}$, show that the problem is equivalent to solving the Volterra integral equation

$$\int_a^x K(x,y)v(y)\,dy - v(x) = g(x),$$

where

$$K(x,y) = -\sum_{m=1}^{n} a_m(x)\frac{(x-y)^{m-1}}{(m-1)!}$$

and

$$\begin{aligned} g(x) &= c_{n-1}a_1(x) + (c_{n-1}(x-a) + c_{n-2}) \\ &\quad + \cdots + (c_{n-1}\frac{(x-a)^{n-1}}{(n-1)!} + \cdots + c_1(x-a) + c_0)a_n(x) \\ &\quad - f(x). \end{aligned}$$

Here Exercise 2.19 may be useful.

(b) Show that it follows from (a) that the Cauchy problem has a uniquely determined solution $u \in \mathcal{C}^n[a,b]$.

Exercise 2.21. Suppose f and g are nonnegative integrable functions on $\mathbf{R}^2$ such that

$$\operatorname{supp} f = \{(x,y);\ x > 0,\ 1/x \le y \le 1 + 1/x\}$$

and

$$\operatorname{supp} g = \{(x,y);\ y \ge 0\}.$$

Prove that $\operatorname{supp} f * g \not\subset \operatorname{supp} f + \operatorname{supp} g$. Why is this not in contradiction with (2.19)?

Exercise 2.22. If $f \in L^p(\mathbf{R}^n)$ and $g \in L^{p'}$, prove that $f*g$ is then uniformly continuous.

Exercise 2.23. On $L^1(\mathbf{R})$, prove that the convolution is associative but check that $(f_1 * f_2) * f_3$, $f_1 * (f_2 * f_3)$ are well-defined and they are different if $f_1 = 1$, $f_2(x) = \sin(\pi x)\chi_{(-1,1)}(x)$, and $f_3 = \chi_{[0,\infty)}$.

Exercise 2.24. Let $\{K_\lambda\}_{\lambda\in\Lambda}$ be a summability kernel on $\mathbf{R}$ and let $f \in L^\infty(\mathbf{R})$. Prove that, if f is continuous on $[a,b]$, then $\lim_{\lambda\to 0} K_\lambda * f = f$ uniformly on $[a,b]$.

Exercise 2.25. Prove that the sequence of **de la Vallée-Poussin sums**, defined as the averages

$$V_{2N} := \frac{1}{N}\sum_{n=N}^{2N-1} D_n = 2F_{2N} - F_N,$$

is a summability kernel.

Exercise 2.26. Let c_0 be the closed subspace of $\ell^\infty(\mathbf{Z})$ of all sequences $x = \{x^k\}_{k=-\infty}^{\infty}$ such that $\lim_{|k|\to\infty} c_k = 0$. Calculate the norm of the Fourier transform $c : f \in L^1(\mathbf{T}) \mapsto c(f) \in c_0$.

Exercise 2.27. On $\mathbf{R}$, let $W(x) := e^{-\pi x^2}$ and, for n variables, let $W(x) := e^{-\pi|x|^2} = W(x_1)\cdots W(x_n)$.

Prove that

$$W_t(x) := \frac{1}{t^n}e^{-\pi|x|^2/t^2} \qquad (t > 0)$$

is a C^∞ summability kernel on $\mathbf{R}^n$. It is called the **Gauss-Weierstrass kernel**.

Exercise 2.28. If $\{e_n\}_{n\geq 1}$ is an orthonormal basis of F, a closed subspace of a Hilbert space H, prove that

$$P_F(x) = \sum_{n\geq 1}(x, e_n)_H e_n.$$

Exercise 2.29. Prove Theorem 2.40 in the case $1 < p < \infty$ by showing first that $T_K f(x)$ is defined a.e. by an application of Hölder's inequality in

$$\int_Y |K(x,y)||f(y)|\,dy = \int_Y |K(x,y)|^{1/p'}|K(x,y)|^{1/p}|f(y)|\,dy.$$

Exercise 2.30. Prove Theorem 2.40(c) in the case $1 < p < \infty$ as an application of the Riesz-Thorin theorem.

Exercise 2.31. The Riesz-Thorin Theorem 2.45 with the convexity estimate $M(\vartheta) \leq M_0^{1-\vartheta}M_1^{\vartheta}$ was proved for complex L^p-spaces. Prove a corresponding result for real spaces but with the estimate $M(\vartheta) \leq 2M_0^{1-\vartheta}M_1^{\vartheta}$, by extending the real linear operator T to the complex linear operator $\tilde{T}$ defined by

$$\tilde{T}(f + ig) := Tf + iTg.$$

Exercise 2.32. Let $\ell^p(2) = \mathbf{R}^2$ with the norm

$$\|(x,y)\|_p = (|x|^p + |y|^p)^{1/p} \qquad (\max(|x|,|y|) \text{ if } p = \infty).$$

Show that the operator $T(x,y) := (x+y, x-y)$ is of type $(\infty, 1)$ with constant $M_0 = 2$ and of type $(2,2)$ with constant $M_1 = 2^{1/2}$ and that T does not satisfy the convexity estimate $M(\vartheta) \leq M_0^{1-\vartheta}M_1^{\vartheta}$.

Exercise 2.33. If $f \in L^p(\mathbf{T})$ and $1 \leq p \leq 2$ and $c(f) = \{c_k(f)\}_{k=-\infty}^{\infty}$ is the sequence of Fourier coefficients of f, defined as $c_k(f) = \int_{\mathbf{T}} f(t)e_{-k}(t)\,dt$, show that $c(f) \in \ell^{p'}$ and

$$\|c(f)\|_{p'} \leq \|f\|_p$$

first if $p = 1$ or $p = 2$ and then for every $1 < p < 2$.

References for further reading:

N. I. Akhiezer and I. M. Glazman, *Theory of linear operators in Hilbert space.*

S. Banach, *Théorie des opérations linéaires.*

S. K. Berberian, *Lectures in Functional Analysis and Operator Theory.*

R. Courant and D. Hilbert, *Methods of Mathematical Physics.*

J. Dieudonné, *Foundations of Modern Analysis.*

L. Kantorovitch and G. Akilov, *Analyse fonctionnelle.*

A. N. Kolmogorov and S. V. Fomin, *Elements of the Theory of Functions and Functional Analysis.*

F. Riesz and B. Sz. Nagy, *Leçons d'analyse fonctionelle.*

W. Rudin, *Real and Complex Analysis.*

Chapter 3

Fréchet spaces and Banach theorems

It will also be useful to consider topological vector spaces which are not normable, but the class of general topological vector spaces proves to be too wide for our needs. Usually it is sufficient to consider spaces with a vector topology which is still metrizable and complete and which can be defined by a sequence of norms or semi-norms instead of a single norm. They are called Fréchet spaces and some fundamental aspects of the theory of Banach spaces still hold on them.

An example of nonnormable Fréchet space is the vector space $\mathcal{C}(\mathbf{R})$ of all continuous functions on $\mathbf{R}$ with the uniform convergence on compact subsets of $\mathbf{R}$, defined by

$$\|f - f_n\|_{[-N,N]} := \max_{-N \le t \le N} |f(t) - f_n(t)| \to 0$$

for all $N > 0$.

The sequence of semi-norms $\|f\|_{[-N,N]}$ defines a vector topology, which can also be described by the distance associated to the Fréchet norm,

$$\|f\| = \sum_{N=1}^{\infty} \frac{1}{2^N} \frac{\|f\|_{[-N,N]}}{1 + \|f\|_{[-N,N]}},$$

which retains many of the properties of a norm.

This Fréchet norm is used to prove the basic Banach theorems concerning the continuity of linear operators.

3.1. Fréchet spaces

A **semi-norm** on the real or complex vector space E is a nonnegative function $p : E \to [0, \infty)$ with the following properties:

1. $p(\lambda x) = |\lambda| p(x)$ and
2. $p(x + y) \leq p(x) + p(y)$.

Then $p(0) = 0$, but it may happen that $p(x) = 0$ for some $x \neq 0$.

It is shown as in the case (2.1) of a norm that

$$|p(x) - p(y)| \leq p(x - y) \tag{3.1}$$

and the p-balls

$$U_p(\varepsilon) := \{x;\, p(x) < \varepsilon\} = \varepsilon U_p(1) = p^{-1}\big((-\varepsilon, \varepsilon)\big) \qquad (p \in \mathcal{P},\ \varepsilon > 0)$$

are convex sets such that $\lambda U_p(\varepsilon) \subset U_p(\varepsilon)$ if $|\lambda| \leq 1$ (it is said that they are **balanced**) and $\bigcup_{t>0} tU_p(\varepsilon) = E$ (they are **absorbing**). Note that, if $p(x) = 0$ and $x \neq 0$, the ball $U_p(\varepsilon)$ contains the whole line $[x]$.

3.1.1. Locally convex spaces. A family $\mathcal{P}$ of semi-norms on E is called **sufficient** if $p(x) = 0$ for every $p \in \mathcal{P}$ implies $x = 0$.

Theorem 3.1. *If $\mathcal{P}$ is a sufficient family of semi-norms on E, then the collection of all the finite intersections of the balls $U_p(\varepsilon)$ ($p \in \mathcal{P}$, $\varepsilon > 0$) is a local basis $\mathcal{U}$ of a vector topology $\mathcal{T}_\mathcal{P}$ on E.*

On the topological vector space $(E, \mathcal{T})$, a semi-norm p is continuous if and only if the ball $U_p(1)$ is an open set.

Proof. We say that $z \in E$ is said to be an interior point of $A \subset E$ if $z + U \subset A$ for some $U \in \mathcal{U}$ and that A is open if every $a \in A$ is an interior point of A.

It is trivial to check that the collection $\mathcal{T}_\mathcal{P}$ of these sets satisfies all the properties of a topology, which is Hausdorff since, if $x \neq y$, there exists $p \in \mathcal{P}$ such that $p(x - y) = \varepsilon > 0$, and $x + U_p(\varepsilon/2)$ and $y + U_p(\varepsilon/2)$ are disjoint, because $x + u = y + v$ with $p(u) < \varepsilon/2$ and $p(v) < \varepsilon/2$ would imply $p(x - y) = p(v - u) < \varepsilon$.

It is also very easy to show that the vector operations are continuous. In the case of the sum, $U_p(\varepsilon/2) + U_p(\varepsilon/2) \subset U_p(\varepsilon)$ and also, if U is a finite intersection of these balls, $(1/2)U + (1/2)U \subset U$ and the sum is continuous at $(0, 0) \in E \times E$. Continuity at any $(x, y) \in E \times E$ follows by translation; if $x+y+U$ is a neighborhood of $x+y$ and $V+V \subset U$, then $(x+V)+(y+V) \subset x + y + U$.

For every continuous semi-norm p, the set $U_p(\varepsilon) = p^{-1}\big((-\varepsilon, \varepsilon)\big)$ is open. Conversely, if q is a semi-norm and $U_p(1)$ is an open set, then $U_p(\varepsilon) =$

$\varepsilon U_p(1) \in \mathcal{T}_\mathcal{P}$ and it follows from (3.1) that, for every $x \in y + U_p(\varepsilon)$,

$$|p(x) - p(y)| \leq p(x - y) < \varepsilon,$$

so $p(x) \in (p(y) - \varepsilon, p(y) + \varepsilon)$ and p is continuous at y. □

A topological vector space $(E, \mathcal{T})$ is said to be a **locally convex space**[1] if there exists a sufficient family $\mathcal{P}$ of semi-norms defining the topology as in Theorem 3.1. In this case, the family of all sets

$$U_{p_1}(\varepsilon) \cap \cdots \cap U_{p_n}(\varepsilon) = \{x;\ \max\{p_1(x), \ldots, p_n(x)\} < \varepsilon\}$$

($\varepsilon > 0$, $p_i \in \mathcal{P}$, $i = 1, \ldots, n$, $n \in \mathbf{N}$) is a local basis for this topology. We will also say that $\{U_p(\varepsilon);\ p \in \mathcal{P},\ \varepsilon > 0\}$ is a **local subbasis** for this topology.

A normable topological vector space is a locally convex space with a sufficient family of semi-norms consisting in a single norm.

Example 3.2. Let X be a nonempty set. Over the vector space

$$\mathbf{C}^X = \prod_{x \in X} \mathbf{C}$$

of all complex functions $f : X \to \mathbf{C}$ ($f = \{f(x)\}_{x \in X} \in \mathbf{C}^X$), the collection of all semi-norms $p_x(f) := |f(x)|$ ($x \in X$) is sufficient and defines the product topology, which is the topology of pointwise convergence, since $p_x(f_n) \to 0$ for every $x \in X$ if and only if $f_n(x) \to 0$ for every $x \in X$.

Recall that the collection of all finite intersections of sets

$$\pi_z^{-1}(D(f_0(z); \varepsilon)) = \Big\{f = \{f(x)\}_{x \in X};\ |f(z) - f_0(z)| < \varepsilon\Big\} = f_0 + U_{p_z}(\varepsilon)$$

is a neighborhood basis of $f_0 = \{f_0(x)\}_{x \in X}$ for the product topology and, according to Theorem 3.1, it is also a neighborhood basis for the topology defined by the semi-norms p_x ($x \in X$).

Example 3.3. Over the vector space $\mathcal{C}(\mathbf{R})$ of all continuous functions on $\mathbf{R}$, the sequence

$$p_n(f) := \|f\|_{[-n,n]} = \sup_{-n \leq t \leq n} |f(t)| \qquad (n \in \mathbf{N})$$

of semi-norms is sufficient. They define the topology of the **local uniform convergence**, or uniform convergence on compact sets.

Every compact set $K \subset \mathbf{R}$ is contained in an interval $[-n, n]$, and then $\|f\|_K \leq p_n(f)$. Hence, $p_n(f - f_k) \to 0$ implies $\|f - f_n\|_K \to 0$, and this means that $f_n \to f$ uniformly on K.

[1] Considered by J. Dieudonné and L. Schwartz in the 1940s to extend the duality theory of normed spaces, locally convex spaces were the basis for the study of the distributions by L. Schwartz.

The continuity condition $\|Tx\|_F \leq C\|x\|_E$ for a linear operator between normed spaces has a natural extension for locally convex spaces:

Theorem 3.4. *Let $\mathcal{P}$ and $\mathcal{Q}$ be two sufficient families of semi-norms for the locally convex space E and F. A linear application $T : E \to F$ is continuous if and only if for every $q \in \mathcal{Q}$ there exist $p_j \in \mathcal{P}$ ($j \in J$ finite) and a constant $C \geq 0$ so that*

$$q(Tx) \leq C \max_{j\in J} p_j(x).$$

A sequence $\{x_n\} \subset E$ is convergent to 0 if and only if $p(x_n) \to 0$ for every $p \in \mathcal{P}$.

Proof. By Theorem 2.2, the continuity of T means that, if $U_q(\varepsilon)$ is a q-ball, $T(U) \subset U_q(\varepsilon)$ for some $U = U_{p_1}(\delta) \cap \cdots \cap U_{p_n}(\delta)$ ($p_j \in \mathcal{P}$). That is,

$$q(Tx) \leq C \max_{j\in J} p_j(x)$$

with $C = \varepsilon/\delta$, since $x/\alpha \in U_{p_1}(1) \cap \cdots \cap U_{p_n}(1)$ if $\alpha > \max_{j\in J} p_j(x)$, and $q(Tx/\alpha) < \varepsilon/\delta$. Hence $(\delta/\alpha)Tx \in U_q(\varepsilon)$ and $q(Tx) < \alpha\varepsilon/\delta$, and we obtain $q(Tx) \leq C \max_{j\in J} p_j(x)$ by allowing $\alpha \downarrow \max_{j\in J} p_j(x)$.

Finally, $x_n \to 0$ if and only if $x_n \in U_p(\varepsilon)$ when $n \geq \nu(p, \varepsilon)$, i.e., eventually $p(x_n) < \varepsilon$, for every $U_p(\varepsilon)$. □

Suppose that E is a locally convex space and that F is a closed subspace of E. The linear **quotient map**

$$\pi : E \to E/F$$

is defined as $\pi(x) = \tilde{x} = x + F$ and, if $\mathcal{P}$ is a sufficient family of semi-norms on E, we consider on E/F the collection $\tilde{\mathcal{P}}$ of semi-norms $\tilde{p}$ defined by

$$\tilde{p}(\tilde{x}) = \inf_{y\in\tilde{x}} p(y) = \inf_{z\in F} p(x - z),$$

for every $p \in \mathcal{P}$.

It is clear that every functional $\tilde{p}$ is a semi-norm on E/F. The family $\tilde{\mathcal{P}}$ is sufficient, since $\tilde{p}(\tilde{x}) = 0$ for all $\tilde{p}$ means that $\inf_{z\in F} p(x - z) = 0$ for all p, every ball $x + U_p(\varepsilon)$ meets F, and then $x \in \bar{F} = F$ and $\tilde{x} = 0$ in E/F.

This new topological vector space E/F is the **quotient locally convex space of E modulus F**, and the quotient map is continuous, since $\tilde{p}(\pi(x)) = \tilde{p}(\tilde{x}) \leq p(x)$ for every $\tilde{p} \in \tilde{\mathcal{P}}$. If E is a normed space, E/F is also a normed space.

In a locally convex space E with the sufficient family of semi-norms $\mathcal{P}$, a subset A is said to be **bounded** if $p(A)$ is a bounded set in $\mathbf{R}$ for every $p \in \mathcal{P}$ or, equivalently, if for every neighborhood U of 0 we have $A \subset rU$ for some $r > 0$.

Indeed, if every p is bounded on A and if U is a neighborhood of 0, we can choose $U_{p_1}(\varepsilon)\cap\cdots\cap U_{p_n}(\varepsilon)\subset U$ and, if $p_j < M_j$ on A $(1\le j\le n)$, then $A\subset rU$ by choosing $r > M_j/\varepsilon$.

Conversely, suppose that A satisfies the condition $A\subset rU_p(1)=U_p(r)$. Then r is an upper bound for $p(A)$.

A compact subset K of the locally convex space E is bounded, since every continuous semi-norm p is bounded on K. If every bounded closed subset of E is compact, it is said that E has the **Heine-Borel property**. As we have seen in Theorem 2.28, a normed space has this property only if its dimension is finite.

3.1.2. Fréchet spaces. In many important examples, the sufficient family of semi-norms will be finite (this is the case of normable spaces, with a single norm) or countable (as in Example 3.3). In this case the locally convex space is said to be **countably semi-normable** and then, if $\mathcal{P}=\{p_1,p_2,\dots,p_n,\dots\}$ is a sufficient sequence of semi-norms, the sequence of all $U_n=\bigcap_{k=1}^n U_{p_k}(1/n)$ forms a decreasing countable local basis of open, convex, and balanced sets, and $\bigcap_{n=1}^\infty U_n=\{0\}$, since the topology is Hausdorff.

We can assume $p_1\le p_2\le\cdots$ since for the increasing sequence of semi-norms $q_1=p_1$, $q_2=\sup(p_1,p_2)$, $q_3=\sup(p_1,p_2,p_3)$, … we obtain $U_{q_n}(\varepsilon)=U_{p_1}(\varepsilon)\cap\cdots\cap U_{p_n}(\varepsilon)$ and both families of semi-norms define the same topology.

The **Fréchet norm** associated to the sequence of semi-norms will be the function

$$\|x\| := \sum_{n=1}^\infty 2^{-n}\frac{p_n(x)}{1+p_n(x)}.$$

It is not a true norm, but it has the following properties:

1. $x=0$ if $\|x\|=0$, since $p_n(x)=0$ $\forall n$ implies $x=0$.
2. $\|-x\|=\|x\|$, since $p_n(-x)=p_n(x)$.
3. $\|x+y\|\le\|x\|+\|y\|$.

To show this last property, note that

$$\frac{a}{1+a}\le\frac{b}{1+b}$$

if $0<a\le b$, and

$$\frac{p_n(x+y)}{1+p_n(x+y)}\le\frac{p_n(x)+p_n(y)}{1+p_n(x)+p_n(y)}\le\frac{p_n(x)}{1+p_n(x)}+\frac{p_n(y)}{1+p_n(y)},$$

since $1+p_n(x)+p_n(y)\ge 1+\max(p_n(x),p_n(y))$.

Then $d(x,y):=\|y-x\|$ is a distance and $d(x+z,y+z)=d(x,y)$.

Theorem 3.5. *Let $\|\cdot\|$ be the Fréchet norm of a countably semi-normable locally convex space E. Then the distance $d(x,y) = \|x-y\|$ defines the topology of E.*

Proof. Let us show that every ball $B_d(0,\delta)$ for the distance d contains some $U \in \mathcal{U}$ and that, conversely, every $U_{p_m}(\varepsilon)$ contains a ball $B_d(0,\delta)$.

(a) $B_d(0, 1/2^k)$ contains $U_{p_{k+1}}(1/2^{k+1})$:

If $p_{k+1}(x) < 1/2^{k+1}$, then $p_1(x) \leq \cdots \leq p_{k+1}(x) < 1/2^{k+1}$ and

$$\|x\| < \sum_{n=1}^{k+1} 2^{-n} \frac{1}{2^n} \frac{1/2^{k+1}}{1+1/2^{k+1}} + \sum_{n=k+2}^{\infty} \frac{1}{2^n} < \frac{1}{2^k}.$$

(b) The p_m-ball $U_{p_m}(1/2^k)$ contains $B_d(0, 1/2^{k+m+1})$ since

$$2^{-m} \frac{p_m(x)}{1+p_m(x)} < 1/2^{k+m+1}$$

if $\|x\| < 1/2^{k+m+1}$; therefore $p_m(x)/(1+p_m(x)) < 1/2^{k+1}$ and it follows that $p_m(x) < 1/2^k$. □

Let E be a countably semi-normable locally convex space with the sufficient family of semi-norms $\mathcal{P}$. We say that $\{x_n\} \subset E$ is a Cauchy sequence if eventually $x_k - x_m \in U$ for every 0-neighborhood U or, equivalently, $p(x_k - x_m) \to 0$ for every $p \in \mathcal{P}$.

By Theorem 3.5, every Cauchy sequence of E is convergent if and only if the metric space E with the distance d defined by the Fréchet norm $\|\cdot\|$ is complete. Then we say that E is a **Fréchet space.**[2]

Every Banach space is a normable Fréchet space, but there are many other important Fréchet spaces that are not normable.

Theorem 3.6. *The countably semi-normable space $\mathcal{C}(\mathbf{R})$ of Example 3.3 with the sufficient sequence of semi-norms $p_n(f) = \|f\|_{[-n,n]}$ is a Fréchet space. It is not normable, since there is no norm $\|\cdot\|$ with the property*

$$p_n(f) \leq C_n\|f\| \qquad (f \in \mathcal{C}(\mathbf{R}),\ C_n > 0 \text{ constant}),$$

for all $n \in \mathbf{N}$.

Proof. If $\mathcal{C}(\mathbf{R})$ were normable, by the norm $\|\cdot\|$, then for f such that $f(n) = nC_n$ we would have $nC_n \leq |f(n)| \leq C_n\|f\|$ and $n \leq \|f\|$ for all $n \in \mathbf{N}$.

[2]Named after the French mathematician Maurice Fréchet, who with his 1906 dissertation titled "Sur quelques points du calcul fonctionnel" is considered one of the founders of modern functional analysis. See footnote 2 in Chapter 1.

If $\{f_k\}$ is a Cauchy sequence, it is uniformly convergent on every interval $[-n,n]$ to a certain continuous function g_n and there is a common extension of all of them to a function g on $\mathbf{R}$, since g_n is the restriction of g_{n+1}. Obviously, $f_k \to g$ uniformly on every $[-n,n]$. □

The construction of Example 3.3 can be extended to the setting of the class $\mathcal{C}(\Omega)$ of all complex continuous functions $f : \Omega \to \mathbf{C}$ on an open subset Ω of $\mathbf{R}^n$, which is the union of an increasing sequence of the compact sets of $\mathbf{R}^n$

$$K_m = \bar{B}(x_0, m) \cap \{x \in \Omega;\ d(x, \Omega^c) \geq \frac{1}{m}\}. \tag{3.2}$$

Every K_m is a subset of the interior G_{m+1} of the next one, K_{m+1}. If $\Omega = \mathbf{R}^n$, $K_m = \bar{B}(x_0, m)$. Every compact set $K \subset \Omega$ is covered by K_N for some N, since $\Omega = \bigcup_{m=1}^{\infty} G_m$ and $K \subset \bigcup_{m=1}^{N} G_m \subset K_N$.

It is also easily shown, as for $\mathcal{C}(\mathbf{R})$, that $\mathcal{C}(\Omega)$, as a complex vector space with the usual operations and with the topology associated to the sufficient increasing sequence of semi-norms $q_m(f) = \|f\|_{K_m}$, is a Fréchet space. Since $\|f\|_K \leq q_N(f)$ if $K \subset K_N$, the family of all semi-norms $\|\cdot\|_K$ (K any compact subset of Ω) defines the same topology on $\mathcal{C}(\Omega)$, which is again the topology of the local uniform convergence.

To define the important example of C^∞ functions, we introduce some terminology that will be useful in the future.

Let us denote $\alpha = (\alpha_1, \ldots, \alpha_n) \in \mathbf{N}^n$ and $D^\alpha = \partial_1^{\alpha_1} \cdots \partial_n^{\alpha_n}$ as in Example 2.7. For every compact subset K of $\mathbf{R}^n$, we define

$$p_{K,\alpha}(f) = \|D^\alpha f\|_K = \sup_{x \in K} |D^\alpha f(x)|$$

if $D^\alpha f$ exists.

Then $\mathcal{E}(\Omega)$ will represent the locally convex space of all C^∞ functions on Ω with the topology defined by the sufficient increasing sequence of semi-norms

$$q_j(f) = \sum_{|\alpha| \leq j} \|D^\alpha f\|_{K_j} = \sum_{|\alpha| \leq j} p_{K_j,\alpha}(f) \qquad (j = 1, 2, \ldots).$$

If K is any compact subset of Ω, then $K \subset K_j$ for all $j \geq N$, for some $N \in \mathbf{N}$, so that $p_{K,\alpha}(f) \leq q_j(f)$ for some j, and the topology of $\mathcal{E}(\Omega)$ is also the topology of the semi-norms $p_{K,\alpha}$. A sequence $\{f_k\} \subset \mathcal{E}(\Omega)$ is convergent to 0 if $D^\alpha f_k \to 0$ uniformly on every compact of Ω, for every $\alpha \in \mathbf{N}^n$.

For every compact subset K of Ω, the collection of functions

$$\mathcal{D}_K(\Omega) = \{f \in \mathcal{E}(\Omega);\ \operatorname{supp} f \subset K\} \tag{3.3}$$

is clearly a closed subspace of $\mathcal{E}(\Omega)$. By extending functions by zeros, we can suppose that $\mathcal{D}_K(\Omega) \subset \mathcal{E}(\mathbf{R}^n)$.

Theorem 3.7. *$\mathcal{E}(\Omega)$ is a Fréchet space.*

Proof. A sequence $\{f_j\}_{j=1}^{\infty} \subset \mathcal{E}(\Omega)$ is a Cauchy sequence if and only if every $\{D^{\alpha} f_j\}_{j=1}^{\infty}$ is uniformly Cauchy over every compact set K and then it is uniformly convergent on K. Hence, $D^{\alpha} f_j \to f^{\alpha}$ uniformly on compact sets and then $D^{\alpha} f = f^{\alpha}$ if $f = f^0$. This means that $f_j \to f$ in $\mathcal{E}(\Omega)$. □

Similarly, for any $m \in \mathbf{N}$ the vector space $\mathcal{E}^m(\Omega)$ of all C^m functions $f : \Omega \to \mathbf{C}$ with the topology defined by the semi-norms

$$p_K(f) := \sum_{|\alpha| \leq m} \|D^{\alpha} f\|_K$$

is also a Fréchet space. A sequence $\{f_k\} \subset \mathcal{E}^m(\Omega)$ is convergent to 0 if $D^{\alpha} f_k \to 0$ uniformly on every compact of Ω, for every $|\alpha| \leq m$.

Also, $\mathcal{D}_K^m(\Omega) = \{f \in \mathcal{E}^m(\Omega);\ \operatorname{supp} f \subset K\}$ is a closed subspace of $\mathcal{E}^m(\Omega)$, for every compact subset K of Ω.

3.2. Banach theorems

We are going to present some profound results that prove to be very useful to show the continuity of linear operators. The ideas are mainly due to Stefan Banach and depend on the following Baire category principle concerning general complete metric spaces.

Theorem 3.8 (Baire). *If $\{G_n\}_{n=1}^{\infty}$ is a sequence of dense and open subsets in a complete metric space M, then $A := \bigcap_{n=1}^{\infty} G_n$ is also dense in M. It is said that A is a dense $\mathcal{G}_{\delta}$-set.*

Proof. We need to show that $A \cap G \neq \emptyset$ for every nonempty open set G in M.

Since G_1 is dense, we can find a ball $\bar{B}(x_1, r_1) \subset G \cap G_1$ with $r_1 < 1$. Similarly, we can choose $\bar{B}(x_2, r_2) \subset B(x_1, r_1) \cap G_2$ and $r_2 < 1/2$, since G_2 is dense. In this way, by induction, we can produce

$$\bar{B}(x_n, r_n) \subset B(x_{n-1}, r_{n-1}) \cap G_n, \qquad r_n < 1/n.$$

Then $d(x_p, x_q) \leq 2/n$ if $p, q \geq n$, since $x_p, x_q \in B(x_n, r_n)$, and $\{x_n\}$ is a Cauchy sequence in M. But M is complete, so we have $\lim_n x_n = x$.

Since x_k lies in the closed set $\bar{B}(x_n, r_n)$ if $k \geq n$, it follows that x lies in each $\bar{B}(x_n, r_n) \subset G_n$ and $x \in A$. Also $x \in \bar{B}(x_1, r_1) \subset G$ and $A \cap G \neq \emptyset$. □

Corollary 3.9. Let F_n ($n \in \mathbf{N}$) be a countable family of closed subsets of a complete metric space M containing no interior points for every $n \in \mathbf{N}$. Then the union $B = \bigcup_{n=1}^{\infty} F_n$ has no interior point either.

Proof. Every open set $G_n := F_n^c$ is dense since, for any nonempty open set G, $G \not\subset F_n$, so that $G \cap G_n \neq \emptyset$. According to Theorem 3.8, $A := \bigcap_{n=1}^{\infty} G_n$ is dense, so that $A^c = B$ does not contain any nonempty open set. □

Theorem 3.10 (Banach-Schauder[3]). *Let E and F be two Fréchet spaces. If $T : E \to F$ is a continuous linear operator and $T(E) = F$, then T is an open mapping, that is, $T(G)$ is open in F if G is an open subset of E.*

In particular, if T is a bijective and continuous linear mapping, then T^{-1} is also continuous.

Proof. Let $\|\cdot\|$ be a Fréchet norm on E and let $\mathcal{U}$ be the local basis of E that contains all the open balls $U_r = \{x \in E;\ \|x\| < r\}$, and let $\mathcal{V}$ be a similar local basis in F.

We want to prove that every $T(U_r)$ is a zero neighborhood in F, since then, if G is an open subset of E and $Ta \in T(G)$ ($a \in G$), there exists $B(a,r) = a + U_r \subset G$ that satisfies $T(a + U_r) \subset T(G)$ and Ta is an interior point in $T(G)$, since $T(a + U_r) = Ta + T(U_r)$ is a neighborhood of Ta.

Let us start by showing that every $\overline{T(U_r)}$ is a zero neighborhood in F. For every $x \in E$, $(1/n)x \to 0$ and $x \in nU_{r/2}$ for some $n \in \mathbf{N}$, so that $E = \bigcup_{n=1}^{\infty} nU_{r/2}$ and $F = T(E) = \bigcup_{n=1}^{\infty} nT(U_{r/2}) = \bigcup_{n=1}^{\infty} n\overline{T(U_{r/2})}$. By the corollary of Baire's theorem, $\overline{T(U_{r/2})}$ has at least one interior point y and we can find $y + V \subset \overline{T(U_{r/2})}$ ($V \in \mathcal{V}$). We have $y \in \overline{T(U_{r/2})} = -\overline{T(U_{r/2})}$, so that $V \subset -y + \overline{T(U_{r/2})} \subset \overline{T(U_{r/2})} + \overline{T(U_{r/2})} \subset \overline{T(U_r)}$ and $\overline{T(U_r)}$ is a neighborhood of $0 \in F$.

If $s > 0$ and $r = s/2$, let us prove now that $V_\sigma \subset \overline{T(U_r)}$ implies $V_\sigma \subset T(U_s)$, and then $T(U_s)$ will be a neighborhood of 0.

Let $y \in V_\sigma$, so that $\|y\| < \sigma$. To prove that $y \in T(U_s)$, we will find $x \in U_s$ so that $y = Tx$.

Write $s_1 = s = \sum_{n=1}^{\infty} r_n$ with $r_n > 0$ and $r_1 = r$. We know that $\overline{T(U_{r_n})}$ is a neighborhood of 0 and, for every $n \geq 2$ there is a ball $V_{\sigma_n} \subset \overline{T(U_{r_n})}$ and we can suppose that $\sigma_n \downarrow 0$.

We have $y \in V_\sigma \subset \overline{T(U_{r_1})}$, so that $\|y - Tz_1\| < \sigma_2$ with $z_1 \in U_{r_1}$, i.e., $\|z_1\| < r_1$.

[3] Also known as the **open mapping theorem** in functional analysis, it was published in 1929 by S. Banach by means of duality in the Banach spaces setting in the first issue of the journal Studia Mathematica and also in 1930 in the same journal by J. Schauder, essentially with the usual direct proof that we have included here.

By induction,

$y - Tz_1 \in V_{\sigma_2} \subset \overline{T(U_{r_2})}$, $\|y - Tz_1 - Tz_2\| < \sigma_3$ with $\|z_2\| < r_2$,

...

$y - Tz_1 - \cdots - Tz_{n-1} \in V_{\sigma_n} \subset \overline{T(U_{r_n})}$, $\|y - \cdots - Tz_n\| < \sigma_{n+1}$ with $\|z_n\| < r_n$,

....

There exists $x = \sum_{n=1}^{\infty} z_n$, since $\|\sum_{n=p}^{q} z_n\| \leq \sum_{n=p}^{\infty} r_n \to 0$ as $p \to \infty$ and partial sums of the series form a Cauchy sequence in the Fréchet space E.

The map T is linear and continuous; thus

$$Tx = T\Big(\lim_{n\to\infty} \sum_{k=1}^{n} z_k\Big) = \lim_{n\to\infty} (Tz_1 + \cdots + Tz_n) = y.$$

Moreover $\|x\| = \lim_{n\to\infty} \|\sum_{k=1}^{n} z_k\| < \sum_{n}^{\infty} r_n = s$, and $x \in U_s$. □

If $T : E \to F$ is a continuous function between two metric spaces, it is obvious that its graph

$$\mathcal{G}(T) := \{(x, y) \in E \times F;\ y = Tx\}$$

is a closed subset of the product space $E \times F$, since if $(x_n, Tx_n) \to (x, y)$ in $E \times F$, then $Tx_n \to Tx$ and $y = Tx$.

That the converse is true if T is a linear operator between two Fréchet spaces is a corollary of the Banach-Schauder theorem:

Theorem 3.11 (Closed graph theorem). *Let E and F be two Fréchet spaces. A linear map $T : E \to F$ is continuous if and only if its graph $\mathcal{G}(T)$ is closed in $E \times F$.*

Proof. First let us show that $E \times F$, with the product topology, is a Fréchet space.

Let $\|\cdot\|_E$ and $\|\cdot\|_F$ be the Fréchet norms defined by the increasing sufficient sequences of semi-norms $\{p_n\}$ and $\{q_n\}$ for E and F, respectively. The corresponding product topology on $E \times F$ is the topology associated to the sufficient sequence of semi-norms $r_n(x, y) := \max(p_n(x), q_n(y))$, since the r_n-balls $U_{r_n}(\varepsilon) = U_{p_n}(\varepsilon) \times U_{q_n}(\varepsilon)$ form a local basis for both topologies. Then $\{(x_n, y_n)\}$ is convergent in $E \times F$ if and only if both $\{x_n\} \subset E$ and $\{y_n\} \subset F$ are convergent, and $E \times F$ is complete if and only if E and F are complete.

If the vector subspace $\mathcal{G}(T)$ is closed in $E \times F$, it is a Fréchet space with the restriction of the product topology in $E \times F$, and we denote $\pi_1(x, Tx) = x$ and $\pi_2(x, Tx) = Tx$ the restrictions of the projections of $E \times F$ on E and

F, respectively. They are two continuous linear maps $\pi_1 : \mathcal{G}(T) \to E$ and $\pi_2 : \mathcal{G}(T) \to F$.

Now π_1 is bijective and, by the Banach-Schauder theorem, the inverse map $\pi_1^{-1} : x \mapsto (x, Tx)$ is continuous from E to $\mathcal{G}(T)$. But $Tx = \pi_2(\pi_1 x)$ and T is continuous as a composition of two continuous maps. □

Remark 3.12. For a map $T : E \to F$ between two metric spaces E and F, the graph $\mathcal{G}(T)$ is closed in $E \times F$ if and only if

$$x_n \to x \text{ and } Tx_n \to y \Rightarrow y = Tx.$$

This condition is clearly weaker than a continuity assumption, which means that

$$x_n \to x \Rightarrow \exists y = \lim Tx_n \text{ and } y = Tx.$$

Note that in the case of a linear map T between normed or Fréchet spaces, if

$$x_n \to 0 \text{ and } Tx_n \to y \Rightarrow y = 0,$$

then the graph is closed.

In a Hilbert space H, an orthogonal projection is a bounded linear operator P such that $P^2 = P$ and $(Px, y)_H = (x, Py)_H$. These last two properties characterize the orthogonal projections:

Theorem 3.13. *Let H be a Hilbert space, and let $P : H \to H$ be a mapping such that $(Px, y)_H = (x, Py)_H$. Then P is linear and bounded, and it is an orthogonal projection if $P^2 = P$.*

Proof. The linearity of P is clear; see (b) in Theorem 2.35. To prove that P is bounded, suppose that $x_x \to 0$ and $Px_n \to y$; then, for any $x \in H$,

$$(x, y)_H = \lim_n (x, Px_n)_H = \lim_n (Px, x_n)_H = 0.$$

Choose $x = y$ and then $y = 0$, so that the graph of P is closed.

If $P^2 = P$, let $F = P(H)$ and $Q = I - P$. Then also $(Qx, y)_H = (x, Qy)_H$, $Q^2 = Q$, and $F = \operatorname{Ker} Q$ is a closed subspace of H. Furthermore $Q(H) = F^\perp$, since $(Px, Qy)_H = (x, P(y - Py))_H = 0$ and $x = Px + Qx = y + z \in F \oplus F^\perp$ as in Theorem 2.35(b). □

It is a well-known elementary fact that a pointwise limit of continuous functions need not be continuous, but with the Banach-Steinhaus theorem[4] we will prove that, for Banach spaces, a pointwise limit of continuous linear operators is always continuous. This theorem is an application of the following **uniform boundedness principle**:

[4]First published in 1927 by S. Banach and H. Steinhaus but also found independently by the Austrian mathematician Hans Hahn, who worked at the Universities of Vienna and Innsbruck.

Theorem 3.14. *Let $T_j : E \to F$ $(j \in J)$ be a family of bounded linear operators between two Banach spaces, E and F. Then*

(a) *either $M := \sup_{j \in J} \|T_j\| < \infty$, and the operators are uniformly bounded on the unit ball of E, or*

(b) *$\psi(x) := \sup_{j \in J} \|T_j(x)\|_F = \infty$ for every x belonging to a $\mathcal{G}_\delta$-dense set $A \subset E$.*

Proof. The level sets $G_n := \{\psi > n\}$ of the function $\psi : E \to [0, \infty]$ are open subsets of E, since $G_n = \bigcup_{j \in J}\{x;\ \|T_j x\|_F > n\}$ and every T_j is continuous. Now we consider two possibilities:

(a) If one of the sets G_n, say G_m, is not dense in E, there is a ball $\bar{B}(a, r)$ contained in G_m^c, so that

$$\psi(a + x) \le m \quad \text{whenever } \|x\|_E \le r,$$

which means that $\|T_j(a + x)\|_F \le m$ for all $j \in J$ if $\|x\|_E \le r$. Then

$$\|T_j(x)\|_F \le \|T_j(a + x)\|_F + \|T_j(a)\|_F \le 2m \qquad (\|x|_E \le r)$$

or, equivalently, $\|T_j(x)\|_F \le 2m/r$ if $\|x\|_E \le 1$, for every $j \in J$, and it follows that $M \le 2m/r < \infty$.

(b) If every G_n is dense, then $\psi(x) = \infty$ for every $x \in A := \bigcap_{n=1}^{\infty} G_n$, since then $\psi(x) > n$ for every $n \in \mathbf{N}$. By Theorem 3.8, A is a dense subset of E. □

Theorem 3.15 (Banach-Steinhaus). *Let $T_n : E \to F$ $(n \in \mathbf{N})$ be a sequence of bounded linear operators between two Banach spaces such that the sequence $\{T_n x\}$ is bounded for every $x \in E$. Suppose further that the limit $\lim_n T_n(x)$ exists in F for every point x belonging to a dense subset D of E.*

Then $T : D \to F$ such that $Tx = \lim_n T_n(x)$ extends to a bounded linear operator $T : E \to F$ such that

$$\|T\| \le \liminf_n \|T_n\|.$$

Thus, every sequence $\{T_n\} \subset \mathcal{L}(E; F)$ such that $Tx = \lim_n T_n(x)$ exists for every $x \in E$ defines a bounded operator $T \in \mathcal{L}(E; F)$.

Proof. By Theorem 3.14, $M := \sup_{n \in \mathbf{N}} \|T_n\| < \infty$.

For every $x \in E$, $\{T_n(x)\}$ is a Cauchy sequence in the Banach space F. Indeed, if $\varepsilon > 0$, there exist $z \in D$ so that $\|x - z\|_E \le \varepsilon$ and $n \in \mathbf{N}$ so that $\|T_p(z) - T_q(z)\|_F \le \varepsilon$ whenever $p, q \ge n$. Then

$$\begin{aligned}\|T_p(x) - T_q(x)\|_F &\le \|T_p(x) - T_p(z)\|_F + \|T_p(z) - T_q(z)\|_F \\ &\quad + \|T_q(z) - T_q(x)\|_F \le 2M\varepsilon + \varepsilon.\end{aligned}$$

We define $T(x) := \lim_n T_n(x)$ and $T : E \to F$ is obviously linear. Moreover,

$$\|T(x)\|_F = \lim_n \|T_n(x)\|_F \leq \liminf_n \|T_n\| \|x\|_E$$

and it follows that $\|T\| \leq \liminf_n \|T_n\|$. □

3.2.1. An application to the convergence problem of Fourier series. If $f \in L^1(\mathbf{T})$ and $c_k(f) = \int_T f(t)e_{-k}(t)\, dt$, recall that

$$c(f) = \{c_k(f)\}_{k=-\infty}^{+\infty} \in c_0$$

and obviously $\|c(f)\|_\infty \leq \|f\|_1$.

The Fourier mapping $f \in L^1(\mathbf{T}) \mapsto c(f) \in c_0$ is continuous and injective, but it cannot be exhaustive, since in this case the inverse map would also be continuous and

$$\|c(f)\|_\infty \geq \delta \|f\|_1$$

for some $\delta > 0$, which leads to a contradiction. Indeed, if D_N is the Dirichlet kernel (see (2.23)), then $\|c(D_N)\|_\infty = 1$ and $\|D_N\|_1 \to \infty$, since

$$\begin{aligned} \|D_N\|_1 &\geq \frac{2}{\pi} \int_0^\pi \left| \frac{\sin[(N+1/2)t]}{t} \right| dt = \frac{2}{\pi} \int_0^{N+1/2} \frac{|\sin t|}{t}\, dt \\ &\geq \frac{2}{\pi} \sum_{k=1}^N \frac{1}{k\pi} \int_{(k-1)\pi}^{k\pi} |\sin t|\, dt = \frac{4}{\pi^2} \sum_{k=1}^N \frac{1}{k}. \end{aligned}$$

The Fourier sums are the operators $S_N = D_N * : \mathcal{C}(\mathbf{T}) \to \mathcal{C}(\mathbf{T})$. Their norms $L_N := \|S_N\|$ are called the **Lebesgue numbers**; obviously $L_N \leq \|D_N\|_1$ and it is shown that

$$L_N = \|D_N\|_1$$

holds by considering a real function g which is a continuous modification of $\operatorname{sgn} D_N$ such that $|g| \leq 1$ and $|S_N g| \geq \|D_N\|_1 - \varepsilon$.

Theorem 3.16. (a) *There are functions $f \in L^1(\mathbf{T})$ such that $S_N f \not\to f$ in $L^1(\mathbf{T})$ as $N \to \infty$.*

(b) *There are functions $f \in \mathcal{C}(\mathbf{T})$ such that $S_N f \not\to f$ in $\mathcal{C}(\mathbf{T})$ as $N \to \infty$. In fact, for every $x \in \mathbf{R}$ there is a dense subset A_x of $\mathcal{C}(\mathbf{T})$ such that, for every $f \in A_x$, $\sup_N |S_N(f, x)| = \infty$.*

Proof. (a) Note that $L_N \to \infty$. According to Theorem 3.14,

$$\sup_N \|S_N(f)\|_1 = \infty$$

for every f belonging to a dense set $A \subset \mathcal{C}(\mathbf{T})$.

(b) Again, the linear forms $u_N(f) := S_N(f, x)$ are continuous on $\mathcal{C}(\mathbf{T})$, with $\|u_N\| = L_N$. □

Remark 3.17. By the Fischer-Riesz Theorem 2.37, $S_N f \to f$ in $L^p(\mathbf{T})$ for every $f \in L^p(\mathbf{T})$ if $p = 2$, and it can be shown, though it is much harder, that this is still true when $1 < p < \infty$ and $p \neq 2$.

The convergence of the Fourier series fails in $L^1(\mathbf{T})$, but according to (2.27) the trigonometric polynomials are dense in every $L^p(\mathbf{T})$ if $1 \leq p < \infty$.

3.3. Exercises

Exercise 3.1. Suppose A is a convex absorbing subset of a vector space E, and let

$$q_A(x) := \inf\{t > 0;\ x \in tA\},$$

which is called the **Minkowski functional** of A. Denote $B = \{q_A < 1\}$ and $\bar{B} = \{q_A \leq 1\}$.

(a) Prove that $q_A(x + y) \leq q_A(x) + q_A(y)$, $q_A(tx) = tq_A(x)$ if $t \geq 0$, and that $B \subset A \subset \bar{B}$ and $q_B = q_A = q_{\bar{B}}$.

(b) Suppose A is also balanced and $\bigcup_{t>0} tA = E$. Prove that q_A is a semi-norm.

Exercise 3.2. Suppose A is a subset of a vector space E. Prove that the **convex hull** of A, defined as

$$\mathrm{co}(A) = \{t_1 a_1 + \cdots + t_n a_n;\ n \in \mathbf{N},\ t_j > 0,\ \sum_{j=1}^{n} t_j = 1,\ a_j \in A\,(1 \leq j \leq n)\},$$

is the intersection of all the convex subsets of E that contain A.

If E is a locally convex space and A is bounded, prove that $\mathrm{co}(A)$ is also bounded and that, if A is open, then $\mathrm{co}(A)$ is also open.

Exercise 3.3. Prove that in every topological vector space E the family of all balanced open neighborhoods of zero is a local basis of E.

Exercise 3.4. Prove that the class $\mathcal{H}(\Omega)$ of all holomorphic functions on an open subset Ω of $\mathbf{C}$ is a closed vector subspace of $\mathcal{C}(\Omega)$ and of $\mathcal{E}(\Omega)$. Hence, $\mathcal{H}(\Omega)$ is a Fréchet space with the topology of the semi-norms $\|\cdot\|_K$ ($K \subset \Omega$ compact).

Exercise 3.5. Show that, for every $p \in [1, \infty]$, the sequence of semi-norms

$$q_n(f) := \|f^{(n)}\|_{L^p(\mathbf{R})} \qquad (n = 0, 1, \ldots)$$

defines the topology of $\mathcal{D}_{[a,b]}(\mathbf{R})$.

Exercise 3.6. Every Fréchet norm, $\|\cdot\|$, satisfies the following properties:

(a) $\|x\| = 0 \Rightarrow x = 0$.

(b) $|\lambda| \leq 1 \Rightarrow \|\lambda x\| \leq \|x\|$.

(c) $\|x + y\| \leq \|x\| + \|y\|$.

(d) $\lim_{\lambda \to 0} \|\lambda x\| = 0$.

(e) $\lim_{\|x\| \to 0} \|\lambda x\| = 0$.

(f) $\|\cdot\|$ is not a norm.

Exercise 3.7. If $\|\cdot\|$ is the Fréchet norm on $\mathcal{C}(\mathbf{R})$ associated to the seminorms $p_n(f) = \|f\|_{[-n,n]}$, show that $\|f\| = 1/2$, $\|g\| = 50/101$, and $\|h\| > 1/2$ if $f(x) = (1 - |x|)^+$, $g(x) = 100f(x-2)$, and $h = (f+g)/2$, and prove that the closed ball $\bar{B}_d(0, 1/2)$ for $d(f,g) = \|g - f\|$ is not convex.

Exercise 3.8. If $E = E^1 \times \cdots \times E^m$ is a finite product of Fréchet spaces endowed with the product topology, then prove that E is also a Fréchet space and extend this result to countable products.

Show that $\mathcal{E}(\mathbf{R})$ can be described as a closed subspace of $\prod_{0 \leq k < \infty} \mathcal{C}(\mathbf{R})$.

Exercise 3.9. (a) Prove that $\mathcal{C}(\mathbf{R})$ does not have the Heine-Borel property by showing that the countable set of all "triangle functions"

$$f_n(t) := \frac{d(t, [-1/n, 1/n]^c)}{|t| + d(t, [-1/n, 1/n]^c)}$$

is bounded and its closure is not compact.

(b) Extend this result to $\mathcal{C}(\Omega)$, where Ω is an open subset of $\mathbf{R}^n$.

Exercise 3.10. Suppose that E_1 and E_2 are two Fréchet spaces and that M is a dense subspace of E_1. Prove that every continuous linear mapping $T : M \to E_2$ has a unique continuous linear extension $\tilde{T} : E_1 \to E_2$.

Exercise 3.11. Prove the following statements:

(a) If E/F is a quotient locally convex space and $\pi : E \to E/F$ is the quotient map, then the image by π of an open subset of E is open in E/F.

(b) If E is a Fréchet or Banach space, then E/F is also a Fréchet space or Banach space, respectively.

(c) If F and M are two closed subspaces of a Banach (or Fréchet) space E and M is of finite dimension, then $F + M$ is also closed in E.

Exercise 3.12. Let $\{e_k\}_{k \in \mathbf{Z}}$ be an orthonormal basis of a Hilbert space H. Let $u_n = e_{-n} + ne_n$ and

$$F := \overline{[e_n;\, n \geq 0]},\ M := \overline{[u_n;\, n \geq 1]}.$$

Prove that F and M are two closed subspaces of H such that $F + M$ is dense and not closed in H. Note that $\sum_{n=1}^{\infty} \frac{1}{n} e_{-n} \in \ell^2(\mathbf{Z}) \setminus (F + M)$.

Exercise 3.13. As an application of Corollary 3.9, prove that $[a,b]$, and every compact metric space without isolated points, is uncountable.

Exercise 3.14. Prove that in an infinite-dimensional Banach space there is no countable algebraic basis.

Exercise 3.15. Find a noncontinuous function $f : \mathbf{R} \to \mathbf{R}$ with a closed graph in $\mathbf{R}^2$.

Exercise 3.16. If $T : \mathcal{C}[0,1] \to \mathcal{C}[0,1]$ is a linear map such that $Tf_n(t) \to Tf(t)$ at every $t \in [0,1]$ whenever $f_n \to f$ in $\mathcal{C}[0,1]$ (that is, uniformly), show that T is continuous.

Exercise 3.17. Assume that $\mathcal{C}[0,1]$ and $\mathcal{C}^1[0,1]$ are endowed with the sup norm $\|\cdot\|_{[0,1]}$. Show that the graph of the derivative operator $D : \mathcal{C}^1[0,1] \to \mathcal{C}[0,1]$ $(Df = f')$ is closed but the operator is unbounded. Why does this not contradict Theorem 3.11?

Exercise 3.18. Let $T : E \to F$ be a linear map between two Fréchet spaces. Prove that, if $y = 0$ whenever $x_n \to 0$ in E and $Tx_n \to y$ in F, then T is continuous.

Exercise 3.19 (Uniform Boundedness Principle for metric spaces). Let M be a complete metric space and $f_j : M \to \mathbf{R}$ $(j \in J)$ a family of continuous functions which is bounded at each point $x \in M$, $|f_j(x)| \le C(x) < \infty$ for all $j \in J$. Prove that the functions f_j are uniformly bounded on a nonempty open subset G of M.

Exercise 3.20 (Approximate quadrature). Let $\{J_n\}$ be a sequence of linear forms on $\mathcal{C}[0,1]$ of the type

$$J_n(f) = \sum_{k=0}^{N(n)} A_n^k f(t_k^n),$$

where, for each n, $\{t_k^n\}_{k=1}^{N(n)}$ is a given finite sequence of points in $[0,1]$ that are called the nodes of J_n.

The sequence is called a **quadrature method** if

$$\int_0^1 f(t)\,dt = \lim_{n\to\infty} J_n(f)$$

holds for every $f \in \mathcal{C}[0,1]$.

Prove that $\|J_n\|_{\mathcal{C}[0,1]'} = \sum_{k=0}^{N(n)} |A_n^k|$ and that $\{J_n\}$ is a quadrature method if and only if the following conditions are satisfied:

(i) $\lim_{n\to\infty} J_n(x^{k-1}) = 1/k$ for every $k \in \mathbf{N}$ and

(ii) $\sup_n \sum_{k=0}^{N(n)} |A_n^k| < \infty$.

If $A_n^k \ge 0$ for all n and k, then (i) implies (ii).

Exercise 3.21. (a) Suppose E_1, E_2, and F are three Banach spaces. Prove that a bilinear map $B : E_1 \times E_2 \to F$ is continuous if and only if it is separately continuous; that is, the linear maps $B(\cdot, y)$ and $B(x, \cdot)$ are all continuous.

(b) If E is the normed space of all real polynomial functions on $[0, 1]$ with the norm $\|f\|_1 = \int_0^1 |f(t)|\,dt$, prove that $B(f, g) := \int_0^1 f(t)g(t)\,dt$ defines a separately continuous bilinear map $B : E \times E \to \mathbf{R}$ which is not continuous.

Exercise 3.22. Let E_1, E_2, and F be three Banach spaces. Prove that a bilinear map $B : E_1 \times E_2 \to F$ is continuous if and only if its graph

$$\mathcal{G}(B) := \{(x_1, x_2, y) \in E_1 \times E_2 \times F;\ B(x_1, x_2) = y\}$$

is a closed subset of the product space $E_1 \times E_2 \times F$.

Exercise 3.23. Suppose $T : L^1(0, 1) \to L^1(0, 1)$ is a bounded linear operator and $p, q \geq 1$. If $T(L^p(0, 1)) \subset L^q(0, 1)$, is it necessarily true that the restriction $T : L^p(0, 1) \to L^q(0, 1)$ will also be continuous?

References for further reading:

S. Banach, *Théorie des opérations linéaires.*

S. K. Berberian, *Lectures in Functional Analysis and Operator Theory.*

B. A. Conway, *A Course in Functional Analysis.*

R. E. Edwards, *Functional Analysis, Theory and Applications.*

G. Köthe, *Topological Vector Spaces I.*

R. Meise and D. Vogdt, *Introduction to Functional Analysis.*

W. Rudin, *Functional Analysis.*

Chapter 4

Duality

An essential aspect of functional analysis is the study and applications of duality, which deals with continuous linear forms on functional spaces.

In the case of a Hilbert space, the projection theorem will allow us to prove the description of the dual given by the Riesz representation theorem and by its extension known as the Lax-Milgram theorem, which is useful in the resolution of some boundary value problems, as we will see in some examples in Chapter 7.

But if we are dealing with a more general normed space, or with any locally convex space, to ensure the existence of continuous linear extensions of continuous linear functionals defined on subspaces, we need the Hahn-Banach theorem. This theorem adopts several essentially equivalent versions, and we will start from its analytical or dominated extension form and then the geometric or separation form will follow.

We include in this chapter a number of applications of both the Riesz and the Hahn-Banach theorems, such as the description of the duality of L^p spaces, interpolation of linear operators, von Neumann's proof of the Radon-Nikodym theorem, and an introduction to the spectral theory of compact operators.

4.1. The dual of a Hilbert space

In this section, H will denote a Hilbert space. Its norm $\|\cdot\|_H$ is associated to a scalar product,

$$\|x\|_H^2 = (x, x)_H.$$

Note that, for every $x \in H$, $(\cdot, x)_H$ is a linear form on H. We will see that its norm is equal to $\|x\|_H$ and that every continuous linear form is of this type, as for Euclidean spaces.

4.1.1. Riesz representation and Lax-Milgram theorem. The decomposition $H = F \oplus F^{\perp}$ given by the Projection Theorem 2.35 allows an easy proof of the following fundamental representation result concerning the dual of a Hilbert space.

Theorem 4.1 (Riesz representation). *The map* $J : H \to H'$ *such that* $J(x) = (\cdot, x)_H$ *is a bijective skew linear isometry, that is,*

(1) $\|(\cdot, x)_H\|_{H'} = \|x\|_H$,

(2) $J(x_1 + x_2) = J(x_1) + J(x_2)$,

(3) $J(\lambda x) = \bar{\lambda} J(x)$, *and*

(4) *if* $u \in H'$, *then* $u = (\cdot, x)_H$ *for some* $x \in H$.

Proof. The Schwarz inequality (2.11) means that $\|(\cdot, x)_H\|_{H'} \leq \|x\|_H$, and the linear form $(\cdot, x)_H$ reaches this value $\|x\|_H$ at $x_0 = x/\|x\|_H$, which proves (1).

The identities (2) and (3) are obvious.

Finally, if $0 \neq u \in H'$ and $F = \operatorname{Ker} u$, then $F \neq H = F \oplus F^{\perp}$ and there exists $z \in F^{\perp}$, $\|z\|_H = 1$. Since $u(z)x - u(x)z \in F$ for every $x \in H$, we have $0 = (u(z)x - u(x)z, z)_H$, which is equivalent to $u(x) = (x, \overline{u(z)}z)_H$. Thus, $u = (\cdot, \overline{u(z)}z)_H$. □

Example 4.2. By an application of the Riesz representation theorem to the Hilbert space $L^2 = L^2(\mu)$,

$$(L^2)' = \{(\cdot, g)_2;\ g \in L^2\},$$

and $g \mapsto (\cdot, g)_2$ is a skew linear isometric bijection from L^2 onto $(L^2)'$. Since $g \in L^2 \mapsto \bar{g} \in L^2$ is also a skew linear isometric bijection, $g \mapsto (\cdot, \bar{g})_2$ is a bijective linear isometry from L^2 onto $(L^2)'$ that allows us to consider L^2 as its own dual. The notation

$$\langle f, g \rangle := \int f g \, d\mu \qquad (f, g \in L^2)$$

is a usual one, $u = \langle \cdot, g \rangle$ $(g \in L^2)$ are all the continuous linear forms on L^2, and $\|u\| = \|g\|_2$.

It is customary to identify $\langle \cdot, g \rangle$ with g and to say that the dual $L^2(\mu)'$ of $L^2(\mu)$ is $L^2(\mu)$.

The following extension of the Riesz theorem was given by P. D. Lax[1] and A. Milgram when studying parabolic partial differential equations:

Theorem 4.3 (Lax-Milgram). *Let H be a Hilbert space and suppose that $B : H \times H \to \mathbf{K}$ is a bounded sesquilinear form; that is, B satisfies the conditions*

(1) *$B(\cdot, a)$ is linear and $B(a, \cdot)$ skew linear for every $a \in H$ and*

(2) *$|B(x, y)| \leq C\|x\|_H\|y\|_H$ for some constant C, for all $x, y \in H$.*

If B is **coercive**, *meaning that*

$$|B(x, x)| \geq c\|x\|^2 \qquad (x \in H)$$

for some constant $c > 0$, then for every $u \in H'$ there exists a uniquely determined element $y \in H$ such that $u = B(\cdot, y)$. Hence $y \in H \mapsto B(\cdot, y) \in H'$ is a bijective skew linear map.

Proof. By virtue of the Riesz Theorem 4.1, for every $y \in H$, there is a unique $T(y) \in H$ such that $B(x, y) = (x, T(y))_H$ for all $x \in H$, and it is readily seen that the mapping $T : H \to H$ is linear. For instance,

$$(x, T(\lambda y))_H = B(x, \lambda y) = B(\bar{\lambda} x, y) = (\bar{\lambda} x, T(y))_H = (x, \lambda T(y))_H,$$

and then $T(\lambda y) = \lambda T(y)$.

It follows from the assumption $|(x, T(y))_H| = |B(x, y)| \leq C\|x\|_H\|y\|_H$ that

$$\|T(y)\|_H \leq C\|y\|_H$$

and T is bounded. Similarly, by the coercivity assumption,

$$c\|y\|_H^2 \leq |(y, T(y))_H| \leq \|y\|_H\|T(y)\|_H,$$

so that

$$c\|y\|_H \leq \|T(y)\|_H \leq C\|y\|_H.$$

Obviously T is one-to-one, and these estimates imply that $T(H)$ is closed since, if $T(y_n) \to z_0$, then

$$c\|y_n - y_m\|_H \leq \|T(y_n) - T(y_m)\|_H,$$

and we can find $y_0 = \lim y_n$, so that $z_0 = T(y_0) \in z(H)$.

To prove that T is onto, suppose that $x \in T(H)^\perp$, so that $B(x, y) = 0$ for all $y \in H$ and then $0 = B(x, x) \geq b\|x\|_H$; thus $x = 0$ and $T(H)^\perp = \{0\}$, which proves that $T(H) = H$, since $T(H)$ is closed.

This shows that for every $u \in H'$ there is an element $T(y) \in H$ such that $u(x) = (x, T(y))_H$, and then $u(x) = B(x, y)$ for all $x \in H$.

---

[1]Peter David Lax, while holding a position at the Courant Institute, was awarded the Abel Prize (2005) "for his groundbreaking contributions to the theory and application of partial differential equations and to the computation of their solutions".

Note that y is unique, since it follows from $B(x, y_1) = B(x, y_2)$, or $B(x, y) = 0$ for $y = y_1 - y_2$, that $0 = |(y, T(y))_H| = |B(y, y)| \geq c\|y\|_H^2$ and then $y = 0$, and hence $y_1 = y_2$. □

4.1.2. The adjoint. Suppose $T \in \mathcal{L}(H_1; H_2)$, where H_1 and H_2 are Hilbert spaces. The transpose T' acts between the duals H_2' and H_1' by the rule $T'v = v \circ T$; that is,

$$(T'v)(x) = v(Tx) \qquad (x \in H_1,\, v \in H_2'). \tag{4.1}$$

By the Riesz representation Theorem 4.1, $v = (\cdot, y)_{H_2}$ and $T'v = (\cdot, T^*y)_{H_1}$ for some $T^*y \in H_1$, and (4.1) becomes

$$(x, T^*y)_{H_1} = (Tx, y)_{H_2} \qquad (x \in H_1,\, y \in H_2). \tag{4.2}$$

The **adjoint** of T is the linear operator $T^* : H_2 \to H_1$ characterized by the identity (4.2), and its linearity follows from this relation. For instance,

$$(x, T^*(y_1 + y_2))_{H_1} = (Tx, y_1)_{H_2} + (Tx, y_2)_{H_2} = (x, T^*y_1 + T^*y_2)_{H_1}$$

for every $x \in H_1$, and then $T^*(y_1 + y_2) = T^*y_1 + T^*y_2$.

Clearly $T^{**} = T$ and $(ST)^* = T^*S^*$ if the composition is defined.

Theorem 4.4. *The map $T \in \mathcal{L}(H_1; H_2) \mapsto T^* \in \mathcal{L}(H_2; H_1)$ is a skew linear isometry such that $\|T^*T\| = \|T\|^2 = \|TT^*\|$, and for every $T \in \mathcal{L}(H_1; H_2)$ the following properties hold:*

(a) $(\operatorname{Im} T)^\perp = \operatorname{Ker} T^*$,

(b) $(\operatorname{Ker} T^*)^\perp = \overline{\operatorname{Im} T}$,

(c) $(\operatorname{Im} T^*)^\perp = \operatorname{Ker} T$, *and*

(d) $(\operatorname{Ker} T)^\perp = \overline{\operatorname{Im} T^*}$.

Proof. Since $(x, (\lambda T)^*y)_{H_1} = (\lambda Tx, y)_{H_2} = \lambda(x, T^*y)_{H_1} = (x, \bar{\lambda}T^*y)_{H_1}$, we obtain $(\lambda T)^* = \bar{\lambda}T^*$. It is also clear that $(S + T)^* = S^* + T^*$.

By the Riesz representation Theorem 4.1,

$$\|T\| = \sup_{\|x\|_{H_1} \leq 1} \|Tx\|_{H_2} = \sup_{\|x\|_{H_1}, \|y\|_{H_2} \leq 1} |(y, Tx)_{H_2}|. \tag{4.3}$$

Hence,

$$\|T^*\| = \sup_{\|x\|_{H_1}, \|y\|_{H_1} \leq 1} |(x, T^*y)_{H_2}| = \sup_{\|x\|_{H_1}, \|y\|_{H_2} \leq 1} |(y, Tx)_{H_2}| = \|T\|$$

and also

$$\|T\|^2 = \sup_{\|x\|_{H_1} \leq 1} \|Tx\|_{H_2}^2 = \sup_{\|x\|_{H_1} \leq 1} (x, T^*Tx)_{H_1} \leq \|T^*T\| \leq \|T\|^2.$$

To prove (a), note that $(y, Tx)_{H_2} = 0$ if and only if $(T^*y, x)_{H_1} = 0$, which holds for every $x \in H$ when $T^*y = 0$. The remaining properties follow very easily from the identity $T^{**} = T$ and from the relation $F^{\perp\perp} = \bar{F}$, if F is a vector subspace of H_1 (or H_2), contained in Theorem 2.36. □

Remark 4.5. It is worth observing that, in the complex case, $(\lambda T)^* = \bar{\lambda} T^*$, but $(\lambda T)' = \lambda T'$, since $(\lambda T)'(v) = v \circ (\lambda T) = \lambda v \circ T$.

An operator $T \in \mathcal{L}(H)$ is said to be **self-adjoint** if $T^* = T$. By Theorem 2.35(b), every orthogonal projection is self-adjoint.

The **Hilbert-Schmidt operators** T_K, defined by integral kernels $K \in L^2(X \times Y)$ as

$$T_K f(x) := \int_Y K(x,y) f(y)\, dy,$$

form another important family of bounded linear operators between Hilbert spaces. Here X and Y are assumed to be two σ-finite measure spaces.

Theorem 4.6. *If $K \in L^2(X \times Y)$, then $T_K : L^2(Y) \to L^2(X)$ and $\|T_K\| \leq \|K\|_2$. The adjoint $T_K^* : L^2(X) \to L^2(Y)$ is the Hilbert-Schmidt operator T_{K^*} defined by the kernel $K^*(y,x) = \overline{K(x,y)}$. Hence, if $X = Y$, T_K is self-adjoint when $K(y,x) = \overline{K(x,y)}$ a.e. on $X \times X$.*

Proof. By Schwarz inequality,

$$\int_Y |K(x,y)||f(y)|\, dy \leq \Big(\int_Y |K(x,y)|^2 dy\Big)^{1/2} \Big(\int_Y |f(y)|^2 dy\Big)^{1/2}$$

with $\int_Y |K(x,y)|^2 dy < \infty$ a.e., since $\int_X \int_Y |K(x,y)|^2 dy\, dx < \infty$. Then $T_K f(x)$ is defined a.e. and

$$|T_K f(x)|^2 \leq \int_Y |K(x,y)|^2 dy \int_Y |f(y)|^2 dy.$$

By integrating both sides, $\|T_K f\|_2 \leq \|K\|_2 \|f\|_2$.

Fubini's theorem shows that $(T_K f, g)_2 = (f, T_{K^*} g)_2$. □

The description (4.3) of the norm of an operator by duality has a useful variant for self-adjoint operators:

Theorem 4.7. *If $T \in \mathcal{L}(H)$ is self-adjoint, then $(Tx, x)_H \in \mathbf{R}$ for every $x \in H$ and*

$$\|T\| = \sup_{\|x\|_H = 1} |(Tx, x)_H|.$$

Proof. We call $S := \sup_{\|x\|_H=1} |(Tx,x)_H|$, so that $|(Tx,x)_H| \leq \|x\|_H^2 S$. Since $S \leq \|T\|$, we only need to show that $\|Tx\|_H \leq S$ if $\|x\|_H = 1$, and we can assume that $Tx \neq 0$.

We will use the **polarization identities**, which extend (2.15) and (2.16),

$$(Tx,y)_H = \frac{1}{4}\big((T(x+y),x+y)_H - (T(x-y),x-y)_H\big)$$

if $\mathbf{K} = \mathbf{R}$ and

$$\begin{aligned}(Tx,y)_H &= \frac{1}{4}((T(x+y),x+y)_H - (T(x-y),x-y)_H \\ &\qquad +i(T(x+iy),x+iy)_H - i(T(x-iy),x-iy)_H\big)\end{aligned}$$

in the complex case. They are checked by expanding the inner products.

Since $(Tx,x)_H = (x,Tx)_H \in \mathbf{R}$,

$$\Re(Tx,y)_H = \frac{1}{4}\big((T(x+y),x+y)_H - (T(x-y),x-y)_H\big).$$

Let $\|x\|_H = 1$ and $y = (1/\|Tx\|_H)Tx$ so that also $\|y\|_H = 1$ and

$$\|Tx\|_H = \Re(Tx,y)_H \leq \frac{S}{4}(\|x+y\|_H^2 + \|x-y\|_H^2) = \frac{S}{4}(\|x\|_H^2 + \|y\|_H^2);$$

hence $\|Tx\|_H \leq S$. □

4.2. Applications of the Riesz representation theorem

Throughout this section, $(\Omega, \mathcal{B}, \mu)$ will be a σ-finite measure space. Our aim is to obtain some consequences of the duality result of Example 4.2 for $L^2(\mu)$.

4.2.1. Radon-Nikodym theorem. Suppose $0 \leq h \in L^1(\mu)$. The finite measure

$$\nu(A) := \int_A h\,d\mu$$

on $(\Omega, \mathcal{B})$ is such that $\nu(A) = 0$ if $\mu(A) = 0$. When this happens, we say that ν is **absolutely continuous** (relatively to μ) and write $\nu \ll \mu$.

Theorem 4.8 (Radon-Nikodym). *Let μ and ν be two σ-finite measures on $(\Omega, \mathcal{B})$. If $\nu \ll \mu$, there exists a unique a.e. determined nonnegative measurable function h such that*

$$\nu(A) = \int_A h\,d\mu \qquad (A \in \mathcal{B}). \tag{4.4}$$

If ν is finite, $h \in L^1(\mu)$.

Proof. First assume that μ and ν are finite and define $\lambda = \mu + \nu$. Then $\lambda(A) = 0$ if and only if $\mu(A) = 0$.

The linear form $u(f) := \int f\,d\mu$ is bounded on the real Hilbert space $L^2(\mu)$ and also on $L^2(\lambda)$, since $\mu \leq \lambda$. Then there exists a unique function $g \in L^2(\lambda)$ such that

$$\int f\,d\mu = \int fg\,d\lambda \qquad (f \in L^2(\lambda)).$$

We rewrite this identity as

$$\int f(1-g)\,d\mu = \int fg\,d\nu. \tag{4.5}$$

Formally $(1-g)\,d\mu = g\,d\nu$ and we will try

$$h := \frac{1-g}{g}.$$

We need to prove that $0 < g \leq 1$ μ-a.e.

First $F := \{g \leq 0\}$ is μ-null, since for $f = \chi_F$ we obtain from (4.5) that $\mu(F) \leq \int \chi_F(1-g)\,d\mu = \int \chi_F g\,d\nu \leq 0$.

To prove that $G := \{g > 1\}$ is also a μ-null set, assume that $\mu(G) > 0$. Then, by taking $f = \chi_G$, again from (4.5) we obtain that $0 > \int_G (1-g)\,d\mu = \int_G g\,d\nu \geq 0$, which is impossible.

By changing g on a μ-null set if necessary, $0 < g \leq 1$ everywhere. The function $k = fg$ is λ-integrable, and equation (4.5) reads

$$\int kh\,d\mu = \int k\,d\nu.$$

If $k = \chi_A$ and $f = \chi_A/g \in L^2(\lambda)$, we obtain $\int_A h\,d\mu = \nu(A)$.

For any $A \in \mathcal{B}$, by denoting $A_n = A \cap \{g > 1/n\}$, we obtain $\chi_{A_n}/g \in L^2(\lambda)$ and $\int_{A_n} h\,d\mu = \nu(A_n)$. Then, by monotone convergence, $\int_A h\,d\mu = \nu(A)$.

Observe that $0 \leq h \in L^1(\mu)$ and it is unique since, if $\int_A f\,d\mu = 0$ for every $A \in \mathcal{B}$, then $f = 0$ μ-a.e.

Now let μ and ν be σ-finite. We can find $\Omega_n \in \mathcal{B}$ $(n \in \mathbf{N})$ with $\Omega_n \uparrow \Omega$ and $\mu(\Omega_n), \nu(\Omega_n) < \infty$. If $A \subset \Omega_n$, then $\nu(A) = \int_A h_n\,d\mu$, with $h_n(x) = h_{n+1}(x)$ a.e. on Ω_n, and we can take $h_n = (h_{n+1})_{|\Omega_n}$. We define $h_n = 0$ on Ω_n^c and the nonnegative function $h(x) := \lim_n h_n(x)$ is measurable; then, by monotone convergence,

$$\nu(A) = \lim_n \nu(A \cap \Omega_n) = \lim_n \int_{A \cap \Omega_n} h\,d\mu = \int_A h\,d\mu.$$

The uniqueness of h follows from the uniqueness on every Ω_n. □

Formula (4.4) is often represented by the notation $d\nu = h\,d\mu$, and h is called the **Radon-Nikodym derivative of** ν.

4.2.2. The dual of L^p. Recall that, if $1 \leq p < \infty$, $L^p = L^p(\mu)$ is the Banach space of all μ-measurable (real or complex) functions f defined by the condition

$$\|f\|_p := \left(\int |f|^p d\mu\right)^{1/p} < \infty,$$

modulo the subspace of functions vanishing a.e., with the usual modification for $p = \infty$.

By denoting

$$\langle f, g\rangle := \int fg\, d\mu$$

with $p' = p/(p-1)$ ($p' = \infty$ if $p = 1$, and $p' = 1$ if $p = \infty$) if the integral exists, then, by Hölder's inequality,

$$|\langle f, g\rangle| \leq \|f\|_p \|g\|_{p'} \qquad (f \in L^p(\mu),\ g \in L^{p'}(\mu)).$$

Note that, for every $g \in L^{p'}$, $\langle \cdot, g\rangle \in (L^p)'$ and $\|\langle \cdot, g\rangle\|_{(L^p)'} \leq \|g\|_{p'}$ for any $p \in [1, \infty]$, and we say that $g \in L^{p'} \mapsto \langle \cdot, g\rangle \in (L^p)'$ is the natural map from $L^{p'}$ into the dual of L^p.

We know from Example 4.2 that $L^2(\mu)'$ can be identified with $L^2(\mu)$. As an application of the Radon-Nikodym theorem, we will show that $L^p(\mu)'$ can also be identified with $L^{p'}(\mu)$, if $1 \leq p < \infty$.

To prepare the proof, we will consider **positive linear forms** on the real L^p spaces. We say that $v \in (L^p)'$ is positive if $v(f) \geq 0$ when $0 \leq f \in L^p$, and then we write $v \in (L^p)'_+$. Also, $f \in (L^p)^+$ if $f \in L^p$ and $f \geq 0$ a.e.

Lemma 4.9. $(L^p)' = (L^p)'_+ - (L^p)'_+$.

Proof. Let $v \in (L^p)'$ and $f \in (L^p)^+$ and define

$$v_+(f) := \sup_{0 \leq g \leq f} v(g).$$

Obviously, $v_+(\alpha f) = \alpha v_+(f)$ if $\alpha \in \mathbf{R}^+$ and $f \in (L^p)^+$. To show that v_+ is also additive, we observe that, if $f_1, f_2 \in (L^p)^+$, then

$$\begin{aligned} v_+(f_1) + v_+(f_2) &= \sup_{0 \leq g_1 \leq f_1,\, 0 \leq g_2 \leq f_2} v(g_1 + g_2) \\ &\leq \sup_{0 \leq g \leq f_1 + f_2} v(g) = v_+(f_1 + f_2). \end{aligned}$$

For the reversed inequality, we claim that

$$\{g \in (L^p)^+;\ g \leq f_1 + f_2\} \subset \{g \in (L^p)^+;\ g \leq f_1\} + \{g \in (L^p)^+;\ g \leq f_2\}.$$

To prove this, if $0 \leq g \leq f_1 + f_2$ and $g_1 := \inf(g, f_1)$, then the function $g_2 := g - g_1 \geq 0$ satisfies $g_2 \leq f_2$ since, when $g_1(x) = f_1(x)$,

$$g_2(x) = g(x) - f_1(x) \leq f_2(x)$$

and $g_2(x) = 0 \le f_2(x)$ when $g_1(x) = g(x)$. Hence, also

$$v_+(f_1 + f_2) \le v_+(f_1) + v_+(f_2).$$

Every real function $f \in L^p$ admits a decomposition $f = f_1 - f_2$ with $f_1, f_2 \in (L^p)^+$, for instance by taking $f_1 = f^+ = \sup(f, 0)$ and $f_2 = f^- = \sup(-f, 0)$, and

$$v_+(f) := v_+(f_1) - v_+(f_2)$$

does not depend on the decomposition, since $f_1 - f_2 = f^+ - f^-$ implies $v_+(f_1) + v_+(f^-) = v_+(f^+) + v_+(f_2)$.

It is clear that $v_+(\lambda f) = \lambda v_+(f)$ in both cases $\lambda \ge 0$ and $\lambda < 0$ ($v_+(-f) = v_+(f^-) - v_+(f^+) = -v_+(f)$). Additivity also holds for v_+, since

$$\begin{aligned} v_+(f_1 + f_2) &= v_+((f_1^+ + f_2^+) - (f_1^- + f_2^-)) \\ &= (v_+(f_1^+) + v_-(f_2^+)) - (v_+(f_1^-) + v_+(f_2^-)) \\ &= v_+(f_1) + v_+(f_2), \end{aligned}$$

and it is continuous, since

$$|v_+(f)| \le v_+(f^+) + v_+(f^-) \le \|v\|_{(L^p)'}(\|f^+\|_p + \|f^-\|_p) \le 2\|v\|_{(L^p)'}\|f\|_p.$$

Finally, $v_- := v_+ - v$ is also linear and continuous, and

$$v_-(f) = \sup_{0 \le g \le f} v(g) - v(f) \ge 0$$

if $f \ge 0$. □

Theorem 4.10 (Riesz representation theorem for $(L^p)'$). *Suppose* $1 \le p < \infty$. *For every* $v \in L^p(\mu)'$ *there is a uniquely determined function* $g \in L^{p'}(\mu)$ *such that*

$$v(f) = \int_\Omega fg \, d\mu \qquad (f \in L^p),$$

and $\|g\|_{p'} = \|v\|_{(L^p)'}$.

Proof. (a) First let $\mu(\Omega) < \infty$ and $v \in (L^p)'_+$.

By $\nu(A) := v(\chi_A)$ ($\|\chi_A\|_p = \mu(A)^{1/p} < \infty$) we obtain a finite measure ν on $(\Omega, \mathcal{B})$, since $\nu(\emptyset) = 0$ and

$$\nu\big(\biguplus_n A_n\big) = v\big(\lim_N \sum_{n=1}^N \chi_{A_n}\big) = \lim_N \sum_{n=1}^N v(\chi_{A_n}) = \sum_n \nu(A_n).$$

It is clear that $\nu \ll \mu$, since $u(\chi_A) = u(0) = 0$ if $\mu(A) = 0$.

By the Radon-Nikodym theorem for finite measures, $v(\chi_A) = \nu(A) = \int_A h \, d\mu$ for a unique integrable function $h \ge 0$, and also $v(s) = \int sh \, d\mu$ if s is a simple function.

If $0 \le f \in L^p$, we choose simple functions s_n so that $0 \le s_n \uparrow f$. Then $s_n \to f$ in L^p by dominated convergence; hence

$$v(f) = \lim_n v(s_n) = \lim_n \int s_n h\, d\mu = \int fh\, d\mu$$

by monotone convergence, and $\int fh\, d\mu \le \|v\|_{L^p(\mu)'}\|f\|_p$.

(b) In the general σ-finite case and for $v \in L^p(\mu)'_+$, let $\Omega = \biguplus_{n=1}^\infty \Omega_n$ with $\mu(\Omega_n) < \infty$.

We obtain h_n associated to the restriction v_n of v to $L^p(\mu_n)$ as in (a), and we extend h_n with zeros. Then, if $h := \sum_n h_n \ge 0$ and $f \in L^p(\mu)$, we can write

$$\langle f, h\rangle = \int \sum_n fh_n\, d\mu = \sum_n \int_{\Omega_n} fh\, d\mu = \sum_n v(f\chi_{\Omega_n}) = v(f).$$

Note that fh is integrable, since $\sum_n f\chi_{\Omega_n} = f$ in $L^p(\mu)$ by dominated convergence, the last equality follows, and

$$\int |fh|\, d\mu = \sum_n \int |fh_n|\, d\mu = v(|f|) < \infty.$$

(c) If $v = v_+ - v_- \in (L^p)'$, by (b) there exists h so that $fh \in L^1$ for every real function $f \in L^p$ and $v(f) = \int fh\, d\mu$.

(d) In the complex case, if $v \in (L^p)'$, we obtain $h = h_1 + ih_2$ so that $fh_1, fh_2 \in L^1$ for every $f \in L^p$, $\Re v(f) = \int fh_1\, d\mu$, $\Im v(f) = \int fh_1\, d\mu$. Then $v(f) = \int fh\, d\mu$ if $f \in L^p$ is a real function. By linearity, also $v(f) = \int fh\, d\mu$ for any complex function $f \in L^p$.

(e) The function h obtained in (c) and (d) belongs to $L^{p'}$ and $\|h\|_{p'} = \|v\|_{(L^p)'}$:

Suppose first that $p > 1$ and let $B_n = \Omega_n \cap \{|h| \le n\}$, with $\Omega_n \uparrow \Omega$. Then the functions $f_n = |h|^{p'-1}\operatorname{sgn}(h)\chi_{B_n}$ ($\operatorname{sgn} z = |z|/z$, with $0/0 := 0$) belong to L^p, $v(f_n) = \int |h|^{p'}\chi_{B_n}\, d\mu$, and

$$\int_{B_n} h^{p'}\, d\mu \le \|v\|_{(L^p)'}\|f_n\|_p = \|v\|_{(L^p)'}\Big(\int_{B_n} h^{p'}\, d\mu\Big)^{1/p}.$$

By monotone convergence, $\|h\|_{p'}^{p'} \le \|v\|_{(L^p)'}\|h\|_{p'}^{p'/p}$, and $\|h\|_{p'} \le \|v\|_{(L^p)'}$. Since $v = \langle \cdot, h\rangle$, by Hölder's inequality $\|v\|_{(L^p)'} \le \|h\|_{p'}$.

When $p = 1$, we also have $\|v\|_{(L^1)'} \le \|h\|_\infty$. Suppose $\|v\|_{(L^1)'} + \varepsilon < \|h\|_\infty$; then, since the measure is assumed to be σ-finite, there exists some

$A \in \mathcal{B}$ such that $0 < \mu(A) < \infty$ and $|h(x)| \geq \|v\|_{(L^1)'} + \varepsilon$ for every $x \in A$. If $f_n := \operatorname{sgn}(h)\chi_{B_n \cap A}$, then

$$v(f_n) = \int_{B_n \cap A} |h| \, d\mu \geq (\|v\|_{(L^1)'} + \varepsilon)\mu(B_n \cap A)$$

and, since also $v(f_n) \leq \|v\|_{(L^1)'}\|f_n\|_1 = \|v\|_{(L^1)'}\mu(B_n \cap A)$, we arrive at a contradiction by allowing $n \uparrow \infty$.

(f) Finally, h is uniquely determined since, if $\int_A h \, d\mu = 0$ for every $A \in \mathcal{B}$ with finite measure, then $h = 0$ a.e. □

Note that if $1 \leq p < \infty$, by Theorem 4.10, the natural linear mapping $g \mapsto \langle \cdot, g \rangle$ is a bijective isometry from $L^{p'}$ onto $(L^p)'$ which allows us to represent by $L^{p'}(\mu)$ the dual of $L^p(\mu)$ and to describe the L^p-norm by duality:

$$\|g\|_{p'} = \sup_{\|f\|_p \leq 1} |\langle f, g \rangle| = \sup_{\|s\|_p = 1;\, s \in S} |\langle f, \cdot \rangle|, \tag{4.6}$$

where S denotes the vector space of all integrable simple functions.

Indeed, S is a dense subspace of L^p, so that the norm of $u = \langle f, \cdot \rangle$ does not change when we restrict u to this subspace.

Remark 4.11. Similarly, if $1 \leq p < \infty$, $\ell^{p'}$ can be regarded as the dual of ℓ^p throughout the mapping $y \in \ell^{p'} \mapsto \langle \cdot, y \rangle \in (\ell^p)'$, now with $\langle x, y \rangle = \sum_{n=1}^{\infty} x_n y_n$, $x = (x_n) \in \ell^p$, and $y = (y_n) \in \ell^{p'}$. Recall that ℓ^p is the space L^p associated to the counting measure on $\mathbf{N}$ (see Exercise 1.11).

4.2.3. The dual of $\mathcal{C}(K)$. Another fundamental theorem of F. Riesz shows that every continuous linear form on $\mathcal{C}(K)$ is represented by a complex measure.[2]

Lemma 4.12. *For every complex measure μ there is a $|\mu|$-a.e. uniquely defined $|\mu|$-integrable function h such that $|h| = 1$ which satisfies*

$$\mu(B) = \int_B h \, d|\mu| \qquad (B \in \mathcal{B}). \tag{4.7}$$

Proof. The existence of $h \in L^1(|\mu|)$ follows from an application of the Radon-Nikodym theorem to $\Re\mu^+$, $\Re\mu^-$, $\Im\mu^+$, and $\Im\mu^-$. By (1.7), the four of them are finite and absolutely continuous with respect to $|\mu|$, so that they have an integrable Radon-Nikodym derivative.

The uniqueness of h follows by noting that $\int_B h \, d|\mu| = 0 \; \forall B \in \mathcal{B}$ implies $h = 0$ a.e.

[2]In 1909, F. Riesz obtained the representation of every $u \in \mathcal{C}[0,1]'$ as a Riemann-Stieltjes integral $u(g) = \int_0^1 g(x) \, dF(x)$ when solving a problem posed by Jacques Hadamard in 1903. J. Radon found the extension to compact subsets of $\mathbf{R}^n$ in 1913, S. Banach to metric compact spaces in 1937, and S. Kakutani to general compact spaces in 1941.

If $|\mu|(B) > 0$, then $\left|\int_B h\, d|\mu|\right| = |\mu(B)| \le |\mu|(B)$ and

$$\left|\frac{1}{|\mu|(B)}\int_B h\, d|\mu|\right| \le 1, \tag{4.8}$$

which implies $|h(x)| \le 1$ a.e.

Indeed, $\bar{D}(0,1)^c \subset \mathbf{C}$ is a countable union of discs $D = D(a,r)$ and it is enough to prove that $|\mu|(h^{-1}(D)) = 0$.

If we assume that $|\mu|(B) > 0$ for $B = h^{-1}(D)$, then

$$\left|\frac{1}{|\mu|(B)}\int_B h\, d|\mu| - a\right| \le \frac{1}{|\mu|(B)}\int_B |h-a|\, d|\mu| \le r$$

and

$$\frac{1}{|\mu|(B)}\int_B h\, d|\mu| \notin \bar{D}(0,1),$$

which is in contradiction to (4.8).

To show that also $|h(x)| \ge 1$ a.e., let us check that $B(r) := \{|h| < r\}$ is a $|\mu|$-null set for every $r < 1$. If $B(r) = \biguplus_{k=1}^{\infty} B_k$,

$$\sum_{k=1}^{\infty} |\mu(B_k)| = \sum_{k=1}^{\infty}\left|\int_{B_k} h\, d|\mu|\right| \le \sum_{k=1}^{\infty} r|\mu|(B_k) = r|\mu|(B(r))$$

and $|\mu|(B(r)) \le r|\mu|(B(r))$, so that $|\mu|(B(r)) = 0$, since $r < 1$. □

The identity (4.7) is represented by $d\mu = h\, d|\mu|$, which is called the **polar representation** of μ.

It is natural to define $L^p(\mu) := L^p(|\mu|)$ and $\int f\, d\mu := \int fh\, d|\mu|$ for every $f \in L^1(\mu)$. Obviously $\left|\int f\, d\mu\right| \le \int |f|\, d|\mu|$.

Now we are ready to prove the **Riesz representation theorem**.

Theorem 4.13. *Let K be a compact subset of $\mathbf{R}^n$ and let $\mathcal{C}(K)$ be the real or complex Banach space of all continuous functions on K, with the usual* sup *norm.*

If μ is, respectively, a real or complex Borel measure on K, then

$$u_\mu(g) := \int g\, d\mu$$

defines a continuous linear form on $\mathcal{C}(K)$, and $\mu \mapsto u_\mu$ is a bijective linear map between the vector space $M(K)$ of all Borel real or complex measures on K and the dual space $\mathcal{C}(K)'$.

Proof. Since $|u_\mu(g)| \le \int |g|\, d|\mu| \le \|g\|_K |\mu|(K)$, it follows that

$$\|u_\mu\|_{\mathcal{C}(K)'} \le |\mu|(K)$$

and u_μ is a continuous linear form.[3]

Conversely, if $v \in \mathcal{C}(K)'$, we are going to prove that $v = u_\mu$ for some $\mu \in M(K)$.

In the case $\mathcal{C}(K) = \mathcal{C}(K;\mathbf{R})$ of real functions, and with the same proof of Lemma 4.9, every $v \in \mathcal{C}(K;\mathbf{R})'$ is the difference $v = v_+ - v_-$ of two positive linear forms $v_+, v_- \in \mathcal{C}(K)'_+$. They are defined on any positive function $f \in \mathcal{C}(K)_+$ by

$$v_+(f) := \sup_{0 \le g \le f} v(g),$$

and $v_- = v - v_+$.

Every $f \in \mathcal{C}(K)$ admits a decomposition $f = f_1 - f_2$ with $f_1, f_2 \in \mathcal{C}(K)$, for instance by taking $f_1 = f^+ = \sup(f,0)$ and $f_2 = f^- = \sup(-f,0)$ and, as in Lemma 4.9, $v_+(f) := v_+(f_1) - v_+(f_2)$ does not depend on the descomposition, since $f_1 - f_2 = f^+ - f^-$ implies $v_+(f_1) + v_+(f^-) = v_+(f^+) + v_+(f_2)$.

With this procedure we obtain

$$\mathcal{C}(K)' = \mathcal{C}(K)'_+ - \mathcal{C}(K)'_+.$$

We write

$$v(g) = \int g\,d\mu_+ - \int g\,d\mu_- = \int g\,d\mu \qquad (g \in \mathcal{C}(K)), \tag{4.9}$$

if the Borel measures $\mu_\pm$ represent the positive linear forms $v_\pm$ and $\mu = \mu_+ - \mu_-$, as in the Riesz-Markov representation theorem.

In the complex case, every $v \in \mathcal{C}(K)'$ determines two continuous linear forms $\Re v, \Im v \in \mathcal{C}(K;\mathbf{R})'$. Since

$$v(f) = v(g + ih) = v(g) + iv(h) \qquad (g, h \in \mathcal{C}(K;\mathbf{R})),$$

if $(\Re v)(g) = \int g\,d\mu$ and $(\Im v)(g) = \int g\,d\lambda$ on $\mathcal{C}(K;\mathbf{R})$, we can write

$$v(f) = \int g\,d\mu + i\int h\,d\mu + i\int g\,d\lambda - \int h\,d\lambda = \int f\,d\mu + i\int f\,d\lambda.$$

If μ and λ are two Borel real measures on K, then $\nu = \mu + i\lambda : \mathcal{B} \to \mathbf{C}$ is a complex measure and $v(f) = \int f\,d\nu$.

To prove the uniqueness, let $\int gh\,d|\mu| = 0$ for all $g \in \mathcal{C}(K)$. Since $\mathcal{C}(K)$ is dense in $L^1(|\mu|)$, by Corollary 2.13, we can consider $\|\bar{h} - g_k\|_1 \to 0$, $g_k \in \mathcal{C}(K)$. Then

$$|\mu|(K) = \int (1 - g_k h)\,d|\mu| = \left|\int (\bar{h} - g_k)h\,d|\mu|\right| \le \|\bar{h} - g_k\|_1 \to 0$$

and $\mu = 0$. Thus, the linear map $\mu \mapsto u_\mu$ is injective. □

[3] It can be proved that $\|u_\mu\|_{\mathcal{C}(K)'} = |\mu|(K)$.

4.3. The Hahn-Banach theorem

For a Hilbert space, the projection theorem has been useful to give a complete description of the dual. The duality theory of more general spaces is based on the Hahn-Banach theorem.[4]

4.3.1. Analytic form of Hahn-Banach theorem. The basic form of the Hahn-Banach theorem refers to a **convex functional** on a real vector space E, which is a function $q : E \to \mathbf{R}$ such that $q(x+y) \leq q(x) + q(y)$ and $q(\alpha x) = \alpha q(x)$ $(x, y \in E$ and $\alpha \geq 0)$.

We note that the sets $\{x;\, q(x) \leq 1\}$ and $\{x;\, q(x) < 1\}$ are convex.

Obviously, a semi-norm is a convex functional.

Theorem 4.14 (Hahn-Banach). *Let E be a real vector space, F a vector subspace of E, q a convex functional on E, and u a linear form on F which is dominated by q:*

$$u(y) \leq q(y) \quad (y \in F).$$

Then u can be extended to all E as a linear form v dominated by q:

$$v(x) \leq q(x) \quad (x \in E).$$

Proof. If $F \neq E$, we start with a simple extension of u to the subspace

$$F \oplus [y] = \{z + ty;\, z \in F,\, t \in \mathbf{R}\},$$

the linear span of F and $y \notin F$.

Of course, if $s \in \mathbf{R}$, then $v(z+ty) = u(z) + ts$ is a linear extension of u. Our aim is to obtain the estimate

$$v(z+ty) = u(z) + ts \leq q(z+ty)$$

by choosing a convenient value for $s = v(y)$. Since q is positive homogeneous, if this estimate holds for $t = \pm 1$, then $v(z+ty) = (1/|t|)v(|t|z \pm y) \leq (1/|t|)q(|t|z \pm y) = q(z+ty)$. Hence all we need are the inequalites

$$u(z) + s \leq q(z+y), \quad u(z') - s \leq q(z'-y) \qquad (z, z' \in F);$$

that is, $u(z') - q(z'-y) \leq s \leq q(z+y) - u(z)$. Hence

$$u(z'+z) \leq q(z' - y + z + y) \leq q(z'-y) + q(z+y).$$

It is possible to choose s so that

$$\sup_{z' \in F} (u(z') - q(z'-y)) \leq s \leq \inf_{z \in F} (q(z+y) - u(z))$$

and then v is dominated by q.

---

[4]The first version of the Hahn-Banach theorem dates back to the work of the Austrian mathematician Eduard Helly in 1912, essentially with the same proof given later independently by H. Hahn (1926) and S. Banach (1929). We give the original proof published by S. Banach in 1929; only his transfinite induction has been changed by an application of Zorn's lemma.

Once we know that a one-dimensional extension is always possible, we can continue with a standard application of Zorn's lemma as follows.

Consider the family Φ of all extensions ℓ of u to vector subspaces L of E that are dominated by q, and we order Φ by $(L_1, \ell_1) \le (L_2, \ell_2)$ meaning that $L_1 \subset L_2$ and ℓ_2 agrees with ℓ_1 on L_1.

Every totally ordered subset $\{(L_\alpha, \ell_a)\}$ of Φ has the upper bound (L, ℓ) obtained by defining $L := \bigcup_\alpha L_\alpha$ and $\ell(y) := \ell_\alpha(y)$ if $y \in L_\alpha$. If also $y \in L_{\alpha'}$, then $\ell_\alpha(y) = \ell_{\alpha'}(y)$ by the total ordering of the set $\{(L_\alpha, \ell_a)\}$, and the previous definition is unambiguous. For the same reason, L is a vector subspace of E and ℓ is a linear extension of all the linear forms ℓ_α.

By Zorn's lemma, there is a maximal element $(\tilde{F}, \tilde{\ell})$ in Φ. But according to the first part of this proof, this extension must be the whole space E, since if $\tilde{F} \neq E$, we would obtain an extension v to a strictly larger subspace $\tilde{F} \oplus [y]$ of E, in contradiction to the maximality of $(\tilde{F}, \tilde{\ell})$. □

The following version of the Hahn-Banach theorem for semi-norms holds for real and complex vector spaces:

Theorem 4.15 (Hahn-Banach). *Let F be a vector subspace of the vector space E, q a semi-norm on E, and u a linear form on F that satisfies the estimate*

$$|u(y)| \le q(y) \quad (y \in F).$$

Then there exists an extension of u to a linear form v on E which satisfies

$$|v(x)| \le q(x) \quad (x \in E).$$

Proof. The real case is simple. We know from Theorem 4.14 that there exists an extension v which satisfies $v(x) \le q(x)$ and $-v(x) = v(-x) \le q(-x) = q(x)$, i.e. $|v(x)| \le q(x)$.

Now suppose E is a complex vector space[5] and F is a complex vector subspace of E. In this case we split the complex linear form u into the real and imaginary parts, $u(y) = u_1(y) + iu_2(y)$. Then u_1 and u_2 are real linear forms on F, which is also a real vector subspace of E regarded as a real vector space. Since $u_1(iy) + iu_2(iy) = u(iy) = iu(y)$, u_1 and u_2 are related by $u_1(iy) = -u_2(y)$.

Conversely, if u_1 is a real linear form on F, the additive functional

$$u(y) = u_1(y) - iu_1(iy) \tag{4.10}$$

is a complex linear form, since $u(iy) = iu(y)$ and $u(ry) = ru(y)$ when $r \in \mathbf{R}$.

[5]The complex version of the Hahn-Banach theorem was published simultaneously in 1938 by H. F. Bohnenblust and A. Sobczyk and by G. Buskes; curiously, in his work Banach only considered the real case.

To extend our $u(y) = u_1(y) - iu_1(iy)$, it follows from the real case and $|u_1(y)| \le |u(y)| \le q(y)$ that there exists a real linear extension v_1 of u_1 so that $|v_1(x)| \le q(x)$. Then the complex linear form $v(x) = v_1(x) - iv_1(ix)$ is an extension of u, since $v_1(y) = u_1(y)$.

Finally, we have $|v_1(x)| \le q(x)$ and, if for a given $x \in E$ we write $|v(x)| = \lambda v(x)$ with $|\lambda| = 1$, then it follows that also

$$|v(x)| = v(\lambda x) = v_1(\lambda x) \le q(\lambda x) = |\lambda| q(x) = q(x),$$

since q is a semi-norm. □

4.3.2. The geometric Hahn-Banach theorem. Suppose that K is a convex subset of a real vector space E and that K is **absorbing** in the sense that $\bigcup_{t>0}(tK) = E$. Then the **gauge** or **Minkowski functional** of K is defined as the functional

$$p_K : E \to [0, \infty)$$

such that

$$p_K(x) = \inf \{t > 0;\ \frac{x}{t} \in K\}.$$

It is easily checked that this functional is convex and that

$$K \subset \{x \in E;\ p_K(x) \le 1\}.$$

To show the subadditivity of p_K, note that, if $x/t, y/s \in K$, also

$$\frac{x+y}{t+s} = \frac{t}{t+s}\frac{x}{t} + \frac{s}{t+s}\frac{y}{s} \in K \tag{4.11}$$

and $p_K(x+y) \le t+s$. By taking the infimum with respect to t and then with respect to s, we obtain $p_K(x+y) \le p_K(x)+s$ and $p_K(x+y) \le p_K(x)+p_K(y)$.

It is also worth noticing that $p_K(x) < 1$ if x is an **internal point** of K, in the sense that for any $z \in E$ there is an $\varepsilon = \varepsilon(z) > 0$ such that

$$\{x + tz;\ |t| \le \varepsilon\} \subset K,$$

that is, for every line L through x, $L \cap K$ is a neighborhood of x in L. Indeed, if x is internal, then we have $(1+\varepsilon)x \in K$ for some $\varepsilon > 0$, so that $p_K(x) \le 1/(1+\varepsilon) < 1$.

Sometimes we will use self-explanatory notation such as $f(A) < f(B)$ when $f(a) < f(b)$ for all $a \in A$, $b \in B$.

Theorem 4.16. *Let K be a convex subset of a real vector space E with at least an internal point x_0. For any $y \in E \setminus K$ there is a nonzero linear form $f : E \to \mathbf{R}$ that satisfies*

$$f(K) \le f(y).$$

If all the points of K are internal, then f can be chosen so that

$$f(K) < f(y).$$

Proof. A translation allows us to assume that $x_0 = 0$ is an internal point of K, and then K is absorbing.

Note that $p_K(y) \geq 1$, since $y \neq 0$, and we can define f on $[y]$ so that $f(y) = 1$; that is, $f(ty) = t$. Then $t \leq tp_K(y) = p_K(ty)$ if $t \geq 0$, and $t \leq p_K(ty)$ is obvious if $t < 0$.

Having shown that $f \leq p_K$ on $[y]$, we conclude from Theorem 4.14 that f can be extended to all of E so that $f(x) \leq p_K(x) \leq 1 = f(y)$.

Recall that if x is an internal point of K, then $p_K(x) < 1$. □

Next we consider E equipped with a vector topology over the reals and two disjoint subsets A and B in E. We say that these sets are **separated** if there exists a nonzero $f \in E'$ such that

$$f(A) < f(B).$$

If instead of this we have

$$\sup f(A) < \inf f(B),$$

we say that A and B are **strictly separated**.

Theorem 4.17. *Suppose A and B are two nonempty disjoint convex subsets of a real locally convex space E.*

(a) *If one of them is open, then A and B are separated.*

(b) *If they are closed and one of them is compact, then A and B are strictly separated.*

Proof. We choose $a_0 \in A$, $b_0 \in B$. Then the convex set

$$K = A - B + b_0 - a_0 = \bigcup_{y \in B} (A - y + b_0 - a_0)$$

is an open neighborhood of zero and $y = b_0 - a_0 \notin K$.

An application of Theorem 4.16 provides a linear form f such that $f(y) = 1$ and $f(K) < 1$. Also $f(-K) \geq 1$. Then $U = K \cap -K$ is an open neighborhood of zero such that $|f(U)| \leq 1$, and $f \in E'$.

If $a \in A$ and $b \in B$, then $f(a) < f(b)$, since $a - b + y \in K$ implies

$$f(a) - f(b) = f(a - b + y) - 1 < 0.$$

Then $f(A)$ and $f(B)$ are two disjoint convex subsets in $\mathbf{R}$, and $f(A)$ is open. This is readily shown, since f is a nonzero linear form and, if U is a balanced and convex neighborhood of 0, then $f(U) = I \subset \mathbf{R}$ is a neighborhood of 0 in $\mathbf{R}$ and it follows as in the proof of Theorem 3.10 that $f(G)$ is open if G is open.

Thus, if $r = \inf f(B)$, $f(a) < r$ for every $f(a)$ in the open interval $f(A)$.

To prove (b), since A is compact and B is closed, for every $a \in A$ we can consider $(a+U_a+U_a)\cap B = \emptyset$, with U_a a convex neighborhood of zero; then, by choosing $A \subset \bigcup_{n=1}^N (a_n + U_{a_n})$, $U = \bigcap_{n=1}^N U_{a_n}$ is an open convex set. The set $A+U$ is also open and convex, and it is disjoint with B, since

$$A+U \subset \bigcup_{n=1}^{N} (a_n + U_{a_n} + U) \subset \bigcup_{n=1}^{N} (a_n + U_{a_n} + U_{a_n}).$$

Now we apply (a) to the couple $A+U$, B to obtain $f \in E'$ so that $f(A+U)$ and $f(B)$ are two disjoint intervals in $\mathbf{R}$, $f(A)$ a compact interval contained in the open interval $f(A+U)$, and (b) follows. □

Remark 4.18. For a complex locally convex space E, separation refers to E as a real locally convex space. Note that, if f is a continuous real linear form such that $f(A) < f(B)$, we know from (4.10) that $u(x) = f(y) - if(iy)$ is the uniquely determined complex linear form by the condition $\Re u = f$. It is obviously continuous.

Corollary 4.19. Assume that E is a real or complex locally convex space.

(a) If K is a convex and balanced closed subset of E and $x_0 \in E \setminus K$, then $|u(K)| \leq 1$ and $u(x_0) > 1$, real, for some $u \in E'$.

(b) If M is a closed vector subspace of E and $x_0 \in E \setminus M$, then $u(M) = \{0\}$ and $u(x_0) = 1$ for some $u \in E'$.

Proof. (a) By Theorem 4.17, K and $\{x_0\}$ are two strictly separated sets and $0 \in K$, so that we can choose $f = \Re u$ satisfying $f(x) < 1 < f(x_0)$ for all $x \in K$. Since K is symmetric, $|f(x)| < 1 < f(x_0) = \Re u(x_0)$.

(b) From (a) we obtain $u \in E'$ such that $u(x_0) \notin u(M)$ and M is a proper vector subspace of $\mathbf{K}$, which forces $u(M) = \{0\}$. We can normalize u so that $u(x_0) = 1$. □

4.3.3. Extension properties.

Theorem 4.20. *Suppose F is a subspace of a normed space E.*

(a) *If $u \in F'$, there exists an extension of u to a continuous linear form $v \in E'$ such that $\|v\|_{E'} = \|u\|_{F'}$. If $x_1 \neq x_2$ are two points of E, then there exists $v \in E'$ so that $v(x_1) \neq v(x_2)$.*

(b) *For every $x_0 \in E \setminus F$, there exists $v \in E'$ such that $\|v\|_{E'} = 1$, $v(F) = \{0\}$ and $v(x_0) = d(x_0, F) = \inf_{y\in F} \|y - x_0\|$.*

If $0 \neq x_0 \in E$, then $\|v\|_{E'} = 1$ and $v(x_0) = \|x_0\|_E$ for some $v \in E'$.

Proof. (a) If $q(x) := \|u\|_{F'}\|x\|_E$, by the Hahn-Banach Theorem 4.15, u admits a linear extension v that also satisfies $|v(x)| \leq q(x) = \|u\|_{F'}\|x\|_E$, and obviously $\|v\|_{E'} \geq \|u\|_{F'}$.

If $y = x_1 - x_2 \neq 0$, it is easy to obtain $v \in E'$ such that $v(x_1 - x_2) = \|x_1 - x_2\|_E \neq 0$. Just extend $u(ty) := t\|y\|_E$ as above.

(b) On the subspace $Z = [x_0] \oplus F$ of E we can define the linear form $u(\lambda x_0 + y) = \lambda d(x_0, F)$, so that $\|u\|_{Z'} \leq 1$, since

$$|u(\lambda x_0 + y)| \leq |\lambda| \|x_0 + \frac{1}{\lambda} y\| = \|\lambda x_0 + y\|.$$

Then, by choosing $\|x_0 - y_\varepsilon\|_E < d(x_0, F) + \varepsilon$ ($y_\varepsilon \in F$), we obtain $u(x_0 - y_\varepsilon) = d(x_0, F)$ and

$$\frac{1}{\|x_0 - y_\varepsilon\|_E} |u(x_0 - y_\varepsilon)| > \frac{d(x_0, F)}{d(x_0, F) + \varepsilon}$$

for every $\varepsilon > 0$. Thus, $\|u\|_{F'} = 1$ and u has an extension to $v \in E'$ with $\|v\|_{E'} = 1$.

In the special case $F = \{0\}$, $v(x_0) = d(x_0, 0) = \|x_0\|_E$ and $\|v\|_{E'} = 1$. □

Theorem 4.21. *If F is a subspace of a locally convex space E and $u \in F'$, then there exists an extension of u to a continuous linear form $v \in E'$.*

If x_1 and x_2 are two distinct points of E, there exists $v \in E'$ so that $v(x_1) \neq v(x_2)$.

Proof. The topology of F is defined by the restriction of any sufficient family of semi-norms for the topology of E. By Theorem 3.4, there exists a continuous semi-norm q on E so that $|u(y)| \leq q(y)$ for all $y \in F$. By the Hahn-Banach Theorem 4.15, u admits a linear extension v that also satisfies $|v(x)| \leq q(x)$ and $v \in E'$.

If $y = x_1 - x_2 \neq 0$, $u(\lambda y) := \lambda$ defines a linear form on $F = [y]$ and we can choose a continuous semi-norm p on E such that $p(y) \neq 0$. Then we have $|u(y)| = cp(y)$, $|u(\lambda y)| = cp(\lambda y)$, and u has an extension to some linear form v on E such that $|v(x)| \leq cp(x)$ and $v \in E'$. Since $u(y) \neq 0$, $v(x_1) \neq v(x_2)$. □

Assume that E is a locally convex space and that M a closed subspace of E. If there exists a second closed subspace of E such that $E = M \oplus N$, that is, $E = M + N$ and $M \cap N = \{0\}$, then M and N are said to be **topologically complementary** subspaces of E and M (and N) is a **complemented** subspace.

As an application of Theorem 4.20 and Theorem 4.21, let us show that finite-dimensional subspaces are complemented:

Theorem 4.22. *Suppose that E is a normed space (or any locally convex space) and that N is a finite-dimensional vector subspace of E. Then $E = N \oplus M$ for some closed subspace M of E.*

Proof. Let $\{e_1, \dots, e_n\}$ be a base of N and let π_j $(1 \leq j \leq n)$ be the corresponding projections, so that $y = \sum_{j=1}^n \pi_j(y)e_j$. Since N has finite dimension, every projection is continuous and, by Theorem 4.21, it has a continuous linear extension $\tilde{\pi}_j : E \to \mathbf{K}$. We are going to show that we can take $M = \bigcap_{j=1}^n \operatorname{Ker} \tilde{\pi}_j$, which is a closed vector subspace of E.

Indeed, if $x \in E$, let $y := \sum_{j=1}^n \tilde{\pi}_j(x)e_j \in N$ and $z := x - y$. Then $\tilde{\pi}_j(z) = \tilde{\pi}_j(x) - \pi_j(y) = 0$ for every j and $z \in N$, so that $E = N + M$. This sum is direct, since if $y = \sum_{j=1}^n \pi_j(y)e_j \in N$ is also in M, then $\pi_j(y) = \tilde{\pi}_j(y) = 0$ for every j and $y = 0$. □

Remark 4.23. For any locally convex space E, separation for a point and a closed subspace F can be obtained from Theorem 4.21 by means of the quotient map π from E onto the quotient space E/F.

If $\mathcal{P}$ is a sufficient family of semi-norms on E, recall that E/F is endowed with the topology defined by the family of semi-norms $\tilde{\mathcal{P}}$ defined by

$$\tilde{p}(\tilde{x}) = \inf_{y \in \tilde{x}} p(y) = \inf_{z \in F} p(x - z)$$

and the quotient map is continuous. If $x \in E \setminus F$, then $0 \neq \tilde{x} \in E/F$ and we can find $\tilde{v} \in (E/F)'$ which satisfies $\tilde{v}(\tilde{x}) \neq 0$ so that we only need to define $v = \tilde{v}\pi$.

4.3.4. Proofs by duality: annihilators, total sets, completion, and the transpose. Here we present some duality results that depend on the Hahn-Banach theorem.

Most of these results turn out to be very useful in applications, in spite of the nonconstructive nature of that theorem, in whose proof we have used Zorn's lemma.

Let E' be the dual of a locally convex space E and write

$$\langle x, u \rangle := u(x). \tag{4.12}$$

Then $\langle \cdot, \cdot \rangle : E \times E' \to \mathbf{K}$ is a bilinear form such that, if $u(x) = \langle x, u \rangle = 0$ for all $x \in E$, then $u = 0$, and also $x = 0$ if $\langle x, u \rangle = 0$ for all $u \in E'$, by Theorem 4.21.

The annihilator of $A \subset E$ is the closed subspace of E'

$$A^o := \{v \in E';\ \langle a, v \rangle = 0 \,\forall a \in A\} = \bigcap_{a \in A} \operatorname{Ker} \langle a, \cdot \rangle$$

and the annihilator of $U \subset E'$ is the closed subspace of E

$$U^o := \{x \in E;\ \langle x, u \rangle = 0 \,\forall u \in U\} = \bigcap_{u \in U} \operatorname{Ker} u.$$

Obviously, $A \subset B \Rightarrow B^o \subset A^o$, and $A \subset A^{oo}$.

Annihilators play the role that orthogonality plays in Hilbert spaces. They can be used to characterize by duality the closure of a vector subspace:

Theorem 4.24. *Suppose E is a locally convex space. The closed linear span $\overline{[A]}$ of a subset A of E coincides with A^{oo}, the annihilator in E of $A^o \subset E'$, so that A is total in E if and only if $A^o = \{0\}$.*

Thus, a vector subspace F of E is closed if and only if $F^{oo} = F$.

Proof. It is clear that the annihilator of A coincides with the annihilator of the linear span $[A]$ of A and, by continuity, with the annihilator of the closure of $[A]$; thus, if $F = \overline{[A]}$, we need to prove that $F^{oo} = F$.

Indeed, we have $F \subset F^{oo}$ and, if $x \notin F$, by Theorem 4.20(b), we can choose $v \in E'$ so that $v \in F^o$ and $v(x) \neq 0$. This shows that also $x \notin F^{oo}$.

Note that, if $F \neq E$, there exists $v \in F^o$, $v \neq 0$, so that $F^o \neq \{0\}$. □

Theorems 4.24 and 2.36 are in the basis of certain approximation results. To prove that a point x of a locally convex space E lies in the closure of a subspace F, all we need is to show that $u(x) = 0$ for every $u \in E'$ that vanishes on F.

Theorem 4.25. *For any normed space E the mapping $J : E \to E''$ such that $J(x) = \widehat{x}$, where*

$$\widehat{x}(u) = \langle x, u \rangle = u(x),$$

is a linear isometry from E into the Banach space E'', endowed with the norm $\|w\|_{E''} = \sup_{\|u\|_{E'} \le 1} |w(u)|$.

Hence, the closure of $J(E)$ in E'' is a completion of E.

Proof. The function $\widehat{x} = \langle x, u \rangle$ is clearly linear on E', and $\|\widehat{x}\|_{E''} \le \|x\|_E$, since $|\widehat{x}(u)| = |u(x)| \le \|u\|_{E'} \|x\|_E$. According to Theorem 4.20(b), we can find some $v \in E'$ such that $\|v\|_{E'} \le 1$ and $\widehat{x}(v) = \|x\|_E$; thus $\|\widehat{x}\|_{E''} = \|x\|_E$.

Finally, J is linear:

$$J(x_1 + x_2)(u) = u(x_1 + x_2) = J(x_1)(u) + J(x_2)(u) = (J(x_1) + J(x_2))(u)$$

and also $J(\lambda x)(u) = \lambda J(x)(u)$ for every $u \in E'$. □

Assume now that $T : E \to F$ is a continuous linear operator between two normed spaces. The **transpose** T' of T is defined on every $v \in F'$ by $T'v = v \circ T$, which is obviously a linear and continuous functional on E, and clearly it depends linearly on v. Then

$$T' : F' \to E'$$

is continuous, since

$$\|T'v\|_{E'} = \|vT\|_{E'} \le \|T\| \|v\|_{F'}. \tag{4.13}$$

It is useful to rewrite the definition of T' as

$$\langle Tx, v\rangle = \langle x, T'v\rangle.$$

Theorem 4.26. *The transposition map $T \in \mathcal{L}(E;F) \mapsto T' \in \mathcal{L}(F';E')$ is a linear isometry and for every $T \in \mathcal{L}(E;F)$ the following properties hold:*

(a) $(\operatorname{Im} T)^o = \operatorname{Ker} T'$,

(b) $(\operatorname{Ker} T')^o = \overline{\operatorname{Im} T}$, *and*

(c) $(\operatorname{Im} T')^o = \operatorname{Ker} T$.

Proof. It is clear that $(T+S)' = T' + S'$ and $(\lambda T)' = \lambda T'$. Moreover, as in the Hilbert space case for the adjoint,

$$\|T\| = \sup_{\|x\|_E \le 1,\, \|v\|_{F'} \le 1} |\langle Tx, v\rangle| = \sup_{\|x\|_E \le 1,\, \|v\|_{F'} \le 1} |\langle x, T'v\rangle| = \|T'\|.$$

(a) Note that $\langle Tx, v\rangle = \langle x, T'v\rangle$ and $v \in (\operatorname{Im} T)^o$ if and only if $\langle x, T'v\rangle = 0$ for all $x \in E$, that is, if and only if $v \in \operatorname{Ker} T'$.

(b) By (a) and Theorem 4.24, $(\operatorname{Ker} T')^o = (\operatorname{Im} T)^{oo} = \overline{\operatorname{Im} T}$.

(c) As in (a), since $\langle x, T'v\rangle = \langle Tx, v\rangle$, $x \in \operatorname{Ker} T$ if and only if $\langle x, T'v\rangle = 0$ for all $T'v \in \operatorname{Im} T'$. □

Remark 4.27. It is easily checked that also $\overline{\operatorname{Im} T'} \subset (\operatorname{Ker} T)^o$, but the reverse inclusion is not always true, as shown in Exercise 4.20.

4.4. Spectral theory of compact operators

The principal axes theorem of analytical geometry asserts that any symmetric quadratic form on $\mathbf{R}^n$

$$(Ax, x) = \sum_{i,j=1}^{n} \alpha_{ij} x_i x_j$$

can be rewritten in the normal form $\sum_{i=1}^n \lambda_i x_i^2$ by means of an orthogonal transform. The general form of this theorem, in the language of matrices or operators, says that each real symmetric matrix A is orthogonally equivalent to a diagonal matrix whose diagonal entries are the roots of the equations $\det(A - \lambda I) = 0$, that is, the eigenvalues of A.[6]

The earliest extensions to an infinite-dimensional theory were achieved after a construction of determinants of infinite systems. They were applied to integral equations around 1900, defined by operators with properties that are close to those of the finite-dimensional case.

[6]This was obtained in 1852 by the English mathematician J. Sylvester in terms of the quadratic form (Ax, x), and A. Cayley inaugurated the calculus of matrices in which the reduction to normal form corresponds to a diagonalization process.

Here, for a first version of this spectral theory in the infinite-dimensional case, we are going to consider the class of compact operators in Banach spaces.

Let $T : E \to F$ be a linear map between Banach spaces and let B_E be the closed unit ball of E. Then T is said to be **compact** if $\overline{T(B_E)}$ is compact in F, that is, if every sequence in $T(B_E)$ has a Cauchy subsequence (see Exercise 1.3).

Such an operator is bounded, since $\overline{T(B_E)} \subset \bigcup_{n=0}^{\infty} B_F(0,n)$ and, by compactness, $\overline{T(B_E)} \subset B_F(0,N)$ for some N.

Every bounded linear operator T with finite-dimensional range $T(E)$ is compact, since in $T(E)$ the closure of the bounded set $T(B_E)$ is compact, by Theorem 2.25.

We will use the notation $\mathcal{L}_c(E;F)$ to represent the collection of all compact linear operators between E and F, and $\mathcal{L}_c(E) = \mathcal{L}_c(E;E)$.

4.4.1. Elementary properties.

The following theorem is useful to prove the compactness of certain operators:

Theorem 4.28 (Ascoli-Arzelà[7]). *Let K be a compact metric space and assume that $\Phi \subset \mathcal{C}(K)$ satifies the following two conditions:*

1. $\sup_{f\in\Phi} |f(x)| < \infty$ *for every $x \in K$ ($\Phi(x)$ is bounded for every $x \in K$).*
2. *For every $\varepsilon > 0$ there is some $\delta > 0$ such that* $\sup_{f\in\Phi} |f(x) - f(y)| \leq \varepsilon$ *if $d(x,y) \leq \delta$ (we say that Φ is "equicontinuous").*

Then $\bar{\Phi}$ is compact in $\mathcal{C}(K)$.

Proof. If $\delta = 1/m$, the compact set K has a finite covering by balls $B_K(c_{m,j},\delta)$, and the collection of all the centers for $m = 1,2,\ldots$ is a countable dense set $C = \{c_k\}$ in K.

Let $\{f_n\} \subset \Phi$. By the first condition, we can select a convergent subsequence $\{f_{n,1}(c_1)\}$, then we obtain a subsequence $\{f_{n,2}\} \subset \{f_{n,1}\}$ so that $\{f_{n,2}(c_2)\}$ is also convergent, and so on. Let $\{f_{n'}\}$ be the diagonal sequence $\{f_{m,m}\}$, which is convergent at every $c \in C$.

[7] In 1884 Giulio Ascoli (at the Politecnico di Milano) needed the assumption of equicontinuity to prove that a sequence of uniformly bounded functions possesses a convergent subsequence. In 1889 Cesare Arzelà (in Bologna) considered the case of continuous functions and proved what is nowadays usually called the theorem of Ascoli-Arzelà and in 1896 he published a paper in which he applied his results to prove, under certain extra assumptions, the Dirichlet principle.

For $\varepsilon > 0$, let $\delta = 1/m$ be as in condition 2. For every $x \in K$, we choose $c_j = c_{j(x)} \in C$ so that $x \in \bar{B}_K(c_j, \delta)$ with $j \leq n(\varepsilon)$ and then

$$\begin{aligned}|f_{p'}(x) - f_{q'}(x)| \\ &\leq |f_{p'}(x) - f_{p'}(c_j)| + |f_{p'}(c_j) - f_{q'}(c_j)| + |f_{q'}(c_j) - f_{q'}(x)| \\ &\leq 2\varepsilon + |f_{p'}(c_j) - f_{q'}(c_j)|\end{aligned}$$

where

$$|f_{p'}(c_j) - f_{q'}(c_j)| \leq \max_{k \leq n(\varepsilon)} |f_{p'}(c_k) - f_{q'}(c_k)| \to 0,$$

so that $\|f_{p'} - f_{q'}\|_K \to 0$. □

The following result gathers together some of the basic properties of compact linear operators:

Theorem 4.29. *The collection $\mathcal{L}_c(E;F)$ of all compact linear operators between two Banach spaces is a closed subspace of $\mathcal{L}(E;F)$ and the right or left composition of a compact linear operator with a bounded linear operator is compact.*

If $T \in \mathcal{L}_c(E;F)$, then also $T' \in \mathcal{L}_c(F';E')$; if E and F are Hilbert spaces, $T^* \in \mathcal{L}_c(F;E)$ (**Schauder theorem**).[8]

Proof. It is clear that $\mathcal{L}_c(E;F)$ is a linear subspace of $\mathcal{L}(E;F)$.

If $T = \lim_n T_n$ with $T_n \in \mathcal{L}_c(E;F)$, to prove that T is compact, let $\varepsilon > 0$ and let N be such that $\|T - T_N\| \leq \varepsilon/2$. Since $\overline{T_N(B_E)}$ is compact, it is covered by a finite collection of balls $B_F(y_i, \varepsilon/2)$ $(i \in I)$, and then $T(B_E) \subset \bigcup_{i \in I} B_F(y_i, \varepsilon)$, since for every $x \in B_E$ we have $\|Tx - T_N x\|_F \leq \varepsilon/2$ and $T_N x \in B_F(y_i, \varepsilon/2)$ for some $i \in I$.

Now, by considering $\varepsilon_m = 1/m$, it is easy to check that every sequence $\{Tx_n\}$ with $x_n \in B_E$ has a partial Cauchy sequence, since we obtain successive subsequences $\{Tx_{n,m}\} \subset \{Tx_{n,m-1}\}$ contained in some $B_F(y, \varepsilon_m)$ and the diagonal sequence $\{Tx_{m,m}\}$ is a Cauchy subsequence of $\{Tx_n\}$.

Let $S = TR$ and assume first that $T : F \to G$ is compact and $R : E \to F$ is bounded. Then $R(B_E) \subset \|R\| B_F$ and $S(B_E) \subset \|R\| T(B_F)$, whose closure is compact; hence, S is compact. If T is bounded and R compact, then $\overline{R(B_E)}$ is compact and its image by the continuous map T is also compact, so that $S(B_E)$ is contained in a compact set.

Suppose T is compact and consider $\{v_n\}_{n \in \mathbf{N}} \subset B_{F'}$. To obtain a Cauchy subsequence of $\{T'v_n\}$, let $K := \overline{T(B_E)}$ and $\Phi := \{f_n = v_{n|K};\, n \in \mathbf{N}\} \subset$

[8]In 1930, the Polish mathematician Julius Pawel Schauder proved this result that allowed the use of duality in the Riesz-Fredholm theory for general Banach spaces. Schauder is well known for his fixed point theorem, for the Schauder bases in Banach spaces, and for the Leray-Schauder principle on partial differential equations. See also footnote 3 in Chapter 3.

$\mathcal{C}(K)$. Then Φ is equicontinuous because

$$|f_n(Tx) - f_n(Ty)| = |v_n(Tx) - v_n(Ty)| \leq \|Tx - Ty\|,$$

and it is also uniformly bounded, since $\|f_n\|_K \leq \|v_n\|_{F'} \leq 1$. According the Ascoli-Arzelà Theorem 4.28, $\bar{\Phi}$ is compact in $\mathcal{C}(K)$ and $\{f_n\}$ has a Cauchy subsequence $\{f_{n_k}\}$, so that

$$\|T'v_{n_p} - T'v_{n_q}\|_{E'} = \sup_{x \in B_E} |v_{n_p}(Tx) - v_{n_q}(Tx)| \leq \sup_{a \in K} |f_{n_p}(a) - f_{n_q}(a)| \to 0$$

as $p, q \to \infty$. □

Example 4.30. Every Volterra operator

$$Tf(x) = \int_a^x K(x,y) f(y)\, dy,$$

where $K(x,y)$ is continuous on $\Delta = \{(x,y) \in [a,b] \times [a,b]; a \leq y \leq x \leq b\}$, is a compact operator $T : \mathcal{C}[a,b] \to \mathcal{C}[a,b]$, since $\Phi = T(B_{\mathcal{C}[a,b]})$ satisfies the conditions of the Ascoli-Arzelà Theorem 4.28.

Indeed, if $a \leq t \leq s \leq b$ and if for a given $\varepsilon > 0$ we choose $\delta > 0$ so that $|K(s,y) - K(t,y)| \leq \varepsilon$ if $|s-t| \leq \delta$ and $(s,y), (t,y) \in \Delta$ (K is uniformly continuous on Δ), then

$$\begin{aligned} |Tf(s) - Tf(t)| &\leq \int_a^t |K(s,y) - K(t,y)||f(y)|\, dy \\ &\quad + \int_t^s |K(s,y)||f(y)|\, dy \\ &\leq (b-a)\varepsilon + \|K\|_\Delta |s-t| \end{aligned}$$

for every $f \in B_{\mathcal{C}[a,b]}$ if $|s-t| \leq \delta$, and Φ is equicontinuous. Obviously it is uniformly bounded, since $\|Tf\|_{[a,b]} \leq \|K\|_\Delta (b-a) \|f\|_{[a,b]}$.

Example 4.31. It is shown in a similar way that every Fredholm operator T_K, defined as in (2.22) by a continuous integral kernel K, is compact.

Note that $|T_K f(s) - T_K f(t)| \leq \int_c^d |K(s,y) - K(t,y)||f(y)|\, dy \leq (d-c)\varepsilon$ if $\|f\|_{[c,d]} \leq 1$ and $|s-t| \leq \delta$ small, by the uniform continuity of K.

Example 4.32. The Hilbert-Schmidt operator $T_K : L^2(Y) \to L^2(X)$, defined as in Theorem 4.6 by a kernel $K \in L^2(X \times Y)$, is also compact.

This is proved by choosing a couple of orthogonal bases $\{u_n\} \subset L^2(X)$ and $\{v_m\} \subset L^2(Y)$, so that an application of Fubini's theorem shows that the products $w_{m,n}(x,y) = u_n(x)v_m(y)$ form an orthogonal basis in $L^2(X \times Y)$. By the Fischer-Riesz Theorem 2.37,

$$K = \sum_{n,m} c_{n,m} w_{n,m},$$

with convergence in $L^2(X \times Y)$.

The Hilbert-Schmidt operator defined by the kernel

$$K_N = \sum_{n,m \le N} c_{n,m} w_{n,m}$$

is compact, since it is continuous by Theorem 4.6 and its range is finite-dimensional, contained in $[u_n;\, n \le N]$. Now, again by Theorem 4.6,

$$\|T_K - T_{K_N}\| \le \|K - K_N\|_2 \to 0$$

and, according to Theorem 4.29, T_K is compact.

4.4.2. The Riesz-Fredholm theory. For compact operators there is a complete spectral theory.[9]

Recall that $\lambda \in \mathbf{K}$ is an **eigenvalue** of $T \in \mathcal{L}(E)$ if $\mathcal{N}_T(\lambda) = \operatorname{Ker}(T - \lambda I)$ is nonzero, that is, if $T - \lambda I$ is not one-to-one. Every nonzero $x \in \mathcal{N}_T(\lambda)$ is called an **eigenvector**, and $\mathcal{N}_T(\lambda)$ is an **eigenspace**. Obviously $T = \lambda I$ on $\mathcal{N}_T(\lambda)$, which is a closed subspace of E and it is invariant for T. The **multiplicity** of an eigenvalue λ is the dimension of $\mathcal{N}_T(\lambda)$.

The **spectrum**[10] of T is the set $\sigma(T)$ of all scalars λ such that $T - \lambda I$ is not invertible. That is, $\lambda \in \sigma(T)$ if either λ is an eigenvalue of T or the range $\mathcal{R}_T(\lambda) := \operatorname{Im}(T - \lambda I)$ is not all of E. This subspace of E need not be closed, but it is also invariant for T, since $T(Tx - \lambda x) = T(Tx) - \lambda Tx$ belongs to $\mathcal{R}_T(\lambda)$.

Note that if T is compact and $0 \notin \sigma(T)$, then $\dim(E) < \infty$, since in this case T^{-1} is continuous by the open mapping theorem and $I = T^{-1}T$ is compact, so that the unit ball B_E and the unit sphere $S_E = \{x;\, \|x\|_E = 1\}$ are compact, and Theorem 2.28 applies.

Theorem 4.33 (The Fredholm alternative). *Suppose $T \in \mathcal{L}_c(E)$ and $\lambda \ne 0$. Then:*

(a) $\dim \mathcal{N}_T(\lambda) < \infty$.

(b) $\mathcal{R}_T(\lambda) = \mathcal{N}_{T'}(\lambda)^o$, *and it is closed in E. If E is a Hilbert space, then $\mathcal{R}_T(\lambda) = \mathcal{N}_{T^*}(\bar{\lambda})^\perp$.*

(c) $\mathcal{N}_T(\lambda) = \{0\}$ *if and only if $\mathcal{R}_T(\lambda) = E$; thus, if $0 \ne \lambda \in \sigma(T)$, then λ is an eigenvalue of T.*

---

[9]David Hilbert constructed his spectral theory for ℓ^2 essentially with Fredholm's method (see footnote 5 in Chapter 2), and F. Riesz followed him to develop the theory on L^2. In 1918, Riesz extended Hilbert's notion of a compact operator to complex functional spaces, a few years before the introduction of general Banach spaces.

[10]In this context, the term spectrum was coined by D. Hilbert when dealing with integral equations with quadratic forms, or, equivalently, with linear operators on ℓ^2.

Proof. Since $T-\lambda I = \lambda^{-1}(\lambda T - I)$ and $\lambda T \in \mathcal{L}_c(E)$, we can assume without loss of generality that $\lambda = 1$. Denote $N = \mathcal{N}_T(1)$ and $R = \mathcal{R}_T(1)$.

(a) Since $T_N = I : N \to N$ is compact, $\dim N < \infty$.

(b) According to Theorem 4.22, $E = N \oplus M$, M a closed subspace of E. Then $S := (T-I)_{|M} : M \to R$ is a continuous isomorphism and we only need to show that S^{-1} is also continuous, which means that $C\|x\|_E \leq \|Sx\|_E$ for some constant $C > 0$ and for all $x \in M$.

If this were not the case, then we would find $x_n \in M$ so that $\|x_n\|_E = 1$ and $\|Sx_n\|_E \leq 1/n$, and, since T is compact, passing to a subsequence if necessary, $Tx_n \to z$ and $Sx_n \to 0$, with $z \in M$, since $x_n = Tx_n - Sx_n \to z$ and M is closed. Moreover, $Sz = \lim_n Sx_n = 0$ and $z = 0$ since S is one-to-one, which is in contradiction to $\|z\|_E = \lim_n \|x_n\|_E = 1$.

By the properties of the transpose, $R = \bar{R} = (\operatorname{Ker}(T' - I))^o$.

(c) Suppose $N = \{0\}$ and $R(1) := R \neq E$. Then $T : R(1) \to R(1)$ is compact and, according to (b), $R(2) = (T-I)(R(1))$ is a closed subspace of $R(1)$ and $R(2) \neq R(1)$, since $T - I$ is one-to-one. In this way, by denoting $R(n) = (T-I)^n(E)$, we obtain a strictly decreasing sequence of closed subspaces.

As in Remark 2.27, we choose $u_n \in R(n)$ so that $d(u_n, R(n+1)) \geq 1/2$ and $\|u_n\|_E = 1$. Then, if $p > q$,

$$Tu_p - Tu_q = (T-I)u_p - (T-I)u_q + u_p - u_q = z - u_q$$

with $z \in E(p+1) + E(q+1) + E(p) \subset E(q+1)$, so that $\|Tu_p - Tu_q\|_E \geq 1/2$, which is impossible, since T is compact.

This shows that $R = E$ if $N = \{0\}$. For the converse suppose that $R = E$, so that $\operatorname{Ker}(T' - I) = \operatorname{Im}(T-I)^o = R^o = \{0\}$ and we can apply the previous result to T', which is compact by Schauder's theorem. Hence $N = \operatorname{Im}(T' - I)^o = (E')^o = \{0\}$. □

Theorem 4.34. *Let $T \in \mathcal{L}_c(E)$. Then $\sigma(T) \subset \{\lambda;\, |\lambda| \leq \|T\|\}$ and, for every $\delta > 0$ there are only a finite number of eigenvalues λ of T such that $|\lambda| \geq \delta$.*

Proof. If $Tx = \lambda x$ and $\|x\|_E = 1$, then $|\lambda| = \|Tx\|_E \leq \|T\|$.

Suppose that, for some $\delta > 0$, there are infinitely many different eigenvalues λ_n such that $|1/\lambda_n| \leq 1/\delta$, and let $Tx_n = \lambda_n x_n$ with $\|x_n\|_E = 1$. These eigenvectors are linearly independent, since $x_n = \beta_1 x_1 + \cdots + \beta_{n-1}x_{n-1}$ with $x_1, \ldots, x_{n-1}$ linearly independent would imply, after an application of T, that $\beta_j = \beta_j \lambda_j / \lambda_n$ $(1 \leq j < n)$ and then necessarily $\lambda_n = \lambda_j$.

We can apply Remark 2.27 to the spaces $M_n = [x_1, \ldots, x_n]$ to obtain $u_n = \beta_1 x_1 + \cdots + \beta_n x_n \in M_n$ such that $\|u_n\|_E = 1$ and $d(u_n, M_{n-1}) \geq 1/2$.

Then the sequence $\{u_n/\lambda_n\}$ is bounded and we arrive at a contradiction by showing that it has no Cauchy subsequence:

It is easily checked that $u_n - Tu_n/\lambda_n \in [x_1, \ldots, x_{n-1}] = M_{n-1}$ and, if $p > q$,

$$\left\|\frac{1}{\lambda_p}Tu_p - \frac{1}{\lambda_q}Tu_q\right\|_E = \left\|u_p - \left\{u_p - \frac{1}{\lambda_p}Tu_p + \frac{1}{\lambda_q}Tu_q\right\}\right\|_E \geq 1/2.$$

□

Remark 4.35. If E is any complex Banach space and $T \in \mathcal{L}(E)$, it will be proved in Theorem 8.10 that $\sigma(T)$ is always a nonempty subset of **C** which is contained in the disc $\{\lambda;\ |\lambda| \leq r(T)\}$, where

$$r(T) = \lim_{n\to\infty} \|T^n\|^{1/n} = \inf_n \|T^n\|^{1/n} \leq \|T\|.$$

If $T \in \mathcal{L}_c(E)$, Theorems 4.33 and 4.34 show that $\sigma(T) \setminus \{0\}$ is a finite or countable set of eigenvalues with finite multiplicity.

These nonzero eigenvalues will be repeated according to their multiplicity in a sequence $\{\lambda_n\}$ so that $\{|\lambda_n|\}$ is decreasing. If this sequence is infinite, then $\lambda_n \to 0$, since for every $\varepsilon > 0$, according to Theorem 4.34, only a finite number of them satisfy $|\lambda_n| > \varepsilon$.

It may happen that $\sigma(T) \setminus \{0\} = \emptyset$, as shown by Theorem 2.30 for Volterra operators $T \in \mathcal{L}_c(\mathcal{C}[a,b])$. For these operators $\sigma(T) = \{0\}$, since $\mathcal{C}[a,b]$ is infinite dimensional.

In the special case of a self-adjoint compact operator of a Hilbert space, H, the spectral theorem will show the existence of eigenvalues and will give a diagonal representation for the operator.

Assume that $0 \neq A = A^* \in \mathcal{L}_c(H)$.

Note that eigenvectors of different eigenvalues are orthogonal, since it follows from $Ax = \alpha x$ and $Ay = \beta y$ that $(\alpha - \beta)(x,y)_H = (Ax,y)_H - (x,Ay)_H = 0$.

If $M(A) := \sup_{\|x\|_H=1}(Ax,x)_H$ and $m(A) := \inf_{\|x\|_H=1}(Ax,x)_H$, then

$$\|A\| = \sup_{\|x\|_H=1} |(Ax,x)| = \max(M(A), -m(A)),$$

by Theorem 4.7, and every eigenvalue λ is in the interval $[m(A), M(A)]$, since $Ax = \lambda x$ for some $x \in H$ with $\|x\|_H = 1$, and then $\lambda = (Ax,x)_H$.

Theorem 4.36 (Hilbert-Schmidt spectral theorem[11]). *The self-adjoint compact operator $A \neq 0$ has the eigenvalue α such that $|\alpha| = \|A\|$, and either $\alpha = M(A)$ if $\|A\| = M(A)$ or $\alpha = m(A)$ if $\|A\| = -m(A)$.*

Moreover, if $\{u_n\}$ is an orthonormal sequence of eigenvectors associated to the sequence $\{\lambda_n\}$ of nonzero eigenvalues, then

$$Ax = \sum_{n\geq 1} \lambda_n (x, u_n)_H u_n \qquad (x \in H)$$

in H and, if there are infinitely many eigenvalues, then $A_N \to A$ in $\mathcal{L}(H)$ as $N \to \infty$, where

$$A_N x := \sum_{n=1}^{N} \lambda_n (x, u_n)_H u_n.$$

Proof. If $\|A\| = M(A)$, then $M(A) = \lim_n (Ax_n, x_n)_H$ with $\|x_n\|_H = 1$ and $\|Ax_n\|_H \leq M(A)$. Then it follows from

$$\begin{aligned} \|Ax_n - M(A)x_n\|_H^2 &= \|Ax_n\|_H^2 - 2M(A)(Ax_n, x_n)_H + M(A)^2 \\ &\leq 2M(A)^2 - 2M(A)(Ax_n, x_n)_H \end{aligned}$$

that $\lim_n (Ax_n - M(A)x_n) = 0$ and $M(A) \in \sigma(A)$, since if $\lim_n (Ax_n - \lambda x_n) = 0$ and $\lambda \notin \sigma(A)$, then $x_n = (A - \lambda I)^{-1}(Ax_n - \lambda x_n) = 0$ by continuity. But $M(A) = \|A\| \neq 0$ and $M(A)$ is an eigenvalue.

Similarly, if $\|A\| = -m(A)$, then $m(A) = \lim_n (Ax_n, x_n)$ with $\|x_n\|_H = 1$ and $\|Ax_n\|_H \leq -m(A)$, so that, with the same proof as before,

$$\|Ax_n + m(A)x_n\|_H^2 \to 0$$

and $m(A) \in \sigma(A) \setminus \{0\}$.

To prove the second part of the theorem, note first that, if $N = \operatorname{Ker} A$ and $F = \overline{[u_1, u_2, \ldots]}$, then $F = N^\perp$, so that $H = F \oplus N$.

Indeed, if $Ax = 0$, then $(x, u_n)_H = 0$ for all $n \geq 0$, and $N \subset F^\perp$. It follows from $A(F) \subset F$ that also $A(F^\perp) \subset F^\perp$: if $z \in F^\perp$, then $(Az, x)_H = (z, Ax)_H = 0$ for all $x \in F$, since $Ax \in F$. But necessarily $A(F^\perp) = \{0\}$, and also $F^\perp \subset N$, since the restriction $A : F^\perp \to F^\perp$ is a self-adjoint compact operator, and if we suppose that it is nonzero, then, according to the first part of this theorem, A would have a nonzero eigenvalue α which should be one of the eigenvalues λ_n of $A : H \to H$, so that $u_n \in F \cap F^\perp$, a contradiction.

[11]D. Hilbert first developed his spectral theory for a large class of operators with a spectrum containing only eigenvalues, and E. Schmidt identified them as the compact operators through a "complete continuity condition". See Exercise 5.13.

Now let $x = y + z \in F \oplus N$ with $y \in F$ and $z \in N$. By the Fischer-Riesz theorem,

$$y = \sum_{n\geq 1}(y, u_n)_H u_n = \sum_{n\geq 1}(x, u_n)_H u_n$$

and

$$Ax = Ay = \sum_{n\geq 1}(x, u_n)_H Au_n = \sum_{n\geq 1}\lambda_n(x, u_n)_H u_n.$$

To show that $A_N \to A$, let $\|x\|_H \leq 1$. Then, using Bessel estimates,

$$\|(A - A_N)x\|_H^2 = \sum_{n>N} |\lambda_n(x, u_n)_H|^2 \leq |\lambda_N| \sum_{n>N} |(x, u_n)_H|^2 \leq |\lambda_N|$$

and $\lambda_N \to 0$. □

An **approximate eigenvalue** of a linear operator $A \in \mathcal{L}(H)$ is a number λ such that $\lim_n \|Ax_n - \lambda x_n\|_H = 0$ for some sequence of vectors $x_n \in H$ such that $\|x_n\|_H = 1$. In this case $(Ax_n - \lambda x_n, x_n)_H \to 0$ and $\lambda = \lim_n(Ax_n, x_n)$, so that $\lambda \in [m(A), M(A)]$ if A is self-adjoint.

Approximate eigenvalues belong to $\sigma(A)$ since, if $\lim_n \|Ax_n - \lambda x_n\|_H = 0$ and $\lambda \in \sigma(A)^c$, then $x_n = (A - \lambda I)(Ax_n - \lambda x_n) \to 0$ as $n \to \infty$ would be in contradiction to the condition $\|x_n\|_H = 1$ for all n.

Remark 4.37. If A is a bounded self-adjoint operator, it follows from the proof of Theorem 4.36 that both $m(A)$ and $M(A)$ are approximate eigenvalues.[12]

Indeed, we may assume without loss of generality that $0 \leq m(A) \leq M(A) = \|A\|$, since $M(A)$ is an approximate eigenvalue of A if and only if $M(A) + t$ is an approximate eigenvalue of $A + tI$. The case $\lambda = m(A)$ is similar.

4.5. Exercises

Exercise 4.1 (Banach limits). Consider the delay operator

$$\tau x(n) = x(n + 1)$$

acting on real sequences $x = \{x(n)\}_{n=1}^{\infty} \in \ell^\infty$ and the averages

$$\Lambda_n x = \frac{x(1) + \cdots + x(n)}{n}.$$

Prove that

$$p(x) := \limsup_{n\to\infty} \Lambda_n x$$

[12]It will be proved in Theorem 9.9 that every spectral value of A is an approximate eigenvalue, so that $[m(A), M(A)]$ is the least interval which contains $\sigma(A)$.

defines a convex functional p on $E = \ell^\infty$ and show that there exists a linear functional $\Lambda : \ell^\infty \to \mathbf{R}$ such that, for every $x \in \ell^\infty$,

$$\Lambda(\tau x) = \Lambda(x) \text{ and } \liminf_{n\to\infty} x(n) \le \Lambda(x) \le \limsup_{n\to\infty} x(n).$$

Exercise 4.2. Let x_0 be a point in a real normed space E. If $\|x_0\|_E = 1$, show that there exists $u \in E'$ such that $u(x_0) = 1$ and so that the ball $\bar{B}_E(0,1)$ lies in the half-space $\{u \le 1\}$.

Exercise 4.3. Let M be a closed subspace of a locally convex space E. Prove that if M is of finite codimension (that is, $\dim(E/M) < \infty$), then M is complemented in E.

Exercise 4.4. Suppose $T : F \to \ell^\infty$ is a bounded linear operator on a subspace F of a normed space E. Prove that T can be extended to a bounded linear map $\tilde{T} : E \to \ell^\infty$ with the same norm, $\|\tilde{T}\| = \|T\|$.

Exercise 4.5. If E is a topological vector space, prove that a linear form u on E is continuous if and only if $\operatorname{Ker} u$ is closed.

Exercise 4.6. Suppose that E is a locally convex space and that $A = \{e_n;\, n \in \mathbf{N}\}$ satisfies the following properties:

$$e_n \to 0, \quad E = [A], \text{ and } e_n \notin \overline{[e_j;\, j \ne n]} \quad \forall n \in \mathbf{N}.$$

If $x = \sum_{n=1}^{N(x)} \pi_n(x) e_n \in E$, then prove that the projections π_n are continuous linear forms on E and that the convex hull $\operatorname{co}(K)$ of $K = A \cup \{0\}$ is a closed subset of E which is not compact, but K is compact.

Find a concrete example for E and A.

Exercise 4.7. Prove that the completion $\tilde{H}$ of a normed space H with a norm defined by a scalar product, $\|x\|_H^2 = (x,x)_H$, is a Hilbert space.

Exercise 4.8. Let H be a Hilbert space. When identifying every $x \in H$ with $(\cdot, x) \in H'$, show that $A^\perp = A^o$ for any subset A of H.

Exercise 4.9. We must be careful when identifying $(L^2)' = L^2$ or $(\ell^2)' = \ell^2$, if we are dealing simultaneously with several spaces. Consider the example $H = \ell^2$ and the weighted ℓ^2 space

$$V = \ell^2(\{n^2\}_{n=1}^\infty) := \{x = \{x_n\}_{n=1}^\infty;\, \sum_{n=1}^\infty n^2 |x_n|^2 < \infty\}$$

with the scalar product $(x,y)_V := \sum_{n=1}^\infty n^2 x_n \bar{y}_n$.

Prove that V is a Hilbert space with a continuous inclusion $V \hookrightarrow \ell^2$ and that every $u \in (\ell^2)'$ is uniquely determined by its restriction u_V to V, which is a bounded linear form on V, so that a continuous inclusion $(\ell^2)' \hookrightarrow V'$ is defined and, by considering $(\ell^2)' = \ell^2$, we obtain $V \subset \ell^2 \subset V'$.

It would be nonsense to also consider $V' = V$, and one must choose $(\ell^2)' = \ell^2$ or $V' = V$ when dealing with both V and ℓ^2.

Exercise 4.10. Let E, F, and G be three normed spaces. Show that a bilinear or sesquilinear map $B : E \times F \to G$ is continuous if and only if there exists a constant $C \geq 0$ such that

$$\|B(x,y)\|_G \leq C\|x\|_E\|y\|_F \qquad (x \in E,\ y \in F).$$

Exercise 4.11. Prove that the inclusions $\mathcal{C}^{m+1}[a,b] \hookrightarrow \mathcal{C}^m[a,b]$ are compact.

Exercise 4.12. If $\{f_k\}$ is a bounded sequence of $\mathcal{E}(\mathbf{R})$, then show that for every $m \in \mathbf{N}$ there is a subsequence of $\{f_k^{(m)}\}_{k=1}^\infty$ which is uniformly convergent on compact subsets of $\mathbf{R}$ and $\mathcal{E}(\mathbf{R})$ has the Heine-Borel property. Extend this to every $\mathcal{E}(\Omega)$, Ω an open subset of $\mathbf{R}^n$, and prove that these spaces (and also $\mathcal{D}_K(\Omega)$ if K has nonempty interior) are not normable.

Exercise 4.13. Suppose $0 < c < 1$, m is the Lebesgue measure on $[0,1]$, and μ is another Borel measure on this interval. If $\mu(B) = c$ whenever $m(B) = c$, show first that necessarily μ is absolutely continuous with respect to m and then prove that $\mu = m$.

Exercise 4.14. Let m be the Lebesgue measure on $\mathbf{R}$, consider two Borel subsets E and F of $\mathbf{R}$, and define the Borel measures $\mu_E(B) := m(B \cap E)$, $\mu_F(B) := m(B \cap F)$. Find when μ_E is absolutely continuous with respect to μ_F and, in this case, describe the corresponding Radon-Nikodym derivative.

Exercise 4.15. Let $1 \leq p < \infty$ and $x = \{x_k\} \in \ell^p$. Show that $\|x\|_p = \sum_k x_k y_k$ for some $y = \{y_k\} \in \ell^{p'}$ such that $\|y\|_{p'} = 1$; that is, if $\langle y, x\rangle = \sum_k x_k y_k$ and $u_x = \langle \cdot, x\rangle$, the norm of $u_x \in (\ell^{p'})' = \ell^p$ is attained on the closed unit ball of $\ell^{p'}$. Find a similar result for functions $f \in L^p(\mathbf{R})$.

Exercise 4.16. Find some $x = \{x_k\} \in \ell^\infty$ such that we cannot find any $y = \{y_k\} \in \ell^1$ so that $\|y\|_1 = 1$ and $\|x\|_\infty = \sum_k x_k y_k$. That is, if $\langle y, x\rangle = \sum_k x_k y_k$ and $u_x = \langle \cdot, x\rangle$, the norm of $u_x \in (\ell^1)' = \ell^\infty$ is not attained on the closed unit ball of ℓ^1.

Exercise 4.17. Let c be the subspace of ℓ^∞ which contains all the convergent sequences $x = \{x^n\} \in \ell^\infty$. Prove that c is complete and that $v(x) := \lim x^n$ defines a continuous linear form on c with norm 1 such that there is no $y \in \ell^1$ so that $v(x) = \langle x, y\rangle$ for all $x \in c$ ($\langle x, y\rangle = \sum x^n y^n$). Show that the natural mapping $\ell^1 \to (\ell^\infty)'$ is not exhaustive.

Exercise 4.18. Prove that the natural isometry $J : L^1(a,b) \to L^\infty(a,b)'$, such that $Jf = \langle \cdot, f\rangle$ with $\langle g, f\rangle := \int_a^b g(t)f(t)\,dt$, is not exhaustive.

Remark: It can be proved that $L^1(0,1)$ is not isomorphic to any dual.

Exercise 4.19. Prove that ℓ^1 is isometrically isomorphic to the dual of c_0, the Banach subspace of ℓ^∞ of all the sequences with limit 0.

Exercise 4.20. Prove that if $T : \ell^1 \to \ell^1$ is defined by
$$T(x_n) = T(\frac{x_n}{n}),$$
then $\overline{\operatorname{Im} T'} \neq (\operatorname{Ker} T)^o$.

Exercise 4.21. Prove that the set of all the characteristic functions χ_I of intervals $I \subset (a, b)$ is total in $L^p(a, b)$, for every $1 \leq p < \infty$.

Exercise 4.22 (Minkowski integral inequality)**.** Let $K(x, y)$ be a measurable function on R^2 and let $1 \leq p \leq \infty$. Using the duality properties of L^p, prove that
$$\Big\| \int_{-\infty}^{+\infty} K(\cdot, y)\, dy \Big\|_p \leq \int_{-\infty}^{+\infty} \|K(\cdot, y)\|_p\, dy$$
first if $K \geq 0$ and then when $K(\cdot, y) \in L^p(\mathbf{R})$ for every $y \in \mathbf{R}$.

Exercise 4.23. Let μ be the Borel measure on $(0, 1)$ defined through the Riesz-Markov theorem by the linear form
$$u(g) := \int_0^1 \frac{g(x)}{x}\, dx$$
on $\mathcal{C}_c(0, 1)$. Is μ the restriction of a real Borel measure $\tilde{\mu}$ on $\mathbf{R}$?

Exercise 4.24. Let u be a linear form on the real vector space $\mathcal{C}(K)$ of all real-valued functions on a compact set K of $\mathbf{R}^n$. Prove that u is positive if and only if $u(1) = \sup_{\{|g| \leq 1\}} |u(g)|$ and that in this case $\|u\|_{\mathcal{C}(K)'} = u(1)$.

Exercise 4.25 (The dual of $\mathcal{H}(D)$)**.** In the disk $D = \{|z| < 1\} \subset \mathbf{C}$ consider the circles $\gamma_r(t) = re^{it}$ $(0 \leq t \leq 2\pi)$, $0 < r < 1$, and denote by $\mathcal{H}_0(D^c)$ the vector space of all continuous functions g on D^c that have a holomorphic extension to a neighborhood $U_g = \{z;\, |z| > \varrho\}$ of D^c in $\mathbf{C}$ and such that $g(\infty) = \lim_{z \to \infty} g(z) = 0$. Prove the following statements:

(a) If $g \in \mathcal{H}_0(D^c)$ and $\gamma_r \subset U_g$, then $u_g(f) := \frac{1}{2\pi i} \int_{\gamma_r} f(z) g(z)\, dz$ defines $u_g \in \mathcal{H}(D)'$ which does not depend on r.

(b) If μ is a complex Borel measure on $\varrho \bar{D}$ with $0 < \varrho < 1$, then $u_\mu(f) := \int_{\varrho \bar{D}} f\, d\mu$ also defines $u_\mu \in \mathcal{H}(D)'$.

(c) If μ is as in (b) and $g_\mu(z) := \int_{\varrho \bar{D}} \frac{1}{z - \omega}\, d\mu(\omega)$, then $g_\mu \in \mathcal{H}_0(D^c)$ and $u_{g_\mu} = u_\mu$ (use the Cauchy integral formula $f(\omega) = \frac{1}{2\pi i} \int_{\gamma_r} \frac{f(z)}{z - \omega}\, dz$).

(d) The map $g \in \mathcal{H}_0(D^c) \mapsto u_g \in \mathcal{H}(D)'$ is bijective.

Exercise 4.26. Prove the easy converse of the Schauder theorem: If E and F are two Banach spaces and $T' \in \mathcal{L}(F'; E')$ is compact, then $T \in \mathcal{L}(E; F)$ is also compact.

Exercise 4.27. Find a concrete Volterra operator $T \in \mathcal{L}_c(\mathcal{C}[a,b])$ with no eigenvalues.

Exercise 4.28. Prove that every nonempty compact subset K of $\mathbf{C}$ is the spectrum of a bounded operator of ℓ^2.

Exercise 4.29. Show that if $K = \{\lambda_n; n \in \mathbf{N}\} \cup \{0\}$ with $\lambda_n \to 0$, then $K = \sigma(T)$ for some $T \in \mathcal{L}_c(\ell^2)$.

Exercise 4.30. By Theorem 4.29, if T is the limit in $\mathcal{L}(E;F)$ of a sequence $\{T_n\}$ of continuous linear operators of finite rank, T is compact.

Prove that the converse is true if F is a Hilbert space by associating to every $T \in \mathcal{L}_c(E;F)$ and to every $\varepsilon > 0$ an orthogonal projection P_ε of F on a finite-dimensional subspace such that

$$\|T - P_\varepsilon T\| \leq \varepsilon.$$

P. Enflo (1973) proved that this converse is not true for general Banach spaces by giving a counterexample in the setting of separable reflexive spaces.

References for further reading:

N. I. Akhiezer and I. M. Glazman, *Theory of Linear Operators in Hilbert Space.*

S. Banach, *Théorie des opérations linéaires.*

S. K. Berberian, *Lectures in Functional Analysis and Operator Theory.*

B. A. Conway, *A Course in Functional Analysis.*

R. Courant and D. Hilbert, *Methods of Mathematical Physics.*

L. Kantorovitch and G. Akilov, *Analyse fonctionnelle.*

G. Köthe, *Topological Vector Spaces I.*

P. D. Lax, *Functional Analysis.*

F. Riesz and B. Sz. Nagy, *Leçons d'analyse fonctionelle.*

W. Rudin, *Real and Complex Analysis.*

W. Rudin, *Functional Analysis.*

K. Yosida, *Functional Analysis.*

Chapter 5

Weak topologies

This short chapter is devoted to the introduction of the weak topologies, the only locally convex space topologies that we are considering in this book which can be nonmetrizable.

We are mainly interested in the weak* topology on a dual E', such that the weak* convergence $u_n \to u$ of a sequence in E' means that $u_n(x) \to u(x)$ for all $x \in E$.

For any normed space E, the Alaoglu theorem shows that the closed unit ball in the dual space E' is weak* compact. Moreover, if E is separable, then this closed unit ball equipped with the weak* topology is metrizable. These facts make it easier to use the weak topology on bounded sets of E'.

As an application to the Dirichlet problem for the disc, we include a proof of the Fatou and Herglotz theorems concerning harmonic functions which are the Poisson integrals of functions or measures on the unit circle $\mathbf{T} \subset \mathbf{C}$.

The weak convergence and the weak* topology will appear again when studying distributions and with the Gelfand transform of commutative Banach algebras.

5.1. Weak convergence

A sequence $\{x_n\}$ in a normed space E is said to converge **weakly** to $x \in E$ if $u(x_n) \to u(x)$ for every $u \in E'$.

The usual convergence $x_n \to x$, meaning that $\|x_n - x\|_E \to 0$, is also called the strong convergence. It is stronger than weak convergence, since $|u(x_n) - u(x)| \leq \|u\|_{E'} \|x_n - x\|_E$. The converse will not be true in general (cf. Exercises 5.3 and 5.7).

Similarly, a sequence $\{u_n\}$ in the dual E' of the normed space is said to converge **weakly*** to $u \in E'$ if $u_n(x) \to u(x)$ for every $x \in E$. Again, $u_n \to u$ in E' implies weak* convergence, since $|u_n(x) - u(x)| \leq \|u_n - u\|_{E'} \|x\|_E$.

This weak* convergence is weaker than the weak convergence of $\{u_n\}$ since, if $\omega(u_n) \to \omega(u)$ for every $\omega \in E''$, then also $\widehat{x}(u_n) = u_n(x) \to \widehat{x}(u)$ for every $x \in E$.

We have a similar situation with the pointwise convergence of a sequence of functions $f_n = \{f_n(x)\}_{x\in X} \in \mathbf{C}^X$, since

$$f_n(x) \to f(x) \qquad (x \in X)$$

means that $\delta_x(f_n) \to \delta_x(f)$ for every evaluation functional δ_x, which is a linear form on the vector space $\mathbf{C}^X$ of all functions $f : X \to \mathbf{C}$.

We know from Example 3.2 that this pointwise convergence is the convergence associated to the product topology on $\mathbf{C}^X$.

These weak limits will be limits with respect to certain locally convex space topologies.

5.2. Weak and weak* topologies

Let E be any real or complex vector space and let $\mathcal{E}$ be a vector subspace of the algebraic dual of E, which is the vector space of all linear forms on E. We say that $(E, \mathcal{E})$ is a **dual couple** if $\mathcal{E}$ separates points of E, that is, if $u(x) = u(y)$ for all $u \in \mathcal{E}$ implies $x = y$.

A typical dual couple is a locally convex space E with the dual E', and we will see that every dual couple is of this form, for a convenient topology on E.

Theorem 5.1. *Suppose $(E, \mathcal{E})$ is a dual couple, $u_1, \ldots, u_n \in \mathcal{E}$, and $u \in \mathcal{E}$. Then $u = \sum_{k=1}^n \lambda_k u_k$ if and only if*

$$u_1(x) = \cdots = u_n(x) = 0 \Rightarrow u(x) = 0.$$

Proof. Suppose $u(x) = 0$ whenever $u_1(x) = \cdots = u_n(x) = 0$. Define the one-to-one linear map $\Phi : E \to \mathbf{K}^n$ such that $\Phi(x) = (u_1(x), \ldots, u_n(x))$. There exists $\tilde{u} \in (\mathbf{K}^n)'$ such that $u = \tilde{u} \circ \Phi$ that can be defined on $\Phi(E)$ as $\tilde{u}(u_1(x), \ldots, u_n(x)) := u(x)$ since, if $(u_1(x), \ldots, u_n(x)) = (u_1(y), \ldots, u_n(y))$, it follows from our assumption that $u(x) = u(y)$.

We can write

$$\tilde{u}(\alpha_1, \ldots, \alpha_n) = \lambda_1\alpha_1 + \cdots + \lambda_n\alpha_n,$$

so that

$$u(x) = \tilde{u}(\Phi(x)) = \lambda_1 u_1(x) + \cdots + \lambda_n u_n(x) \qquad (x \in E)$$

and $u = \lambda_1 u_1 + \cdots + \lambda_n u_n$.

The converse is obvious: if $u = \sum_{k=1}^n \lambda_k u_k$, then $u_1(x) = \cdots = u_n(x) = 0 \Rightarrow u(x) = 0$. □

Note that, if E is a locally convex space, $u \in E'$, and $x, x_1, \dots, x_n \in E$, then, according to Theorem 5.1, $x = \sum_{k=1}^n \lambda_k x_k$ if and only if $u(x) = 0$ $(u \in E')$ whenever $u(x_1) = \cdots = u(x_n) = 0$.

We assume that $(E, \mathcal{E})$ is a dual couple.

The **weak topology** $\sigma(E, \mathcal{E})$ is the locally convex topology on E defined by the sufficient family of semi-norms $p_u(x) = |u(x)|$ $(u \in \mathcal{E})$. Similarly, $\sigma(\mathcal{E}, E)$ is defined by the sufficient family of semi-norms $p_x(u) = |u(x)|$ $(x \in E)$. It is the restriction to $\mathcal{E} \subset \mathbf{K}^E$ of the product topology or the topology of the pointwise convergence on $\mathbf{K}^E$ of Example 3.2.

We use the prefix $\sigma(\mathcal{E}, E)-$ to indicate that we are considering the topology $\sigma(\mathcal{E}, E)$ on $\mathcal{E}$.

Theorem 5.2. *With the notation $\widehat{x}(u) = u(x)$, $\sigma(\mathcal{E}, E)$ is the weakest topology on $\mathcal{E}$ that makes every function $\widehat{x}$ continuous.*

The dual of $(\mathcal{E}, \sigma(\mathcal{E}, E))$ is E, as a vector subspace of the algebraic dual of $\mathcal{E}$.

Proof. Every functional $\widehat{x}$ is $\sigma(\mathcal{E}, E)$-continuous, since $|\widehat{x}(u)| = p_x(u)$ and $|\widehat{x}(u)| < \varepsilon$ if $u \in U_{p_x}(\varepsilon)$.

If every function $\widehat{x}$ is $\mathcal{T}$-continuous for a topology $\mathcal{T}$ on $\mathcal{E}$, then every set

$$V(u_0) := \{u \in \mathcal{E};\ |\widehat{x}(u) - \widehat{x}(u_0)| < \varepsilon\} = \{u;\ p_x(u - u_0) < \varepsilon\}$$

is an $\mathcal{T}$-neighborhood of u_0, and it is also a $\sigma(\mathcal{E}, E)$-neighborhood of u_0. Thus, every point u_0 of a weakly open set G is a $\mathcal{T}$-interior point of G and $\mathcal{T}$ is finer than $\sigma(\mathcal{E}, E)$.

Let $\mathcal{E}'$ be the $\sigma(\mathcal{E}, E)$-dual of $\mathcal{E}$. By construction, $E \subset \mathcal{E}'$.

Reciprocally, if $\omega \in \mathcal{E}'$, then there exist $x_1, \dots, x_n \in E$ and a constant $C \geq 0$ such that

$$|\omega(u)| \leq C \max(p_{x_1}(u), \dots, p_{x_n}(u)).$$

Hence, $\omega(u) = 0$ if $u(x_1) = \cdots = u(x_n) = 0$, and then $\omega = \sum_{k=1}^n \lambda_k \widehat{x}_k$ by Theorem 5.1. □

In a locally convex space, the closed convex sets and the closed subspaces are closed for the original topology of the space:

Theorem 5.3. *Suppose C is a convex subset of the locally convex space E. Then the weak closure of C is equal to the closure $\bar{C}$ of C for the topology of E.*

Proof. The original topology is finer than $w := \sigma(E, E')$ and the weak closure $\bar{C}^w$ of C is closed in E, so that $\bar{C} \subset \bar{C}^w$. Conversely, suppose that $x_0 \notin \bar{C}$; according to Theorem 4.17(b), we can choose $u \in E'$ so that

$$\sup \Re u(x_0) < r < \inf \Re u(B)$$

and $\Re u$ is weakly continuous. Hence, $\{\Re u < r\}$ is a weak neighborhood of x_0 which is disjoint with C, and $x_0 \notin \bar{C}^w$. This shows that $\bar{C}^w \subset \bar{C}$. □

Corollary 5.4. For any subset A of a locally convex space E, the weak closure of $[A]$ and its closure for the original topology are the same. Hence, A is total if and only if it is weakly total.

In the special case of a normed space E, we call $w^* = \sigma(E', E)$ the **weak* topology** of the dual space E', and $w = \sigma(E, E')$ is the **weak topology** of E. These topologies are weaker than the corresponding norm topologies on E' and E.

Note that a sequence $\{u_n\} \subset E'$ is convergent to u if $u_n \to u$ for the topology w^* if and only if it is weakly* convergent to u, since we have $|u_n(x) - u(x)| = p_x(u_n - u) \to 0$ for every $x \in E$.

Similarly $x_n \to x$ for the weak topology w if and only if $u(x_n) \to u(x)$ for every $u \in E'$, and $\{x_n\}$ is weakly convergent to x in E.

The most important facts concerning the topology w^* are contained in the following compactness and metrizability result.

Theorem 5.5 (Alaoglu[1]). *The closed unit ball $B_{E'} = \{u \in E';\ \|u\|_{E'} \leq 1\}$ of the dual E' of a normed space E is w^*-compact. If E is separable, the w^*-topology restricted to $B_{E'}$ is metrizable.*

Proof. Recall that w^* is the restriction to E' of the product topology on $\mathbf{K}^E$ and observe that $B_{E'}$ is contained in $K := \Pi_{x \in E} \bar{D}(0, \|x\|)$ since, if $\|u\|_{E'} \leq 1$, then $u = \{u(x)\}_{x \in E}$ satisfies $|u(x)| \leq \|x\|$ for every $x \in E$. By the Tychonoff theorem, K is a compact subset of $\mathbf{K}^E$, and

$$B_{E'} = \bigcap_{x,y} \{f \in K;\ f(x+y) = f(x) + f(y)\} \cap \bigcap_{\lambda, x} \{f \in K;\ f(\lambda x) = \lambda f(x)\}$$

is the intersection of a family of subsets, all of them being closed as defined by the equalities with continuous functions

$$\pi_{x+y}(f) = \pi_x(f) + \pi_y(f) \text{ and } \pi_{\lambda x}(f) = \lambda \pi_x(f).$$

Thus, $B_{E'}$ is a closed subset of K and it is compact.

[1]This is the best known result of the Canadian-American mathematician Leonidas Alaoglu (1938), contained in his thesis (Chicago, 1937). For separable spaces, it was first published by S. Banach (1932).

Suppose now that the sequence $\{x_n\}_{n=1}^{\infty}$ is dense in E. Then the family of semi-norms $p_{x_n}(u) = |u(x_n)|$ is sufficient on E' since, by continuity, $u(x_n) = 0$ for all x_n implies $u(x) = 0$ for all $x \in E$. This sequence of semi-norms defines on E' a locally convex metrizable topology $\mathcal{T}$ which is clearly weaker than the w^*-topology. On $B_{E'}$ these topologies coincide, since Id : $(B_{E'}, w^*) \to (B_{E'}, \mathcal{T})$ is continuous and the image of a compact (or closed) set is $\mathcal{T}$-closed, since it is $\mathcal{T}$-compact. □

Note that Theorem 5.5 states that, if E is separable, the w^*-topology is metrizable when restricted to a bounded set, but this is far from being true in general for w^* on the whole E'. This only happens if E is finite dimensional (Exercise 5.12).

Example 5.6. Suppose $\{K_\lambda\}_{\lambda\in\Lambda}$ is a summability kernel on $\mathbf{R}^n$ such that $\lim_{\lambda\to 0}\sup_{|y|\geq M}|K_\lambda(x)| = 0$ for all $M > 0$. If $f \in L^\infty(\mathbf{R}^n)$, then $\lim_{\lambda\to 0} K_\lambda * f = f$ in the w^*-topology on $L^\infty(\mathbf{R}^n) = L^1(\mathbf{R}^n)'$.

A similar result holds for the periodic summability kernels.

This is shown as in the proof of Theorem 2.41 by considering, for every $g \in L^1$,

$$\begin{aligned}\left|\int (K_\lambda * f - f)g\right| &\leq \int\left|\int (f(x-y) - f(x))g(x)\,dx\right| |K_\lambda(y)|\,dy \\ &\leq \sup_{|y|\leq M}\left|\int (\tau_y f(x) - f(x))g(x)\,dx\right| \\ &\qquad +2\|g\|_1\|f\|_\infty \sup_{|y|\geq M}|K_\lambda(y)|,\end{aligned}$$

where

$$\left|\int (\tau_y f(x) - f(x))g(x)\,dx\right| = \left|\int (\tau_y g(x) - g(x))f(x)\,dx\right| \leq \|\tau_y g - g\|_1\|f\|_\infty,$$

and we know that $\lim_{M\to 0}\sup_{|y|\leq M}\|\tau_y g - g\|_1 = 0$ by Theorem 2.14.

It is worth noticing that the analogue of this last example holds for measures as well:

If μ is a complex Borel measure on $\mathbf{R}^n$ (or on $\mathbf{T}$), which has a polar representation $d\mu = h\,d|\mu|$, and g is a bounded Borel measurable function, then the convolution

$$(\mu * g)(x) := \int g(x-y)\,d\mu(y) = \int g(x-y)h(y)\,d\mu(y)$$

is well-defined.

If every function K_λ of the summability kernel is bounded and continuous, then

$$\mu * K_\lambda - \mu \to 0 \tag{5.1}$$

for the weak topology with respect to $\mathcal{C}_c(\mathbf{R}^n)$ (or $\mathcal{C}(\mathbf{T})$ in the periodic case). Similarly, $f * K_\lambda - f \to 0$ in the w^*-topology on L^∞ if $f \in L^\infty$.

Note also that we can define the Fourier series for any complex measure μ on $\mathbf{T}$ (or on $(-\pi, \pi]$)

$$\mu \sim \sum_{k=-\infty}^{\infty} c_k(\mu) e^{ikt}$$

by

$$c_k(\mu) = \frac{1}{2\pi} \int_{(-\pi,\pi]} e^{-ikt}\, d\mu(t) = \int_{\mathbf{T}} z^{-k}\, d\mu(z).$$

Then the corresponding Cesàro sums are $\sigma_N(\mu) = \mu * F_N$, where F_N is the Cesàro summability kernel (2.25), and it is a special case of these remarks that $\sigma_N(\mu) \to \mu$ in the above w^*-topology.

5.3. An application to the Dirichlet problem in the disc

Our next aim is to show how the previous results are related to the classical Dirichlet problem.

Let U denote the open unit disc $|z| < 1$ in the plane domain, and let

$$\partial = \frac{1}{2}(\partial_x - i\partial_y), \quad \bar{\partial} = \frac{1}{2}(\partial_x + i\partial_y)$$

so that

$$4\partial\bar{\partial} = \partial_x^2 + \partial_y^2 = \triangle.$$

With this notation, the Cauchy-Riemann equations $\partial_x u = \partial_y v$, $\partial_x v = -\partial_y u$ for $f = u + iv$ read $\bar{\partial} f = 0$, and every holomorphic function $f = u + iv$ on U is C^∞ as a two-variables function which is harmonic; this is,

$$\triangle f = 0.$$

Obviously, $u = \Re f$ and $v = \Im f$ are also harmonic.

For a real-valued harmonic function u, any real-valued v such that $f = u + iv$ is holomorphic on U is called a **harmonic conjugate** of u and, according to the Cauchy-Riemann equations, the harmonic conjugate of u is unique up to an additive constant, since $\partial_x v = \partial_y v = 0$ leads to $v = C$. When v is chosen with the condition $v(0) = 0$, v is called "the" harmonic conjugate of u.

Our aim is to study the extension of a function f defined on $\mathbf{T} = \partial U$ to a harmonic function on U, that is, to study the Dirichlet problem for the disc

$$\triangle F = 0, \quad F = f \text{ on } \mathbf{T}. \tag{5.2}$$

The condition $F = f$ on the boundary has to be understood in an appropriate sense.

If $f \in \mathcal{C}(\mathbf{T})$, we look for a classical solution, which is a continuous function F on $\bar{U}$ which is harmonic in U. This problem is completely solved by the Poisson integral that can be obtained by an application of the Hahn-Banach theorem, as follows.

Let E be the subspace of the complex Banach space $\mathcal{C}(\mathbf{T})$ which contains all the complex polynomial functions $g(z) = \sum_{n=1}^{N} c_n z^n$ restricted to $\mathbf{T}$. By the maximum modulus property, $\|g\|_{\mathbf{T}} = \|g\|_{\bar{U}}$, and the evaluation map $g \mapsto g(z_0)$ at a fixed point $z_0 \in \bar{U}$ is a continuous linear form with norm 1 which is extended by the Hahn-Banach theorem to $u_{z_0} \in \mathcal{C}(\mathbf{T})'$ so that $\|u_{z_0}\| = 1$.

Note that if E is any linear subspace of $\mathcal{C}(\mathbf{T})$ with the maximum modulus property, it can, and will, also be assumed to be contained in $\mathcal{C}(\bar{U})$, and the evaluation map is defined on this subspace of $\mathcal{C}(\mathbf{T})$.

By the Riesz representation Theorem 4.13,

$$u_{z_0}(f) = \int_{\mathbf{T}} f \, d\mu_{z_0} \qquad (f \in \mathcal{C}(\mathbf{T}))$$

for a complex Borel measure μ_{z_0} on $\mathbf{T}$, and then we say that μ_{z_0} represents u_{z_0} or z_0. We will see that this measure is uniquely determined by z_0.

Lemma 5.7. *If $u \in \mathcal{C}(\mathbf{T})'$, $\|u\| = 1$, and $u(1) = 1$, then the representing measure of u is positive.*

Proof. We need to prove that $u(f) \geq 0$ if $f \geq 0$ and we can also asume that $f \leq 1$. Let $g = 2f - 1$, so that $-1 \leq g \leq 1$ and, if $u(g) = a + ib$ $(a, b \in \mathbf{R})$, it follows from the hypothesis that for every $x \in \mathbf{R}$

$$1 + x^2 \geq |u(g + ix)|^2 = |a + ib + ix|^2 = a^2 + (b + x)^2 \geq (b + x)^2,$$

so that $b^2 + 2xb \leq 0$ and then $b = 0$, $u(g) = a$. Finally we have $|a| \leq \|g\|_{\mathbf{T}} \leq 1$ and then $a = u(g) = 2u(f) - 1$ forces $u(f) \geq 0$. □

Note that for the functions $e_k(z) = z^k$ $(k \in \mathbf{Z})$, if $z_0 = re^{i\vartheta}$,

$$r^n e^{in\vartheta} = \int_{\mathbf{T}} e_n \, d\mu_{z_0} \qquad (n \geq 0)$$

and, since the measure is positive and $e_{-n} = \bar{e}_n$ on $\mathbf{T}$,

$$r^n e^{-in\vartheta} = \int_{\mathbf{T}} e_{-n}\, d\mu_{z_0}.$$

Hence, $\int_{\mathbf{T}} e_k\, d\mu_{z_0} = r^{|k|} e^{ik\vartheta}$.

By addition, for every $r \in [0,1)$ we obtain the 2π-periodic function

$$P_r(s) = \sum_{k=-\infty}^{\infty} r^{|k|} e^{iks} \tag{5.3}$$

such that

$$\int_{\mathbf{T}} f\, d\mu_{z_0} = \frac{1}{2\pi} \int_{-\pi}^{\pi} f(e^{i\vartheta}) P_r(\vartheta - t)\, dt \tag{5.4}$$

for every trigonometric polynomial, since it holds when $f = e_n$. But the Fejér kernel F_N is a summability kernel and, for every $f \in \mathcal{C}(\mathbf{T})$, the Cesàro sums $\sigma_N(f) = F_N * f$ are trigonometric polynomials such that $\sigma_N(f) \to f$ uniformly. By continuity, (5.4) holds for every $f \in \mathcal{C}(\mathbf{T})$ and the Borel measure μ_{z_0} is the uniquely determined absolutely continuous measure

$$d\mu_{z_0} = \frac{1}{2\pi} P_r(\vartheta - t)\, dt \qquad (z_0 = re^{i\vartheta}).$$

Note that $P_r(\vartheta - t)$ is the real part of

$$1 + 2\sum_{n=1}^{\infty} (z_0 e^{-it})^n = \frac{e^{it} + z_0}{e^{it} - z_0} = \frac{1 - r^2 + 2ir\sin(\vartheta - t)}{|1 - z_0 e^{-it}|};$$

that is,

$$P_r(\vartheta - t) = \frac{1 - r^2}{1 - 2r\cos(\vartheta - t) + r^2}.$$

The family $\{P_r\}_{0<r\le 1}$ is called the **Poisson kernel of the disc**. Every P_r is a positive continuous and periodic function such that

$$\int_{\mathbf{T}} P_r = \sum_{k=-\infty}^{\infty} r^{|k|} \frac{1}{2\pi} \int_{-\pi}^{\pi} e^{iks}\, ds = 1$$

and $P_r(-t) = P_r(t)$. Moreover,

$$\sup_{0<\delta\le|t|\le\pi} P_r(t) \le P_r(\delta) \to 0 \text{ if } \delta \downarrow 0,$$

so that $\{P_r\}_{0<r<1}$ is a summability kernel on $\mathbf{T}$.

We summarize all these results in the following theorem:

Theorem 5.8. *The Poisson kernel $\{P_r\}_{0<r\le 1}$ is the summability kernel on $\mathbf{T}$ such that*

$$P_r(\vartheta - t) = \frac{1 - r^2}{1 - 2r\cos(\vartheta - t) + r^2} = \Re \frac{e^{it} + z_0}{e^{it} - z_0},$$

and

$$d\mu_{z_0} = \frac{1}{2\pi} P_r(\vartheta - t)\, dt \qquad (z_0 = re^{i\vartheta})$$

represents every point $z_0 \in U$ *in the sense that it is the unique measure on* $\mathbf{T}$ *such that*

$$f(z_0) = \frac{1}{2\pi} \int_{-\pi}^{\pi} f(e^{it}) P_r(\vartheta - t)\, dt = (P_r * f)(\vartheta)$$

for every f *in a vector subspace* E *of* $\mathcal{C}(\bar{U})$ *which contains the polynomials and satisfies the maximum modulus property* $\|f\|_{\bar{U}} = \|f\|_{\mathbf{T}}$.

Now we are ready to solve the Dirichlet problem (5.2) using the Poisson integral of a function

$$(P_r * f)(\vartheta) = \frac{1}{2\pi} \int_{-\pi}^{\pi} f(e^{i\vartheta}) P_r(\vartheta - t)\, dt$$

or of a measure μ,

$$(P_r * \mu)(\vartheta) = \frac{1}{2\pi} \int_{-\pi}^{\pi} P_r(\vartheta - t)\, d\mu(t).$$

Theorem 5.9. *Let* $f \in L^1(\mathbf{T})$ *and let* μ *be a complex Borel measure on* $\mathbf{T}$. *Denote*

$$F(re^{i\vartheta}) := (P_r * f)(\vartheta) \quad \text{or} \quad F(re^{i\vartheta}) := (P_r * \mu)(\vartheta),$$

with $0 \le r < 1$ *and* $\vartheta \in \mathbf{R}$.

Then F *is harmonic on the open unit disc* U *and, as* $r \to 1$, *the functions* $F_r(\vartheta) := F(re^{i\vartheta})$ *satisfy the following convergence results:*

(a) *If* $f \in \mathcal{C}(\mathbf{T})$, *then* $F_r \to f$ *uniformly, so that, by defining* $F(e^{i\vartheta}) := f(e^{i\vartheta})$, F *is a classical solution of the Dirichlet problem (5.2).*

(b) *If* $f \in L^p(\mathbf{T})$ *with* $1 \le p < \infty$, *then* $F_r \to f$ *in* $L^p(\mathbf{T})$.

(c) *If* $f \in L^\infty(\mathbf{T})$, *then* $F_r \to \mu$ *in the* w^**-convergence on* $L^\infty(\mathbf{T}) \equiv L^1(\mathbf{T})'$.

(d) *If* $F_r = P_r * \mu$, *then* $F_r \to \mu$ *in the* w^**-convergence for* $\mu \in M(\mathbf{T}) \equiv \mathcal{C}(\mathbf{T})'$.

Proof. If f is real, then F is the real part of the holomorphic function

$$V(z) := \frac{1}{2\pi} \int_{-\pi}^{\pi} \frac{e^{it} + z}{e^{it} - z} f(e^{it})\, dt, \tag{5.5}$$

and it is harmonic. A similar reasoning shows that $F(re^{i\vartheta}) = (P_r * \mu)(\vartheta)$ defines a harmonic function on U.

Since we are dealing with a summability kernel, the statements (a) and (b) hold, and (c)–(d) follow from (5.1). $\square$

Remark 5.10. A simple change of variables gives

$$f(z) = \frac{1}{2\pi}\int_{-\pi}^{\pi} \frac{R^2 - r^2}{R^2 - 2r\cos(\vartheta - t) + r^2} f(a + Re^{it})\, dt$$

for every $z = a + re^{i\vartheta} \in D(a,R)$ if f is continuous on the closed disc $\bar{D}(a,R) = \{z \in \mathbf{C};\ |z-a| \le R\}$ and harmonic in $D(a,R)$.

For $r = 0$ this is the **mean value property** of the harmonic function f:

$$f(a) = \frac{1}{2\pi}\int_{-\pi}^{\pi} f(a + Re^{it})\, dt.$$

Let us now consider the inverse problem and, given a harmonic function F on U, try to find out whether it is the Poisson integral of some function or measure on $\mathbf{T}$. We still denote $F_r(\vartheta) = F(re^{i\vartheta})$ if $z = re^{i\vartheta} \in U$.

Theorem 5.11. *Suppose F is a complex-valued harmonic function in the open unit disc U.*

(a) *F is the Poisson integral of some $f \in \mathcal{C}(\mathbf{T})$ if and only if F_r is uniformly convergent when $r \uparrow 1$. In this case, F is the unique classical solution of the Dirichlet problem (5.2).*

(b) *F is the Poisson integral of some $f \in L^1(\mathbf{T})$ if and only if F_r is convergent in $L^1(\mathbf{T})$ when $r \uparrow 1$.*

(c) **Fatou's theorem:** *If $1 < p \le \infty$, F is the Poisson integral of some $f \in L^p(\mathbf{T})$ if and only if*

$$\sup_{0 \le r < 1} \|F_r\|_p < \infty.$$

(d) *F is the Poisson integral of some complex Borel measure if and only if*

$$\sup_{0 \le r < 1} \|F_r\|_1 < \infty.$$

(e) **Herglotz's theorem:** *F is the Poisson integral of some Borel measure if and only if $F \ge 0$.*

Proof. We need to prove only the direct parts, the converses being contained in Theorem 5.9.

Let us start with (a) by showing that, if $F_r \to f$ uniformly, then F is the unique solution of the Dirichlet problem for the disc. We can suppose that F is real-valued and we will prove that F has to be the real part of the holomorphic function V defined in (5.5).

The function $V_1 = \Re V$ is a classical solution of the Dirichlet problem with the boundary value f. Then $H = F - V_1 \in \mathcal{C}(\bar{U})$ is harmonic in U and zero on $\mathbf{T}$, and we only need to show that $H = 0$ at every point of U.

Assume that $H(z_0) > 0$ for some $z_0 \in U$, denote $\varepsilon = H(z_0)/2$, and let

$$h(z) := H(z) + \varepsilon |z|^2,$$

a continuous function on $\bar{U}$ such that $h = \varepsilon$ on $\mathbf{T}$ and $h(z_0) > \varepsilon$. Then

$$\max h = h(z_1)$$

for some $z_1 \in U$, so that $\partial_x^2 h(z_1) \leq 0$ and $\partial_y^2 h(z_1) \leq 0$. This is in contradiction to $\triangle h(z_1) = 4\varepsilon$, which follows from the definition of h. The assumption $H(z_0) < 0$ would also lead to a similar contradiction and $H = F - V_1 = 0$.

Proceeding now to the proofs of (d) and (e), let $R := \sup_{0 \leq r < 1} \|F_r\|_1$ and $u_r(g) := \langle g, F_r \rangle$. Then $u_r \in \mathcal{C}(\mathbf{T})'$ and $\|u_r\| \leq R$. According to Alaoglu's theorem, since $\mathcal{C}(\mathbf{T})$ is a separable Banach space, the ball with radius R is a metrizable w^*-compact set and from a sequence $r \to 1$ we can choose a subsequence u_{r_n} which is w^*-convergent to some $u \in \mathcal{C}(\mathbf{T})'$.

By the Riesz representation theorem there is a complex Borel measure μ on $\mathbf{T}$ such that

$$u(g) = \int_{\mathbf{T}} g \, d\mu = \lim_{n \to \infty} \int_{\mathbf{T}} g(t) F_{r_n}(t) \, dt \qquad (g \in \mathcal{C}(\mathbf{T})).$$

Note that every function $h_n(z) := F(r_n z)$ is harmonic in a neighborhood of $\bar{U}$ and on U it is the unique solution of the Dirichlet problem (5.2) for $f(e^{it}) = F(r_n e^{it})$. Thus, if $z = re^{i\vartheta}$,

$$h_n(z) = \frac{1}{2\pi} \int_{-\pi}^{\pi} P_r(\vartheta - t) h_n(e^{it}) \, dt.$$

If for a fixed $z = re^{i\vartheta} \in U$ we consider the continuous function $g = P_z$, where

$$P_z(t) := \Re \frac{e^{it} + z}{e^{it} - z} = P_r(\vartheta - t) \quad (z = re^{i\vartheta}),$$

we obtain

$$F(z) = \lim_n h_n(z) = \int_{\mathbf{T}} P_z \, d\mu = \int_{-\pi}^{\pi} P_r(\vartheta - t) \, d\mu(t).$$

If $F \geq 0$, then $\|F_r\|_1 = F(0)$ for every $r \in [0, 1)$ by the mean value property of harmonic functions, and the complex measure μ obtained in (d) is positive.

The proof of (c) is similar. In this case, $u_r(g)$ is defined for every $g \in L^{p'}(\mathbf{T})$ and $u_r \in L^{p'}(\mathbf{T})'$ with $\|u_r\| \leq R = \sup_{0 \leq r < 1} \|F_r\|_p$. Since $1 \leq p' < \infty$, $L^{p'}(\mathbf{T})$ is separable and, according to Alaoglu's theorem, we have some $u_{r_n} \to u$ in the w^*-topology on $L^{p'}(\mathbf{T})'$. Now by the Riesz representation theorem for the dual of an L^q-space, $u(g) = \int_{\mathbf{T}} gh$ for some $h \in L^p(\mathbf{T})$ such that $\|h\|_p \leq R$. Now the proof continues as in (d).

The proof of (b) is simple. If $F_r \to f$ in $L^1(\mathbf{T})$ and $d\mu = f(t)\,dt$ on $\mathbf{T}$, then as in (d) it follows that

$$F(z) = \lim_n h_n(z) = \int_{\mathbf{T}} P_z\,d\mu = \int_{-\pi}^{\pi} P_r(\vartheta - t) f(t)\,dt,$$

since now $u_r = \langle \cdot, F_r \rangle \to \langle \cdot, f \rangle$. □

5.4. Exercises

Exercise 5.1. Prove that the unit ball of ℓ^1 is not weakly compact.

Exercise 5.2. Prove that every sequence in the closed unit ball of $L^2(\mathbf{R})$ has a weakly convergent subsequence, and find a weakly convergent sequence with no convergent subsequence.

Exercise 5.3. In the Banach space $\mathcal{C}[0,1]$, prove that the sequence

$$f_n(x) := \begin{cases} nx & \text{if } 0 \le x < 1/n, \\ 2 - nx & \text{if } 1/n \le x < 2/n, \\ 0 & \text{if } 2/n \le x \le 1 \end{cases}$$

is weakly convergent to 0 but it is not strongly convergent.

Exercise 5.4. If $\delta \in \mathcal{C}[-1,1]'$ is the linear form

$$\delta(g) := g(0)$$

and $\{h_n\}_{n=1}^{\infty} \subset \mathcal{C}[-1,1]$, prove that $h_n \to \delta$ weakly (in the sense that $\langle g, h_n \rangle = \int_{-1}^{1} g h_n \to g(0)$ for every $g \in \mathcal{C}[-1,1]$) if and only if the following three conditions are satisfied:

(1) $\lim_n \int_{-1}^{1} h_n(t)\,dt = 1$,

(2) $\lim_n \int_{-1}^{1} h_n(t)\varphi(t)\,dt = 0$ if $\varphi \in \mathcal{C}^{\infty}[-1,1]$ vanishes in a neighborhood of 0, and

(3) $\sup_n \int_{-1}^{1} h_n(t)\,dt < \infty$.

Prove also that if $\sup_n \int_{-1}^{1} h_n(t)\,dt = \infty$, then there exists a function $g \in \mathcal{C}[-1,1]$ for which $\int_{-1}^{1} g h_n \to \infty$.

Exercise 5.5. If $x_n \to x$ weakly in a Banach space E, prove that $\|x\|_E \le \liminf \|x_n\|_E$.

Exercise 5.6. For every Banach space E there is a linear isometry from E onto a closed subspace of $\mathcal{C}(K)$, where K is the closed unit ball $B_{E'}$ endowed with the restriction of the weak* topology of E'.

Exercise 5.7. Let E be a locally convex space. If $x_n \to 0$ in E, then also $x_n \to 0$ weakly. By considering an orthonormal system in a Hilbert space, show that the converse is not true in general.

Exercise 5.8. In a Fréchet space E, suppose that $x_n \to x$ weakly. Prove that x is the limit in E of a sequence of convex combinations $\sum_{j=1}^{N} \alpha_j x_{n_j}$ of elements from the sequence $\{x_n\}$.

Exercise 5.9. As an application of the uniform boundedness principle, prove that a subset of a normed space E is bounded if and only if it is weakly bounded. If E is complete, also show that a subset of E' is bounded if and only if it is w^*-bounded.

Exercise 5.10. Let $T : E \to F$ be a linear mapping between two Fréchet spaces and let $T'(v) := v \circ T$ $(v \in F')$, the transpose of T. Prove the equivalence of the following properties:

(a) T is continuous.

(b) T is weakly continuous ($T : E(\sigma(E, E')) \to F(\sigma(F, F')$ continuous).

(c) $T'(F') \subset E'$ (and then $T' : F'(\sigma(F', F)) \to E'(\sigma(E', E)$ continuous).

Exercise 5.11. Let E be an infinite-dimensional locally convex space. Prove that the weak topology $\sigma(E', E)$ on E' is metrizable if and only if E has a countable algebraic basis.

Exercise 5.12. Let E be a normed space. Prove that if the weak topology $\sigma(E, E')$ on E, or the weak* topology on E', is metrizable, then E is finite dimensional.

Exercise 5.13. Let H be a separable Hilbert space and let $T \in \mathcal{L}(H)$. Prove that T is compact if and only if, for any sequence $\{x_n\} \subset H$ such that $(x, x_n)_H \to 0$ as $n \to \infty$ for all $x \in H$ (that is, $x_n \to 0$ weakly), it follows that $\|Tx_n\|_H \to 0$. Operators with this property were called completely continuous by Hilbert and Schmidt.

Exercise 5.14. Suppose E is a real normed space and K_1, K_2 are two disjoint weakly compact convex subsets of E. Then prove that $f(K_1) < r < f(K_2)$ for some $f \in E'$ and $r \in \mathbf{R}$.

Exercise 5.15 (Goldstine's theorem). Recall that $E \subset E''$ or, more precisely, $J(E)$ is a subspace of E'' if J is as in Theorem 4.25. Show that the weak* topology on E'' induces on E the weak topology and that the closed unit ball B_E of E is w^*-dense in the closed unit ball $B_{E''}$ of E''. Thus E is w^*-dense in E''.

Exercise 5.16. A normed space E is said to be **reflexive** if $E'' = E$; that is, if $J(E) = E''$ with J as in Theorem 4.25.

(a) Prove that E is reflexive if and only if the closed unit ball B_E of E is weakly compact.

(b) Prove that every closed subspace F of a reflexive normed space E is also reflexive.

(c) Prove that a Banach space E is reflexive if and only if E' is reflexive.

Exercise 5.17. Prove that every Hilbert space H is reflexive and that $L^p(0,1)$ $(1 \le p \le \infty)$ is reflexive if and only if $1 < p < \infty$.

Hint: Use the representation theorems and Exercise 4.18.

Exercise 5.18. Check the details of the Riesz representation theorem for $(\ell^p)'$ $(1 \le p < \infty)$ which show that the dual space of ℓ^p is isometrically isomorphic to $\ell^{p'}$ throughout the bilinear form $\langle x, y\rangle := \sum_n x(n)y(n)$. If $1 < p < \infty$, find a weakly convergent sequence in ℓ^p which is not strongly convergent.

Show that on ℓ^1 the weak topology is strictly weaker than the topology of the norm. It is true, but harder to prove, that in ℓ^1 every weakly convergent sequence is strongly convergent.

Exercise 5.19. Find a sequence of functions f_n such that, for any $1 < p < \infty$, $f_n \to 0$ weakly in $L^p(-\pi, \pi)$ but not strongly.

Exercise 5.20. Prove that $\mathcal{C}[0,1]$ is w^*-dense in $L^\infty(0,1)$ but it is not dense for the topology of the norm.

Exercise 5.21. If

$$\sum_{k=-\infty}^{\infty} c_k e^{ikt} \tag{5.6}$$

is a given trigonometric series, we denote $S_N := \sum_{k=-N}^{N} c_k e^{ikt}$ and

$$\sigma_N := \frac{1}{N+1}\sum_{n=0}^{N} S_n, \tag{5.7}$$

the Cesàro means.

Prove that (5.6) is the Fourier series of a complex Borel measure on $\mathbf{T}$ if and only if $\sup_N \|\sigma_N\|_1 < \infty$.

Exercise 5.22. Prove that (5.6) is the Fourier series of a function $f \in L^p(\mathbf{T})$ $(1 < p \le \infty)$ if and only if $\sup_N \|\sigma_N\|_p < \infty$.

Exercise 5.23. Prove that (5.6) is the Fourier series of a function $f \in L^1(\mathbf{T})$ if and only if the sequence $\{\sigma_N\}$ of Fourier sums is a convergent sequence in $L^1(\mathbf{T})$.

Exercise 5.24. Prove that (5.6) is the Fourier series of a 2π-periodic continuous function if and only if $\{\sigma_N\}$ is a uniformly convergent sequence.

References for further reading:

S. Banach, *Théorie des opérations linéaires.*

N. Dunford and J. T. Schwartz, *Linear Operators: Part 1.*

G. Köthe, *Topological Vector Spaces I.*

P. D. Lax, *Functional Analysis.*

F. Riesz and B. Sz. Nagy, *Leçons d'analyse fonctionelle.*

W. Rudin, *Real and Complex Analysis.*

W. Rudin, *Functional Analysis.*

K. Yosida, *Functional Analysis.*

Chapter 6

Distributions

For a long time, physicists have been operating with certain "singular functions" that are not true functions in the usual sense.

A typical simple example was Dirac's function δ, assumed to be "supported by $\{0\}$ but so large on this point that $\int_{\mathbf{R}} \delta(t)\,dt = 1$", which is the unit impulse of signal theory.[1]

Previously, in 1883, to solve some physical problems, Heaviside[2] introduced a symbolic calculus that included the derivatives of singular functions. With this calculus it was accepted that δ is the derivative Y' of Heaviside's function $Y = \chi_{[0,\infty)}$, and then a formal partial integration with φ regular enough and with compact support gives

$$\int_{\mathbf{R}} \varphi(t)\delta(t)\,dt = -\int_{[0,\infty)} \varphi'(t)\,dt = \varphi(0).$$

In his 1932 text [**43**] on quantum mechanics, von Neumann warned against the use of these unclear objects. In the preface of the book he says that "The method of Dirac ... in no way satisfies the requirements of

[1]The δ function was introduced by the British physicist, and one of the founders of quantum mechanics, Paul Adrien Maurice Dirac as the continuous form of the discrete Kronecker delta in the formulation of quantum mechanics, which is contained in his 1930 book "The Principles of Quantum Mechanics", a landmark in the history of science.

[2]The English telegraph operator and self-taught engineer, physicist, and mathematician Oliver Heaviside patented the co-axial cable in 1880, reformulated Maxwell's initially cumbersome equations by reducing the original system of 20 differential equations to 4 differential equations in 1884, and between 1880 and 1887 developed a controversial operational calculus that motivated his saying "Mathematics is an experimental science, and definitions do not come first, but later on".

mathematical rigor [with] the introduction of 'improper' functions with self-contradictory properties" and he calls the delta functions and their derivatives "mathematical fictions". Very soon afterwards Bochner introduced in a rigorous way singular functions in Fourier analysis, and Sobolev (1938) used weak derivatives in the study of partial differential equations.

But it was L. Schwartz[3] who defined them as linear forms acting on a family of test functions φ, with the appropriate continuity properties.

The idea was that, as a mater of fact, a singular function such as δ always appears inside an integral formula. In the case of the Dirac function, the useful property is that $\int_{\mathbf{R}} \varphi(t)\delta(t)\, dt = \varphi(0)$, and δ can be directly defined as the linear form $\varphi \mapsto \varphi(0)$ acting on a convenient family of test functions φ.

If we consider any function $0 \leq \varrho \in \mathcal{C}_c(\mathbf{R})$ supported by $[-r, r]$ and such that $\int_{\mathbf{R}} \varrho(t)\, dt = 1$, the summability kernel $\varrho_n(t) := n\varrho(nt)$ is such that

$$\lim_{n\to\infty} \int_{\mathbf{R}} \varphi(t)\varrho_n(t)\, dt = \lim_{n\to\infty} \langle \varphi, \varrho_n \rangle = \varphi(0) \qquad (\varphi \in \mathcal{C}(\mathbf{R})),$$

since $|\int_{\mathbf{R}} \varphi(t)\varrho_n(t)\, dt - \varphi(0)| \leq \int_{-r/n}^{r/n} |\varphi(t) - \varphi(0)|\varrho_n(t)\, dt \to 0$. This means that δ is also a weak limit of the sequence $\{\varrho_n\}$ by associating to ϱ_n the linear form $\langle \cdot, \varrho_n \rangle$.

As a positive linear form on the space of test functions $\mathcal{C}_c(\mathbf{R})$, by the Riesz-Markov theorem δ can also be considered a Borel measure. The class of test functions for distributions will be much smaller than $\mathcal{C}_c(\mathbf{R})$, and operations with distributions will include well-defined generalized derivatives. The derivatives δ', δ'', ... of δ will be distributions but will not longer be measures.

With the description of the general theory of distributions, we will include some basic facts and examples concerning differential equations, a field where the influence of functional analysis has grown continuously precisely due to the use of distributions. In the next chapter we will find more examples.

The results are stated for complex-valued distributions, but the reader can check that they are also valid for real distributions.

6.1. Test functions

Let Ω be a nonempty open subset of $\mathbf{R}^n$.

[3]Laurent Schwartz formalized the mathematical theory of the generalized functions or distributions in 1945, and his 1950–51 book "Théorie des Distributions" [**40**] remains a basic reference for this topic. See footnote 1 in Chapter 3.

In Section 3.1 we introduced the Fréchet space $\mathcal{E}(\Omega)$ of all C^∞ complex functions on Ω, endowed with the locally convex topology of the local uniform convergence of functions and their derivatives.

If $\mathcal{K}(\Omega)$ represents the family of all compact subsets of Ω, the space of **test functions** is the vector space of all C^∞ complex functions on Ω with compact support,

$$\mathcal{D}(\Omega) = \bigcup_{K\in\mathcal{K}(\Omega)} \mathcal{D}_K(\Omega).$$

Recall that, if $K \in \mathcal{K}(\Omega)$, in (3.3) we defined $\mathcal{D}_K(\Omega)$ as the closed subspace of $\mathcal{E}(\Omega)$ that contains all $f \in \mathcal{E}(\Omega)$ whose support lies in K, so that its topology is defined by the increasing sequence of norms

$$q_N(f) = \sum_{|\alpha|\le N} \|D^\alpha f\|_K \qquad (N \in \mathbf{N}). \tag{6.1}$$

Thus, $\varphi_k \to \varphi$ in $\mathcal{D}_K(\Omega)$ means that $D^\alpha\varphi_k \to D^\alpha\varphi$ uniformly, for every $\alpha \in \mathbf{N}^n$.

Example 6.1. Let $g(t) = e^{-1/t}\chi_{(0,+\infty]}(t)$ and denote $\varrho(x) = Cg(1-|x|^2)$ with $C > 0$ such that $\int \varrho(x)\,dx = 1$. Then ϱ is a test function whose graph has the familiar bell shape that satisfies $0 \le \varrho \in \mathcal{D}_{\bar{B}(0,1)}(\mathbf{R}^n)$.

The derivatives are $g^{(n)}(t) = P_n(1/t)e^{-1/t}$ if $t > 0$, where every P_n is a polynomial and $g^{(n)}(t) = 0$ if $t < 0$; thus $g^{(n)}(0) = \lim_{t\to 0} g^{(n)}(t) = 0$ for every $n \in \mathbf{N}$, so that g is C^∞ supported by $[0,\infty)$, and $\varrho \in \mathcal{D}_{\bar{B}(0,1)}(\mathbf{R}^n)$.

From ϱ we can define a C^∞ summability kernel $\{\varrho_\varepsilon\}_{0<\varepsilon<\varepsilon_0}$ on $\mathbf{R}^n$ by $\varrho_\varepsilon(x) = \varepsilon^{-n}\varrho(x/\varepsilon)$ (see (2.22)). Note that $0 \le \varrho_\varepsilon \in \mathcal{D}_{\bar{B}(0,\varepsilon)}(\mathbf{R}^n)$ and that $\int \varrho_\varepsilon(x)\,dx = \int \varrho(x)\,dx = 1$. Such a function is often called a **mollifier**.

Let us define the local analogue of $L^1(\Omega)$ by denoting $L^1_{\text{loc}}(\Omega)$ the vector space of complex measurable functions locally integrable on the open set Ω, i.e. functions integrable on every compact subset K of Ω. As usual, two locally integrable functions are supposed to be equivalent if they coincide a.e.

If $f \in L^1_{\text{loc}}(\mathbf{R}^n)$, $\varrho_\varepsilon * f$ given by

$$(\varrho_\varepsilon * f)(x) = \int_{\mathbf{R}^n} \varrho_\varepsilon(x-y)f(y)\,dy = \int_{\bar{B}(0,\varepsilon)} f(x-y)\varrho_\varepsilon(y)\,dy$$

is a well-defined C^∞ function, since we can differentiate under the integral sign.

New test functions can be constructed from Example 6.1. For every couple $K \subset \Omega$, there is a test function $\varrho \in \mathcal{D}(\Omega)$ which is a smooth Urysohn function for this couple, so that $K \prec \varrho \prec \Omega$. It can be defined as follows:

Let $0 < \delta < d(K, \Omega^c)/2$ and let $0 \le \varphi \in \mathcal{D}(B(0,\delta))$ with $\int \varphi(x)\,dx = 1$, as in Example 6.1. We denote

$$K(\delta) = K + \bar{B}(0,\delta) = \{x \in \mathbf{R}^n;\ d(x,K) \le \delta\}$$

and define $\varrho = \chi_{K(\delta)} * \varphi$.

The properties of convolution (cf. (b) and (c) in Subsection 2.4.1) yield $\varrho \in \mathcal{D}_{K(2\delta)}(\Omega)$, since $f * \varrho \in \mathcal{E}(\mathbf{R}^n)$ and $\operatorname{supp}(f * \varrho) \subset K + \bar{B}(0,\delta) = K(\delta)$. To obtain $\varrho = 1$ in a neighborhood of K, we only need to change K by a compact neighborhood of K.

The C^∞ summability kernel $\{\varrho_\varepsilon\}$ is useful to regularize nonsmooth functions:

Theorem 6.2. (a) *If $f \in \mathcal{C}_K(\mathbf{R}^n)$ and $0 < \varepsilon < \eta$, then $f * \varrho_\varepsilon \in \mathcal{D}_{K(\eta)}(\mathbf{R}^n)$ and $\lim_{\varepsilon\to 0} f * \varrho_\varepsilon = f$ uniformly on $\mathbf{R}^n$.*

(b) *If $f \in L^p(\mathbf{R}^n)$ $(1 \le p < \infty)$, then $\lim_{\varepsilon \to 0} f * \varrho_\varepsilon = f$ in $L^p(\mathbf{R}^n)$.*

Proof. Let $\operatorname{supp} f \subset K$ and $\operatorname{supp} \varrho_\varepsilon \subset \bar{B}(0,\varepsilon)$. Then $f * \varrho_\varepsilon \in \mathcal{E}(\mathbf{R}^n)$ and $\operatorname{supp}(f * \varrho_e) \subset K(\varepsilon) \subset K(\eta)$.

Now (a) and (b) follow from Theorem 2.41. □

The small class $\mathcal{D}(\Omega)$ is large enough to be dense in many spaces, such as $L^p(\Omega)$ if $1 \le p < \infty$ (see Exercise 6.2) and $\mathcal{E}(\Omega)$:

Theorem 6.3. *The set $\mathcal{D}(\Omega)$ of all test functions is dense in $\mathcal{E}(\Omega)$.*

Proof. For every $f \in \mathcal{E}(\Omega)$ we consider $\{\varrho_m f\} \subset \mathcal{D}(\Omega)$, where ϱ_m are the Urysohn functions associated to an increasing sequence of compact subsets K_m of Ω such that every other compact subset K of Ω is contained in one of them, K_N, and $\varrho_m f = f$, so that $D^\alpha(\varrho_m f) = D^\alpha f$ on K if $m > N$. Then $D^\alpha(\varrho_m f) \to D^\alpha f$ uniformly on every compact set $K \subset \Omega$. □

6.2. The distributions

The distributions in a nonempty open subset Ω of $\mathbf{R}^n$ are defined as the linear forms on $\mathcal{D}(\Omega)$ which satisfy a convenient continuity property.

On $\mathcal{D}(\Omega) = \bigcup_{K \in \mathcal{K}(\Omega)} \mathcal{D}_K(\Omega)$, instead of defining a topology, it will be sufficient to consider a notion of convergence.[4]

We say that $\varphi_k \to \varphi$ in $\mathcal{D}(\Omega)$ if $\varphi_k \to \varphi$ in $\mathcal{D}_K(\Omega)$ for some $K \in \mathcal{K}(\Omega)$; that is, φ and φ_k are test functions on Ω that satisfy

- $\operatorname{supp} \varphi_k \subset K$ for some fixed compact set $K \subset \Omega$ and

[4]This notion of convergence follows from the topology defined in Exercise 6.5. With this topology, $\mathcal{D}(\Omega)$ is known as the inductive limit of the spaces $\mathcal{D}_K(\Omega)$.

- $D^\alpha \varphi_k \to D^\alpha \varphi$ uniformly as $k \to \infty$, for every $\alpha \in \mathbf{N}^n$.

A **distribution** on the open set Ω is defined as a linear form

$$u : \mathcal{D}(\Omega) \to \mathbf{C}$$

which is continuous with respect to the above convergence or, equivalently, such that $u_{|\mathcal{D}_K(\Omega)} \in \mathcal{D}_K(\Omega)'$ for every compact set $K \subset \Omega$.

We denote by $\mathcal{D}'(\Omega)$ the complex vector space of all distributions on Ω.

Hence $u \in \mathcal{D}'(\Omega)$ means that $u : \mathcal{D}(\Omega) \to \mathbf{C}$ is a linear form such that $u(\varphi_k) \to 0$ whenever $\varphi_k \to 0$ in some $\mathcal{D}_K(\Omega)$.

Thus, if $u \in \mathcal{D}'(\Omega)$, for every compact set $K \subset \Omega$ there are an integer $N = N_K > 0$ and a constant $C_K > 0$, both depending on K, such that

$$|u(\varphi)| \leq C_K q_N(\varphi) = C_K \sum_{|\alpha| \leq N} \|D^\alpha \varphi\|_K \qquad (\varphi \in \mathcal{D}_K(\Omega)).$$

If there is an N independent of K, then the smallest such N is called the **order** of the distribution u (the constant C_K still can depend on K). If this N does not exist, u is said to be of **infinite order**.

Example 6.4. Suppose f is a locally integrable function on Ω. Then we can define

$$u_f(\varphi) := \langle \varphi, f \rangle = \int_\Omega f(x)\varphi(x)\, dx \qquad (\varphi \in \mathcal{D}(\Omega)).$$

It is clear that u_f is a linear form on $\mathcal{D}(\Omega)$ such that $|u_f(\varphi)| \leq \|f\|_{L^1(K)} \|\varphi\|_K$, that is, $u_f \in \mathcal{D}'(\Omega)$.

Next we prove that the linear mapping $f \in L^1_{\text{loc}}(\Omega) \mapsto \langle \cdot, f \rangle \in \mathcal{D}'(\Omega)$ is one-to-one, so that we can consider $L^1_{\text{loc}}(\Omega) \subset \mathcal{D}'(\Omega)$ and it is said that the distribution u_f is a function. If no confusion is possible, we write f for u_f.

Theorem 6.5. *If $f \in L^1_{\text{loc}}(\Omega)$ and $\int_\Omega f(x)\varphi(x)\, dx = 0$ for every test function $\varphi \in \mathcal{D}(\Omega)$, then $f = 0$ (a.e.).*

Proof. We can assume that f is a real function and we only need to prove that $f = 0$ a.e. on every ball $B(a, r) \subset \Omega$.

If $A = \{x \in B(a, r);\, f(x) \geq 0\}$, choose $K_m \subset A \subset G_m \subset B(a, r)$ so that $|G_m \setminus K_m| \downarrow 0$ (K_m compact and G_m open sets). If φ_m is a Urysohn function for $K_m \subset G_m$, then $\varphi_m \to \chi_A$ a.e. and

$$\int_A f^+(x)\, dx = \lim \int f(x)\varphi_m(x)\, dx = 0.$$

Hence, $f^+(x) = 0$ a.e. on $B(a, r)$. Analogously, $f^-(x) = 0$ a.e. on $B(a, r)$. □

Example 6.6. For every $a \in \Omega$, the **Dirac distribution**, δ_a, is defined as $\delta_a(\varphi) = \varphi(a)$. On $\mathbf{R}^n$, we denote $\delta = \delta_0$.

Example 6.7. If μ is a Borel measure, or a complex Borel measure on Ω, then

$$\langle g, \mu \rangle := \int_\Omega g\, d\mu = \int_\Omega gh\, d|\mu| \qquad (g \in \mathcal{C}_c(\Omega))$$

defines a linear form $\langle \cdot, \mu \rangle$ on $\mathcal{C}_c(\Omega)$, and the restriction to $\mathcal{D}(\Omega)$ is clearly a distribution which is identified with μ. Here $d\mu = hd|\mu|$ is the polar representation of μ.

The pair $(\mathcal{D}(\Omega), \mathcal{D}'(\Omega))$ is a dual couple since, if $u(\varphi) = 0$ for every $u \in \mathcal{D}'(\Omega)$, then $\varphi = 0$ ($\int |\varphi|^2 = \langle \bar{\varphi}, \varphi \rangle = 0$). Thus, if u is a distribution and φ a test function, we also write $\langle \varphi, u \rangle$ instead of $u(\varphi)$.

Eventually we will use the notation $\langle \varphi(x), u(x) \rangle$ if we need to refer to the variable that is used at each moment.

The space $\mathcal{D}'(\Omega)$ is endowed with the weak topology $\sigma(\mathcal{D}'(\Omega), \mathcal{D}(\Omega))$, and the **distributional convergence** $u_k \to u$ means that $u_k(\varphi) \to u(\varphi)$ for every test function φ.

Multiplication by a C^∞ function, $g \in \mathcal{E}(\Omega)$, is naturally extended to distributions: If $f \in L^1_{\mathrm{loc}}(\Omega)$, also $gf \in L^1_{\mathrm{loc}}(\Omega)$ and, as a distribution,

$$\langle \varphi, gf \rangle = \int_\Omega \varphi(x) g(x) f(x)\, dx = \langle g\varphi, f \rangle,$$

which suggests that we can define gu for every $u \in \mathcal{D}'(\Omega)$ by the rule $(gu)(\varphi) = u(g\varphi)$; that is,

$$\langle \varphi, gu \rangle := \langle g\varphi, u \rangle.$$

Since $g\varphi \in \mathcal{D}(\Omega)$ for every $\varphi \in \mathcal{D}(\Omega)$, to prove that $gu \in \mathcal{D}'(\Omega)$, we only need to check the continuity property for this new linear functional gu on $\mathcal{D}(\Omega)$. But multiplication by g is a linear operator $g\cdot : \mathcal{D}(\Omega) \to \mathcal{D}(\Omega)$ such that

$$\varphi_k \to \varphi \text{ in } \mathcal{D}(\Omega) \Rightarrow g\varphi_k \to g\varphi \text{ in } \mathcal{D}(\Omega) \tag{6.2}$$

since, if $\varphi_k \to \varphi$ in $\mathcal{D}_K(\Omega)$, it is easily checked that also $g\varphi_k \to g\varphi$ in $\mathcal{D}_K(\Omega)$ from

$$\|\partial_j(g\varphi)\|_K \leq \|\partial_j g\|_K \|\varphi\|_K + \|g\|_K \|\partial_j \varphi\|_K$$

and from the corresponding estimates for the successive derivatives $D^\alpha(g\varphi)$.

Hence, also $(gu)(\varphi_k) = u(g\varphi_k) \to u(g\varphi) = (gu)(\varphi)$, and $gu \in \mathcal{D}'(\Omega)$.

There is a general procedure for extending some operations on $\mathcal{D}(\Omega)$ to operations on $\mathcal{D}'(\Omega)$ which includes the above multiplication by a C^∞ function as a special case.

If $T : \mathcal{D}(\Omega_1) \to \mathcal{D}(\Omega_2)$ is a linear map such that $T'u = u \circ T \in \mathcal{D}'(\Omega_1)$ for every $u \in \mathcal{D}'(\Omega_2)$, then $T' : \mathcal{D}'(\Omega_2) \to \mathcal{D}'(\Omega_1)$ is the transpose of T:

$$\langle T\varphi, u\rangle = \langle \varphi, T'u\rangle \qquad (\varphi \in \mathcal{D}(\Omega_1),\ u \in \mathcal{D}'(\Omega_2)).$$

Both T and T' are weakly continuous, since

$$p_\varphi(T'u) = |\langle \varphi, T'u\rangle| = |\langle T\varphi, u\rangle| = p_{T\varphi}(u)$$

and $p_u(T\varphi) = p_{T'u}(\varphi)$, where the p_φ and the p_u are, respectively, the seminorms that define the weak topology on $\mathcal{D}$ and on $\mathcal{D}'$.

Sometimes there are two linear operators

$$T : \mathcal{D}(\Omega_1) \to \mathcal{D}(\Omega_2), \quad R : \mathcal{D}(\Omega_2) \to \mathcal{D}(\Omega_1)$$

which are transposes of each other in the sense that

$$\langle \varphi, R\psi\rangle = \langle T\varphi, \psi\rangle \qquad (\varphi \in \mathcal{D}(\Omega_1),\ \psi \in \mathcal{D}(\Omega_2).$$

Then, for every $u \in \mathcal{D}'(\Omega_2)$ we can define $Ru = u \circ T$, which is a linear form on $\mathcal{D}(\Omega_1)$ characterized by the condition

$$\langle \varphi, Ru\rangle = \langle T\varphi, u\rangle.$$

If T satisfies the continuity condition $T\varphi_k \to 0$ in $\mathcal{D}(\Omega_1)$ whenever $\varphi_k \to 0$ in $\mathcal{D}(\Omega_1)$, then every Ru is a distribution on Ω_1, and $R : \mathcal{D}'(\Omega_2) \to \mathcal{D}'(\Omega_1)$ is the transpose T' of $T : \mathcal{D}(\Omega_1) \to \mathcal{D}(\Omega_2)$. This transpose is considered the extension of $R : \mathcal{D}(\Omega_2) \to \mathcal{D}(\Omega_1)$.

Very often, the restriction of $R = T' : \mathcal{D}'(\Omega_2) \to \mathcal{D}'(\Omega_1)$ to $L^1_{\mathrm{loc}}(\Omega_2)$ is also a natural extension of $R : \mathcal{D}(\Omega_2) \to \mathcal{D}(\Omega_1)$.

This is the case of multiplication by a C^∞ function $g \in \mathcal{E}(\Omega)$, $R = g\cdot$ on $\mathcal{D}(\Omega)$. With our definitions, also $T = g\cdot$ on $\mathcal{D}(\Omega)$, since

$$\langle \varphi, g\psi\rangle = \int_\Omega \varphi(x)g(x)\psi(x)\,dx = \langle g\varphi, \psi\rangle,$$

and $R = g\cdot$, extended to $\mathcal{D}'(\Omega)$, coincides with $g\cdot$ when restricted to $L^1_{\mathrm{loc}}(\Omega)$.

Let us now consider the case of a C^∞ change of variables $\psi : \Omega_2 \to \Omega_1$, and let Rf represent the function $f(\psi^{-1}(x))$ if $f \in L^1_{\mathrm{loc}}(\Omega_2)$, so that

$$\langle \varphi, Rf\rangle = \int_{\Omega_2} f(y)\varphi(\psi(y))|J_\psi(y)|\,dy = \langle |J_\psi|\varphi(\psi), f\rangle.$$

The following theorem shows that $\psi : \mathcal{D}(\Omega_1) \to \mathcal{D}(\Omega_2)$ continuously.

Theorem 6.8. *Let $\psi : \Omega_2 \to \Omega_1$ be a C^∞-change of variables, K a compact set in Ω_1, and let $L = \psi^{-1}(K)$. Then the linear map $\varphi \mapsto \varphi(\psi)$ is continuous from $\mathcal{D}_K(\Omega_1)$ to $\mathcal{D}_L(\Omega_2)$.*

Proof. If $\varphi(x) = 0$ when $x \notin K$ and $y \notin L$, then $x = \psi(y) \notin K$ and $\varphi(\psi)(y) = 0$. So $\Psi : \varphi \in \mathcal{D}_K(\Omega_1) \mapsto \varphi(\psi) \in \mathcal{D}_L(\Omega_2)$ is a well-defined linear mapping. From $\partial_j(\varphi(\psi)) = \partial_j\psi(\partial_j\varphi)(\psi)$, it follows as for (6.2) that $\|\partial_j(\varphi(\psi))\|_L \leq \|\partial_j\psi\|_L\|\partial_j\varphi\|_K$. Similarly,

$$\|\partial_k\partial_j(\varphi(\psi))\|_L \leq \|\partial_k\partial_j\psi\|_L\|\partial_j\varphi\|_K + \|\partial_j\psi\|_L\|\partial_k\partial_j\varphi\|_K$$

and, by induction, $\|D^\alpha\varphi(\psi)\|_L \leq C\sum_{\beta\leq\alpha}\|D^\beta\varphi\|_K$. Thus, $\varphi_k \to 0$ in $\mathcal{D}_K(\Omega_1)$ implies $\Psi(\varphi_k) \to 0$ in $\mathcal{D}_L(\Omega_2)$. □

Let $T\varphi := |J_\psi|\varphi(\psi)$ if $\varphi \in \mathcal{D}(\Omega_1)$, and let $u \in \mathcal{D}'(\Omega_2)$. Then the linear form $T'u$ on $\mathcal{D}(\Omega_1)$ defined as the transpose of T by

$$\langle\varphi, T'u\rangle = \langle|J_\psi|\varphi(\psi), u\rangle$$

is a distribution, since it is the composition

$$\varphi \mapsto \varphi(\psi) \mapsto |J_\psi|\varphi(\psi) \mapsto u(|J_\psi|\varphi(\psi))$$

of three linear mappings which have the appropriate continuity properties.

If $u = f \in L^1_{\mathrm{loc}}(\Omega_2)$, $T'u = Rf$. Hence, we write $Ru = T'u$ and

$$\langle\varphi, Ru\rangle = \langle|J_\psi|\varphi(\psi), u\rangle$$

defines the change of variables for distributions. Translations, scaling, and symmetry of distributions on $\Omega = \mathbf{R}^n$ are special cases:

Example 6.9. On $\mathbf{R}^n$ the translated $\tau_a(u)$, the symmetric $\tilde{u}$, and the dilation $u(\alpha x)$ of a distribution u are the distributions defined as follows:

(a) $\langle\varphi, \tau_a(u)\rangle = \langle\tau_{-a}\varphi, u\rangle$, or $\langle\varphi(x), u(x-a)\rangle = \langle\varphi(x+a), u(x)\rangle$.

(b) $\langle\varphi, \tilde{u}\rangle = \langle\tilde{\varphi}, u\rangle$, or $\langle\varphi(x), u(-x)\rangle = \langle\varphi(-x), u(x)\rangle$.

(c) $\langle\varphi(x), u(\alpha^{-1}x)\rangle = |\alpha|^n\langle\varphi(\alpha x), u(x)\rangle$ if $\alpha \neq 0$.

An example is $\tau_a\delta = \delta_a$, since

$$\langle\varphi, \tau_a\delta\rangle = \langle\varphi(x+a), \delta(x)\rangle = \varphi(a) = \langle\varphi, \delta_a\rangle.$$

6.3. Differentiation of distributions

If $f \in \mathcal{E}^1(\mathbf{R})$, or if f is an everywhere differentiable function on $\mathbf{R}$ and $f' \in L^1_{\mathrm{loc}}(\mathbf{R})$ (see e.g. Rudin's "Real and Complex Analysis" [**39**]), integration by parts yields

$$\langle\varphi, f'\rangle = \int_{-r}^{r} \varphi(t)f'(t)\,dt = -\int_{-r}^{r} \varphi'(t)f(t)\,dt = -\langle\varphi', f\rangle$$

if $\varphi \in \mathcal{D}_{[-r,r]}(\mathbf{R})$.

This result suggests that we associate to each distribution u on the open set $\Omega \subset \mathbf{R}^n$ the partial "derivatives" $\partial_j u$ (or $Du = u'$ when $n = 1$) by the formula

$$\langle \varphi, \partial_j u \rangle := \langle -\partial_j \varphi, u \rangle \qquad (1 \le j \le n).$$

That is, $(\partial_j u)(\varphi) = -u(\partial_j \varphi)$ ($u'(\varphi) = -u(\varphi')$ if $n = 1$).

If $\varphi_k \to 0$ in $\mathcal{D}_K(\Omega)$, then

$$-\langle \partial_j \varphi_k, u \rangle \to 0,$$

since $\partial_j \varphi_k \to 0$ in $\mathcal{D}_K(\Omega)$. Thus, $\partial_j u \in \mathcal{D}'(\Omega)$, the operator

$$\partial_j : \mathcal{D}'(\Omega) \to \mathcal{D}'(\Omega)$$

is the adjoint of

$$-\partial_j : \mathcal{D}(\Omega) \to \mathcal{D}(\Omega),$$

and ∂_j on $\mathcal{D}'(\Omega)$ extends the partial derivatives of C^1 functions on Ω.

Note that $\partial_j : L^1_{\mathrm{loc}}(\Omega) \to \mathcal{D}'(\Omega)$ can be defined as the restriction of $\partial_j : \mathcal{D}'(\Omega) \to \mathcal{D}'(\Omega)$. It is a custom, which we shall usually follow, to write $\partial_j f$ instead of $\partial_j u_f$, which is a distribution called the **distributional derivative** or **weak derivative** of $f \in L^1_{\mathrm{loc}}(\Omega)$. It is characterized by the identity

$$\langle \varphi, \partial_j f \rangle = -\int_\Omega f(x) \partial_j \varphi(x)\, dx \qquad (\varphi \in \mathcal{D}(\Omega)).$$

The αth distributional derivatives are defined by induction. If $D^\alpha = \partial_1^{\alpha_1} \circ \cdots \circ \partial_n^{\alpha_n}$, then

$$D^\alpha : \mathcal{D}'(\Omega) \to \mathcal{D}'(\Omega)$$

is a continuous linear operator such that

$$\langle \varphi, D^\alpha u \rangle = (-1)^{|\alpha|} \langle D^\alpha \varphi, u \rangle.$$

We also write $D^\alpha f$ for $D^\alpha u_f$ if $f \in L^1_{\mathrm{loc}}(\Omega)$.

Theorem 6.10. *If $f \in \mathcal{E}^m(\Omega)$ and $|\alpha| \le m$, then the usual pointwise derivative $D^\alpha f$ coincides with the distributional derivative: $D^\alpha u_f = u_{D^\alpha f}$.*

The Leibniz formula for the derivatives of a product of functions[5] *is extended to*

$$\partial_j (fu) = (\partial_j f) u + f \partial_j u \qquad (f \in \mathcal{E}^1(\Omega),\ u \in \mathcal{D}'(\Omega)).$$

Proof. By induction, assume $m = 1$ and consider $D^\alpha = \partial_1$:

$$\langle \varphi, \partial_1 u_f \rangle = -\int_K \partial_1 \varphi(x) f(x)\, dx.$$

[5]See in Exercise 2.1 the extension of the Leibniz formula to higher-order derivatives.

We can suppose that $f(x) = 0$ if $x \in \mathbf{R}^n \setminus G$, where $G \subset \Omega$ is an open set containing $K = \operatorname{supp}\varphi$, so that

$$\langle \varphi, \partial_1 f \rangle = -\int_{\mathbf{R}^n} \partial_1\varphi(x) f(x)\,dx.$$

If $x = (t, \bar{x}) \in \mathbf{R} \times \mathbf{R}^{n-1}$ and $f_{\bar{x}}(t) = f(x)$, integration by parts gives

$$\langle \varphi, \partial_1 u_f \rangle = -\int_{\mathbf{R}^{n-1}} \int_{-r}^{r} \varphi_{\bar{x}}'(t) f_{\bar{x}}(t)\,dt\,d\bar{x} = \langle \varphi, \partial_1 f \rangle.$$

For the product,

$$\langle \varphi, \partial_j(fu) \rangle = -\langle f\partial_j\varphi, u \rangle = \langle \varphi\partial_j f - \partial(f\varphi), u \rangle = \langle \varphi, (\partial_j f)u + f\partial_j u \rangle.$$

□

If f is absolutely continuous, so that f' exists a.e., and if $f' \in L^1_{\text{loc}}(\mathbf{R})$, it is shown in Exercise 6.20 that this a.e. derivative is also the distributional derivative of f. But one has to be careful since the following examples show that there are many functions $f \in L^1_{\text{loc}}(\mathbf{R})$ with an a.e. defined derivative $f' \in L^1_{\text{loc}}(\mathbf{R})$ which is not the distributional derivative of f.

Example 6.11. Let f be a C^1 function on $(a,b) \setminus \{t_1, t_2, \dots, t_n\}$ $(a < t_1 < t_2 < \cdots < t_n < b)$, such that the right and left limits $f(t_j+)$ and $f(t_j-)$ exist and are finite $(1 \le j \le n)$. If f' on $(a,b) \setminus \{t_1, t_2, \dots, t_n\}$ is locally integrable on (a,b), then the distributional derivative of $u_f = \langle \cdot, f \rangle$ is

$$\begin{aligned} u_f' &= f' + \big(f(t_1+) - f(t_1-)\big)\delta_{t_1} + \big(f(t_2+) - f(t_2-)\big)\delta_{t_2} \\ &\quad + \cdots + \big(f(t_n+) - f(t_n-)\big)\delta_{t_n}. \end{aligned}$$

Define $\varphi = 0$ outside of (a,b). If $t_0 = a$ and $t_{n+1} = b$,

$$\langle \varphi, u_f' \rangle = -\sum_{j=1}^{n} \int_{t_j}^{t_{j+1}} \varphi'(t) f(t)\,dt$$

and integration by parts gives

$$\begin{aligned} \langle \varphi, u_f' \rangle &= \sum_{j=1}^{n} \Big(\int_{t_j}^{t_{j+1}} \varphi(t) f'(t)\,dt + f(t_j-)\varphi(t_j) - f(t_{j+1}+)\varphi(t_{j+1}) \Big) \\ &= \int_a^b \varphi(t) f'(t)\,dt + \sum_{j=1}^{n} \big(f(t_j+) - f(t_j-)\big)\varphi(t_j). \end{aligned}$$

Example 6.12. If $f \in \mathcal{E}(\mathbf{R})$ and $Y = \chi_{[0,\infty)}$ (**Heaviside function**), then

$$(fY)' = f'Y + f(0)\delta \ \text{ and } \ (fY)^{(n)} = f^{(n)}Y + \sum_{k=0}^{n-1} f^{(k)}(0)\delta^{(n-k-1)}.$$

The first identity is contained in Example 6.11. Then, by induction,

$$(fY)^{(n+1)} = ((fY)^{(n)})' = f^{(n+1)}Y + f^{(n)}\delta + \sum_{k=0}^{n-1} f^{(k)}(0)\delta^{(n-k)}$$

and $f^{(n)}\delta = f^{(n)}(0)\delta$.

The following results show a couple of instances where the derivatives of distributions behave as the derivatives of functions.

First, on an interval of $\mathbf{R}$, the constant functions are the unique distributions with zero derivative:

Theorem 6.13. *Let $u \in \mathcal{D}'(a,b)$ $(-\infty \le a < b \le +\infty)$. Then $u' = 0$ if and only if u is a constant function.*

Proof. If $\varphi \in \mathcal{D}(a,b)$, let $\psi = \varphi - (\int_a^b \varphi)\varrho$, where we choose $\varrho \in \mathcal{D}(a,b)$ so that $\int_a^b \varrho = 1$. Then $\psi \in \mathcal{D}(a,b)$ and $\Psi(x) = \int_a^x \psi$ is also a test function since, if $\operatorname{supp}\varphi$, $\operatorname{supp}\varrho \subset [c,d] \subset (a,b)$, then $\Psi(x) = 0$ if $x \notin [c,d]$.

Moreover $u'(\Psi) = -u(\Psi') = -u(\psi) = -u(\varphi) + (\int_a^b \varphi)u(\varrho)$ and, if $u' = 0$, also $\int_a^b u(\varrho)\varphi(t)\,dt - u(\varphi) = 0$. Thus, $u(\varphi) = \langle \varphi, u(\varrho)\rangle$ and u is the constant function $u(\varrho)$. □

The derivative of a distribution on $\mathbf{R}^n$ can be defined as a quotient limit of distributions:

Theorem 6.14. *Suppose $\varphi \in \mathcal{D}(\mathbf{R}^n)$ and $u \in \mathcal{D}'(\mathbf{R}^n)$, and let e_j be the jth standard basis vector of $\mathbf{R}^n$. Then*

$$\lim_{h\to 0} \frac{\tau_{-he_j}\varphi - \varphi}{h} = \partial_j\varphi \quad in\ \mathcal{D}(\mathbf{R}^n)$$

and

$$\lim_{h\to 0} \frac{\tau_{-he_j}u - u}{h} = \partial_j u \quad in\ \mathcal{D}'(\mathbf{R}^n).$$

Proof. We can suppose $j = 1$ and we choose a compact set $K \subset \mathbf{R}^n$ which contains $\operatorname{supp}\varphi$ and so that $d = d(\operatorname{supp}\varphi, K^c) > 0$. By the Lagrange mean value theorem,

$$\left|\frac{\varphi(x+he_1) - \varphi(x)}{h} - \partial_1\varphi(x)\right| = |\partial_1\varphi(x + h_x e_1) - \partial_1\varphi(x)|$$

with $|h_x| \le |h|$. Note that $\operatorname{supp}\varphi + he_1 \subset K$ if $|h| < d$ and, by taking the sup in $x \in K$,

$$\left\|\frac{\varphi(x+he_1) - \varphi(x)}{h} - \partial_1\varphi(x)\right\|_K \le \sup_{|t|\le|h|} \|\partial_1\varphi(x+te_1) - \partial_1\varphi(x)\|_K,$$

where the right side tends to 0 as $|h| \to 0$, since $\partial_1\varphi$ is uniformly continuous.

The same holds for every $D^\alpha\varphi$ and, if $h \to 0$,

$$\frac{\varphi(x+he_1)-\varphi(x)}{h} \to \partial_1\varphi(x)$$

in $\mathcal{D}_K(\mathbf{R}^n)$ and in $\mathcal{D}(\mathbf{R}^n)$.

From the continuity of the distribution u on $\mathcal{D}_K(\mathbf{R}^n)$,

$$\frac{\langle \varphi(x-he_1), u(x)\rangle - \langle\varphi, u\rangle}{h} \to \langle -\partial_1\varphi, u\rangle$$

if $h \to 0$, so that $\lim_{h\to 0}\langle \varphi, h^{-1}(\tau_{-he_j}u - u)\rangle = \langle \varphi, \partial_j u\rangle$. □

6.4. Convolution of distributions

The convolution $f * g$ of two integrable functions f and g is a well-defined integrable function and, if one of them has compact support, then

$$\begin{aligned}\langle \varphi, f*g\rangle &= \int_{\mathbf{R}^n} g(y)\int_{\mathbf{R}^n}\varphi(x)f(x-y)\,dx\,dy\\ &= \langle \varphi(y+z), f(z)g(y)\rangle = \langle\langle \varphi(y+z), f(z)\rangle, g(y)\rangle.\end{aligned}$$

If u and v are two distributions, this suggests that we should try

$$\langle \varphi, u*v\rangle = \langle\langle \varphi(y+z), u(z)\rangle, v(y)\rangle$$

as a possible definition, but then we need to apply v to the function $f(y) = \langle \varphi(y+z), u(z)\rangle$. Note that if $\operatorname{supp}\varphi \subset \bar{B}(0,r)$, we can only state that $\varphi(y+z) = 0$ whenever (y,z) lies in $|y+z| \geq r$, and f can no longer have a compact support.

This shows that it is convenient to consider cases where u can be applied to functions with an unbounded support. This will be possible if u belongs to the class of distributions with compact support.

6.4.1. Support of a distribution and distributions with compact support. We recall that the support of a continuous function f on Ω is the closure in Ω of the set of points where $f(x) \neq 0$ or, equivalently, the complement of the largest open subset of Ω on which f is zero.

To define an analogous concept for distributions, it is convenient to show that they are locally determined, and partitions of unity are useful to this end.

Recall that, according to Theorem 1.3 and Remark 1.4, if $\{\Omega_1, \ldots, \Omega_m\}$ is a finite family of open subsets of $\mathbf{R}^n$ which covers a compact set $K \subset \mathbf{R}^n$, then there exists a partition of unity $\varphi_j \in \mathcal{D}(\Omega_j)$ $(1 \leq j \leq m)$ subordinate to this covering.

Note also that if G is an open subset of Ω, we can consider $\mathcal{D}(G) \subset \mathcal{D}(\Omega) \subset \mathcal{D}(\mathbf{R}^n)$ and the restriction of a distribution $u \in \mathcal{D}'(\Omega)$ to $\mathcal{D}(G)$ is also a distribution, $u_G \in \mathcal{D}'(G)$. If the restrictions of $u, v \in \mathcal{D}'(\Omega)$ are the same, we say that these distributions are equal on G, and we write $u = v$ on G.

Theorem 6.15. *For a given distribution $u \in \mathcal{D}'(\Omega)$, there is a largest open subset G of Ω where u is zero.*

Proof. Let G be the union of all open subsets of Ω where u is zero, and let $\varphi \in \mathcal{D}_K(G)$. Then $K \subset \Omega_1 \cup \cdots \cup \Omega_m$ with $u = 0$ on every Ω_j, and we can choose a partition of unity $\{\varphi_j\}_{j=1}^m$ subordinate to this covering of K. Then $u(\varphi) = u(\sum_{j=1}^m \varphi_j \varphi) = \sum_{j=1}^m u(\varphi_j\varphi) = 0$, since $\operatorname{supp} \varphi_j \varphi \subset \Omega_j$. □

The **support of a distribution** $u \in \mathcal{D}'(\Omega)$, $\operatorname{supp} u$, is defined as the complement $\Omega \setminus G$ of the largest open subset G of Ω where u vanishes.

Theorem 6.16. *The support of $u \in \mathcal{D}'(\Omega)$ is compact if and only if u is the restriction to $\mathcal{D}(\Omega)$ of some $v \in \mathcal{E}'(\Omega)$, which is uniquely determined by u. We call $\mathcal{E}'(\Omega)$ the dual of $\mathcal{E}(\Omega)$.*

Proof. Since the inclusions $\mathcal{D}_K(\Omega) \hookrightarrow \mathcal{E}(\Omega)$ are all continuous, the restriction u of every $v \in \mathcal{E}'(\Omega)$ to $\mathcal{D}(\Omega)$ is a distribution. The continuity of v means that

$$|v(f)| \leq C q_{K,m}(f) \qquad (f \in \mathcal{E}(\Omega)) \tag{6.3}$$

for some semi-norm $q_{K,m}(f) = \max_{|\alpha|\leq m} \|D^\alpha f\|_K$ and some constant $C > 0$. Thus $v(f) = 0$ if $\operatorname{supp} f \subset \Omega \setminus K$, since in this case $q_{K,m}(f) = 0$, and then $\operatorname{supp} v \subset K$.

Reciprocally, given $u \in \mathcal{D}'(\Omega)$ with $\operatorname{supp} u = K$, compact, we choose $K(\delta) = K + \bar{B}(0,\delta) \subset \Omega$ and $K(\delta) \prec \varrho \prec \Omega$. We claim that

$$v(f) := u(f\varrho)$$

defines a linear form on $\mathcal{E}(\Omega)$ which is an extension of u which does not depend on ϱ or δ.

Indeed, if also $K(\delta') \prec \varrho_1 \prec \Omega$, then $u(f\varrho) - u(f\varrho_1) = u(f(\varrho - \varrho_1)) = 0$ since $\operatorname{supp}(\varrho - \varrho_1) \cap K = \emptyset$.

Moreover, if $\varphi \in \mathcal{D}_K(\Omega)$, it is shown that $v(\varphi) = u(\varphi\varrho) = u(\varphi)$ by choosing a compact set L such that $K \cup \operatorname{supp}\varphi \subset \operatorname{Int}(L)$ and $L \prec \varrho \prec \Omega$. If $f_k \to 0$ in $\mathcal{E}(\Omega)$, it is easily checked that $f_k\varrho \to 0$ in $\mathcal{D}(\Omega)$ and then $v(f_k) \to 0$. That is, $v \in \mathcal{E}'(\Omega)$.

According to Theorem 6.3, the set $\mathcal{D}(\Omega)$ of all test functions is dense in $\mathcal{E}(\Omega)$, and v is uniquely determined by u. □

Since $v \in \mathcal{E}'(\Omega) \mapsto u = v_{|\mathcal{D}(\Omega)} \in \mathcal{D}'(\Omega)$ is linear and one-to-one onto the class of all distributions on Ω with compact support, $\mathcal{E}'(\Omega)$ is said to be the **space of distributions with compact support** on Ω.

6.4.2. Convolution of distributions with functions. Next we define the convolution of a distribution with a function on the whole $\mathbf{R}^n$ when at least one of them has compact support. Suppose first that $\varphi \in \mathcal{D}(\mathbf{R}^n)$ and $u \in \mathcal{D}'(\mathbf{R}^n)$.

Since $(f * \varphi)(x) = \langle \varphi(x-y), f(y)\rangle$ if $f \in L^1_{\text{loc}}(\mathbf{R}^n)$, we define

$$(u * \varphi)(x) := \langle \varphi(x-y), u(y)\rangle = \langle (\tau_x \tilde{\varphi})(y), u(y)\rangle = \langle \tilde{\varphi}, \tau_{-x} u\rangle = \langle \tau_{-x}\varphi, \tilde{u}\rangle.$$

Note that $(u * \varphi)(x)$ is continuous with respect to u in $\mathcal{D}'(\mathbf{R}^n)$ (when $\varphi \in \mathcal{D}(\mathbf{R}^n)$ and $x \in \mathbf{R}^n$ are fixed), with respect to φ, and with respect to x. That is,

(6.4)

$$(u_k * \varphi)(x) \to (u * \varphi)(x), \ (u * \varphi_k)(x) \to (u * \varphi)(x), \ (u_k * \varphi)(x_k) \to (u * \varphi)(x)$$

when $u_k \to u$ in $\mathcal{D}'(\mathbf{R}^n)$, $\varphi_k \to \varphi$ in $\mathcal{D}_K(\mathbf{R}^n)$, and $x_k \to x$ in $\mathbf{R}^n$, respectively.

Indeed, $(u_k * \varphi)(x) = \langle \tilde{\tau}_x \varphi, u_k\rangle \to \langle \tilde{\tau}_x \varphi, u\rangle$ if $u_k \to u$, since on $\mathcal{D}'(\mathbf{R}^n)$ we are considering the weak convergence.

Also, if $\varphi_k \to \varphi$, then $(u * \varphi_k)(x) = \langle \tilde{\varphi}_k, \tau_{-x} u\rangle \to \langle \varphi, \widetilde{\tau_{-x} u}\rangle$.

Finally, if $x_k \to x$, then $(u * \varphi)(x_k) = \langle \tau_{-x_k}\varphi, \tilde{u}\rangle \to \langle \tau_{-x}\varphi, \tilde{u}\rangle$ from the uniform continuity of φ and of all $D^\alpha \varphi$, since $\|D^\alpha \varphi - \tau_{-x_k} D^\alpha \varphi\|_{\mathbf{R}^n} \le \varepsilon$ if k is large, and there is a single compact set $L \subset \mathbf{R}^n$ so that $\operatorname{supp} \tau_{-x_k}\varphi \subset L$.

Let us gather together the basic properties of this convolution:

Theorem 6.17. *If $\varphi \in \mathcal{D}(\mathbf{R}^n)$ and $u \in \mathcal{D}'(\mathbf{R}^n)$, then $u * \varphi \in \mathcal{E}(\mathbf{R}^n)$ and the following properties are satisfied:*

(a) $\tau_a(u * \varphi) = (\tau_a u) * \varphi = u * (\tau_a \varphi)$.

(b) $D^\alpha(u * \varphi) = (D^\alpha u) * \varphi = u * (D^\alpha \varphi)$.

(c) $\operatorname{supp}(u * \varphi) \subset \operatorname{supp} u + \operatorname{supp} \varphi$.

(d) $(u * \varphi) * \psi = u * (\varphi * \psi)$ *if* $\psi \in \mathcal{C}_c(\mathbf{R}^n)$.

(e) $u * \varphi_k \to u * \varphi$ *in* $\mathcal{E}(\mathbf{R}^n)$ *if* $\varphi_k \to \varphi$ *in* $\mathcal{D}(\mathbf{R}^n)$*; hence,* $u* : \mathcal{D}(\mathbf{R}^n) \to \mathcal{E}(\mathbf{R}^n)$ *is continuous (when restricted to every* $\mathcal{D}_K(\mathbf{R}^n)$*).*

Proof. (a) From the definition,

$$(u * \varphi)(x-a) = \langle \varphi(x-(y+a)), u(y)\rangle = \langle \varphi(x-y), \tau_a u(y)\rangle = ((\tau_a u) * \varphi)(x)$$

and also $\langle \varphi(x-a-y), u(y)\rangle = (u * (\tau_a \varphi))(x)$.

(b) We can assume $D^\alpha = \partial_1$. Then, according to Theorem 6.14, an application of (a) gives

$$\begin{aligned}\partial_1(u*\varphi)(x) &= \lim_{t\to 0}\frac{\tau_{-te_1}(u*\varphi)(x)-(u*\varphi)(x)}{t}\\ &= \lim_{t\to 0}\langle\tilde\varphi, \frac{\tau_{-x-te_1}u-\tau_{-x}u}{t}\rangle\\ &= \langle\tilde\varphi,\tau_{-x}\partial_1 u\rangle = ((\partial_1 u)*\varphi)(x).\end{aligned}$$

Similarly,

$$\partial_1(u*\varphi)(x) = \lim_{t\to 0}\langle\frac{\tau_{-te_1}\tilde\varphi-\tilde\varphi}{t},\tau_{-x}u\rangle = (u*\partial_1\varphi)(x).$$

(c) If $\operatorname{supp}\varphi(x-\cdot)\cap\operatorname{supp}u=\emptyset$, then $\langle\varphi(x-\cdot),u\rangle = 0$. Hence, if $(u*\varphi)(x)\neq 0$, necessarily $\operatorname{supp}\varphi(x-\cdot)\cap\operatorname{supp}u\neq\emptyset$, so that $x-y\in\operatorname{supp}\varphi$ at least for one point $y\in\operatorname{supp}u$ and then $x\in\operatorname{supp}u+\operatorname{supp}\varphi$. Thus, $\{u*\varphi\neq 0\}\subset\operatorname{supp}u+\operatorname{supp}\varphi$, which is closed as a sum of a compact set and a closed set, so that

$$\operatorname{supp}u*\varphi = \overline{\{u*\varphi\neq 0\}}\subset\operatorname{supp}u+\operatorname{supp}\varphi.$$

(d) We claim that, by writing the integral that defines $(\varphi*\psi)(x)$ for a given point $x\in\mathbf{R}^n$ as the limit of Riemann sums with uniform increments $0<h<1$, then

$$\lim_{h\to 0}\sum_{z\in\mathbf{Z}^n}\varphi(x-hz)\psi(hz)h^n = (\varphi*\psi)(x) \tag{6.5}$$

in $\mathcal{D}(\mathbf{R}^n)$. Note that the sums are in fact finite, since φ has a bounded support, and the argument in (c) shows that their supports are contained in $\operatorname{supp}\varphi+\operatorname{supp}\psi$.

To prove this claim, note that $(x,y)\mapsto D^\alpha\varphi(x-y)\psi(y)$ is uniformly continuous on $\mathbf{R}^{2n}$, since it is continuous and it is readily checked that its support is compact, so that $\sum_{z\in\mathbf{Z}^n}D^\alpha\varphi(x-hz)\psi(hz)h^n\to(\varphi*\psi)(x)$ uniformly in x.

Now, since $\langle\varphi(x-y-hz),u(y)\rangle = (u*\varphi(\cdot-hz))(x)$,

$$\begin{aligned}(u*(\varphi*\psi))(x) &= \lim_{h\to 0}\langle(\sum_{z\in\mathbf{Z}^n}\varphi(x-y-hz)\psi(hz)h^n),u(y)\rangle\\ &= \lim_{h\to 0}\sum_{z\in\mathbf{Z}^n}(u*\varphi)(x-hz)\psi(hz)h^n\\ &= \int_{\mathbf{R}^n}(u*\varphi)(x-y)\psi(y)\,dy = ((u*\varphi)*\psi)(x).\end{aligned}$$

(e) The restrictions $u*:\mathcal{D}_K(\mathbf{R}^n)\to\mathcal{E}(\mathbf{R}^n)$ are continuous by the closed graph theorem, since, if $\varphi_k\to\varphi$ in $\mathcal{D}(\mathbf{R}^n)$ and $u*\varphi_k\to f$, then it follows from (6.4) that $(u*\varphi_k)(x) \to (u*\varphi)(x)$ for every $x \in \mathbf{R}^n$. Hence, $f(x) = (u*\varphi)(x)$. □

The convolution that we have defined can be considerably extended.

If $u \in \mathcal{D}'(\mathbf{R}^n)$ is a distribution with compact support and $f \in \mathcal{E}(\mathbf{R}^n)$, then we also may define

$$(u*f)(x) := u(\tau_x \tilde{f})$$

since u is extended to a unique $u \in \mathcal{E}'(\mathbf{R}^n)$, by Theorem 6.16. We have an analogous result to Theorem 6.17:

Theorem 6.18. *If $f \in \mathcal{E}(\mathbf{R}^n)$ and $u \in \mathcal{E}'(\mathbf{R}^n)$, then $u*f \in \mathcal{E}(\mathbf{R}^n)$ and the following properties are satisfied:*

(a) $\tau_a(u*f) = (\tau_a u)*f = u*(\tau_a f)$.

(b) $D^\alpha(u*f) = (D^\alpha u)*f = u*(D^\alpha f)$.

(c) $(u*f)*\psi = u*(f*\psi)$ *if* $\psi \in \mathcal{D}(\mathbf{R}^n)$.

Proof. The statement (a) is proved exactly as in Theorem 6.17 and $(D^\alpha u)*f = u*(D^\alpha f)$, which is one half of (b), follows directly from the definition. To also show that $D^\alpha(u*f)(x) = u*(D^\alpha f)(x)$, we choose $\varrho = \tilde{\varrho} \in \mathcal{D}(\mathbf{R}^n)$ such that $\varrho = 1$ on a neighborhood of $\operatorname{supp} u \cup \tau_x \tilde{f}$ and then

$$(u*D^\alpha f)(x) = u*(D^\alpha(\varrho f))(x) = D^\alpha(u*(\varrho f))(x)$$

with $(u*(\varrho f))(x) = u(\varrho \tau_x \tilde{f}) = (u*f)(x)$. Note also that

$$((D^\alpha u)*f)(x) = ((D^\alpha u)*(\varrho f))(x)$$

is continuous.

To prove (c), let G be a bounded open set that contains $\operatorname{supp} u$ and choose any $\varrho \in \mathcal{D}(\mathbf{R}^n)$ such that $\tilde{\varrho} = \tilde{f}$ on $\operatorname{supp}\psi + G$. Then $\widetilde{f*\psi} = \widetilde{\varrho*\psi}$ on G and it follows from the definition that

$$(u*(f*\psi))(0) = (u*(\varrho*\psi))(0) = (u*(\psi*\varrho))(0).$$

Since $\tau_h\tilde{f} = \tau_h\tilde{\varrho}$ on W, $u*f = u*\varrho$ at every point in $-\operatorname{supp}\psi$ and it follows that

$$((u*f)*\psi)(0) = ((u*\varrho)*\psi)(0).$$

But $\operatorname{supp}(u*\psi) \subset \operatorname{supp} u + \operatorname{supp}\psi$, so that

$$((u*\psi)*f)(0) = ((u*\psi)*\varrho)(0)$$

and $((u*\psi)*\varrho)(0) = (u*(\psi*\varrho))(0)$, which combined with the first identity gives

$$(u*(f*\psi))(0) = ((u*\psi)*f)(0)$$

and we obtain (c) from (a) by applying a convenient translation. □

Example 6.19. $\delta * f = f$ for every $f \in \mathcal{E}(\mathbf{R}^n)$, since

$$(\delta * f)(x) = \langle f(x-y), \delta(y) \rangle = f(x).$$

6.4.3. Convolution of distributions. A typical fact of convolution operators is that they commute with translations. By Theorem 6.17(a), for every distribution $u \in \mathcal{D}'(\mathbf{R}^n)$, the operator $u* : \mathcal{D}(\mathbf{R}^n) \to \mathcal{E}(\mathbf{R}^n)$ is linear, continuous on every $\mathcal{D}_K(\mathbf{R}^n)$, and $u * (\tau_a \varphi) = \tau_a(u * \varphi)$. The converse is also true:

Theorem 6.20. *If $T : \mathcal{D}(\mathbf{R}^n) \to \mathcal{C}(\mathbf{R}^n)$ is linear, continuous on every $\mathcal{D}_K(\mathbf{R}^n)$ (or, simply, $(T\varphi_k)(0) \to 0$ if $\varphi_k \to 0$ in $\mathcal{D}(\mathbf{R}^n)$) and if it commutes with translations, then $T = u*$ for a uniquely determined distribution $u \in \mathcal{D}'(\mathbf{R}^n)$.*

Proof. Since necessarily $(T\varphi)(0) = (u * \varphi)(0) = (\tau_0 u)(\tilde{\varphi}) = u(\tilde{\varphi})$, we must define $u(\varphi) := (T\tilde{\varphi})(0)$. Note that u is linear and, if $\varphi_k \to 0$ in $\mathcal{D}(\mathbf{R}^n)$, then $u(\varphi_k) = (T\tilde{\varphi}_k)(0) \to 0$; hence $u \in \mathcal{D}'(\mathbf{R}^n)$.

But $T(\tau_{-a}\varphi)(0) = (u*\tau_{-a}\varphi)(0)$ and, since T commutes with translations,

$$(T\varphi)(a) = (\tau_{-a}T\varphi)(0) = T(\tau_{-a}\varphi)(0) = (u * \tau_{-a}\varphi)(0) = (u * \varphi)(a).$$

□

The preceding theorem will allow us to extend the previous definitions to a convolution of two distributions u and v on $\mathbf{R}^n$ if at least the support of one of them is compact. In this case we can define

$$T_{u,v}(\varphi) := u * (v * \varphi).$$

Note that, if $v \in \mathcal{E}'(\mathbf{R}^n)$, then $v * \varphi \in \mathcal{D}(\mathbf{R}^n)$ and $u * (v * \varphi)$ is well-defined; similarly, if u has compact support, then $v * \varphi \in \mathcal{E}(\mathbf{R}^n)$ and we also are allowed to consider

$$u * (v * \varphi)(x) = u(\tau_x(\widetilde{v * \varphi})).$$

In both cases, the linear map $T_{u,v} : \mathcal{D}(\mathbf{R}^n) \to \mathcal{E}(\mathbf{R}^n)$ is continuous on every $\mathcal{D}_K(\mathbf{R}^n)$ and commutes with translations. By Theorem 6.20, there exists a unique $w \in \mathcal{D}'(\mathbf{R}^n)$ such that $T_{u,v} = w*$. Then we define $u * v := w$, so that $u * v \in \mathcal{D}'(\mathbf{R}^n)$ is characterized by the identity

$$(u * v) * \varphi = u * (v * \varphi) \qquad (\varphi \in \mathcal{D}(\mathbf{R}^n)).$$

Theorem 6.21. *Let $u_1, u_2, u_3, u, v \in \mathcal{D}'(\mathbf{R}^n)$.*

(a) *If $u * \varphi = 0$ for every $\varphi \in \mathcal{D}(\mathbf{R}^n)$, then $u = 0$.*

(b) *If at least two of $u_1, u_2, u_3 \in \mathcal{D}'(\mathbf{R}^n)$ have compact support, then*

$$u_1 * (u_2 * u_3) = (u_1 * u_2) * u_3.$$

(c) *If at least the support of one of u and v is compact, then $u*v = v*u$,*

$$D^\alpha(u * v) = (D^\alpha u) * v = (u * D^\alpha v),$$

and $\operatorname{supp}(u * v) \subset \operatorname{supp} u + \operatorname{supp} v$.

Proof. (a) $u(\varphi) = u(\tau_0\tilde{\varphi}) = (u * \varphi)(0) = 0$ if $u * \varphi = 0$.

(b) From the definitions, $(u_1 * (u_2 * u_3)) * \varphi = ((u_1 * u_2) * u_3) * \varphi$ since, in any case,

$$(u_1 * (u_2 * u_3)) * \varphi = u_1 * ((u_2 * u_3) * \varphi) = u_1 * (u_2 * (u_3 * \varphi))$$

and

$$((u_1 * u_2) * u_3) * \varphi = (u_1 * u_2) * (u_3 * \varphi) = u_1 * (u_2 * (u_3 * \varphi)).$$

(c) Let $u_1 = u * v$ and $u_2 = v * u$. To show that $u_1 = u_2$, we only need to prove that $u_1 * (\varphi_1 * \varphi_2) = u_2 * (\varphi_1 * \varphi_2)$ for any $\varphi_1, \varphi_2 \in \mathcal{D}(\mathbf{R}^n)$, since then, by an application of (b), $(u_1 * \varphi_1) * \varphi_2 = (u_2 * \varphi_1) * \varphi_2$. Now we apply (a), so that $u_1 * \varphi_1 = u_2 * \varphi_1$ and then $u_1 = u_2$.

But, by using the fact that the convolution of functions is commutative and (b),

$$(u * v) * (\varphi_1 * \varphi_2) = u * ((v * \varphi_1) * \varphi_2) = u * (\varphi_2 * (v * \varphi_1)) = (u * \varphi_2) * (v * \varphi_1)$$

and also

$$(v * u) * (\varphi_1 * \varphi_2) = (v * \varphi_1) * (u * \varphi_2) = (u * \varphi_2) * (v * \varphi_1).$$

Hence, $v * u = u * v$.

To study the supports, let ϱ be a test function supported by $\bar{B}(0,1)$ and with $\int \varrho = 1$, and let $\varrho_k(x) = k^n \varrho(kx)$. Assume that v has compact support. Then $\operatorname{supp}(u * v) \subset \operatorname{supp}(u * v * \varrho_k)$ and

$$\operatorname{supp}(u * v * \varrho_k) \subset \operatorname{supp} u + \operatorname{supp}(v * \varrho_k) \subset (\operatorname{supp} u + \operatorname{supp} v) + \operatorname{supp} \varrho_k$$

where $\operatorname{supp} \varrho_k = \bar{B}(0, 1/k)$; hence

$$\bigcap_k ((\operatorname{supp} u + \operatorname{supp} v) + \bar{B}(0, 1/k)) = \operatorname{supp} u + \operatorname{supp} v,$$

a closed set, since $\operatorname{supp} v$ is compact.

Finally,

$$(D^\alpha(u * v)) * \varphi = (u * v) * D^\alpha \varphi = u * (v * D^\alpha \varphi) = u * (D^\alpha v) * \varphi$$

for every φ, and $D^\alpha(u * v) = u * (D^\alpha v)$. □

Example 6.22. If $u \in \mathcal{D}'(\mathbf{R}^n)$, $u*\delta = u$, since $(u*\delta)*\varphi = u*(\delta*\varphi) = u*\varphi$.

This example shows that the derivatives are convolution operators:

$$D^\alpha u = D^\alpha(\delta * u) = (D^\alpha \delta) * u.$$

6.5. Distributional differential equations

6.5.1. Linear differential equations. A very simple differential equation on $(\alpha, \beta) \subset \mathbf{R}$ is

$$u' = 0$$

and in Theorem 6.13 we proved that its distributional solutions are the classical constant solutions.

This property is easily extended to all ordinary linear equations with smooth coefficients:

Theorem 6.23. *Let $a \in \mathcal{E}(\alpha, \beta)$ and $f \in \mathcal{C}(\alpha, \beta)$. The solutions $u \in \mathcal{D}'(\alpha, \beta)$ of the first order linear differential equation*

$$u' + au = f$$

are the classical solutions $u \in \mathcal{E}^1(\alpha, \beta)$.

This fact still holds for linear systems:

$$\vec{u}' + A\vec{u} = \vec{f}$$

($A = \{a_i^j\}_{i,j=1}^n$ an $n \times n$ matrix, $\vec{f} = (f_1, \ldots, f_n) \in \mathcal{C}(\alpha, \beta)^n$, $\vec{u} \in \mathcal{D}'(\alpha, \beta)^n$).

Proof. If $a = 0$ and $F \in \mathcal{E}^1(\alpha, \beta)$ is a primitive of f, then $(u - F)' = u' - F' = 0$ and $u - F = C$. Hence, $u = F + C$ is a C^1 function.

In the general numerical case, if $\int a$ is a primitive of a, then the function $e = \exp \int a$ is a C^∞ solution of $e' = ea$ and, for any solution u of $u' + au = f$,

$$(eu)' = eu' + e'u = e(u' + au) = ef,$$

which is continuous. Then eu is a C^1 function and so is $u = e^{-1}eu$.

In the vector case we follow the same model with $\int A = \{\int a_i^j\}_{i,j=1}^n$. Then $e = \exp \int A = \sum_{k=0}^\infty (1/k!)(\int A)^k$ is a square matrix with an inverse e^{-1} that allows us to write again $\vec{u} = e^{-1}e\vec{u}$. □

Corollary 6.24. If $f \in \mathcal{C}(\alpha, \beta)$ and $a_0, a_1, \ldots, a_{n-1} \in \mathcal{E}(\alpha, \beta)$, then the distributional solutions of the linear equation

$$u^{(n)} + a_{n-1}u^{(n-1)} + \cdots + a_1 u' + a_0 u = f$$

are the classical solutions $u \in \mathcal{E}^n(\alpha, \beta)$.

Proof. With the usual procedure, we denote $u_k = u^{(k-1)}$ $(k = 1, \ldots, n)$ to obtain the first order linear system

$$u_n' + a_{n-1}u_n + \cdots + a_1 u_2 + a_0 u_1 = f,\ u_{n-1}' - u_n = 0, \ldots,\ u_1' - u_2 = 0$$

and then we apply Theorem 6.23. □

Such a regularity property does not hold in the several variables case, and we have an example in Exercise 6.35. A differential operator with constant coefficients on $\mathbf{R}^n$,

$$P(D) = \sum_{|\alpha| \leq N} a_\alpha D^\alpha,$$

is said to be **hypoelliptic**[6] whenever, for every open set $\Omega \subset \mathbf{R}^n$, if $f \in \mathcal{E}(\Omega)$ and u is a distributional solution $u \in \mathcal{D}'(\Omega)$ of $P(D)u = f$, u must also belong to $\mathcal{E}(\Omega)$.

There is a characterization of all the polynomials P such that $P(D)$ is hypoelliptic due to Hörmander (see, for instance, Yoshida's "Functional Analysis" [**44**]). Here we are going to prove a sufficient condition in terms of what is known as a fundamental solution.

6.5.2. Fundamental solutions. On $\mathbf{R}^n$, a **fundamental solution** of a differential operator with constant coefficients

$$P(D) = \sum_{|\alpha| \leq N} a_\alpha D^\alpha$$

is a distributional solution of the equation $P(D)u = \delta$.

The interest in having a fundamental solution, E, is due to the fact that, if v is any distribution and E or v has a compact support, so that $v * E$ is well-defined, then

$$P(D)(E * v) = (P(D)E) * v = \delta * v = v \tag{6.6}$$

and $E * v$ is a solution of $P(D)u = v$.

By the Malgrange-Ehrenpreis theorem, every differential operator with constant coefficients has a fundamental solution.[7]

In the following pages we are going to show a few important concrete examples.

Theorem 6.25. *On* $\mathbf{R}$, *the ordinary linear operator with constant coefficients*

$$P(D)u = u^{(n)} + a_1 u^{(n-1)} + \cdots + a_{n-1}u' + a_n u$$

has the fundamental solution $E = fY$, *if* $f \in \mathcal{E}(\mathbf{R})$ *is the solution of*

$$P(D)f = 0, \quad f^{(n-1)}(0) = 1, \quad f^{(n-2)}(0) = \cdots = f'(0) = f(0) = 0.$$

---

[6] If the characteristic polynomial $P_N(x) = \sum_{|\alpha|=N} a_\alpha x^\alpha$ does not vanish at any point, $P(D)$ is said to be **elliptic**; an example is the Laplacian. Every elliptic operator is hypoelliptic (see [**14**]).

[7] A proof due to Hörmander can be found in Yosida's "Functional Analysis" [**44**] or in Rudin's "Functional Analysis" [**38**].

Proof. The derivative of $E = fY$ is $E' = f\delta + f'Y = f(0)\delta + f'Y = f'Y$. Then $E'' = f'(0)\delta + f''Y = f''Y$ and, by induction

$$E^{(n)} = f^{(n-1)}(0)\delta + f^{(n)}Y = \delta + f^{(n)}Y.$$

Hence,

$$\begin{aligned} P(D)E &= \delta + f^{(n)}Y + a_1 f^{(n-1)}Y + \cdots + a_{n-1}f'Y + a_n fY \\ &= \delta + YP(D)f = \delta \end{aligned}$$

as announced. □

As a first example in several variables, let us look for a fundamental solution of the Laplacian

$$\triangle \equiv \partial_1^2 + \cdots + \partial_n^2$$

on $\mathbf{R}^n$, which is the most important of all differential operators.

To this end, **Green's identities**[8]

$$\int_\Omega v(x)\triangle u(x)\,dx + \int_\Omega \nabla u(x)\cdot\nabla v(x)\,dx = \int_S v\partial_\nu u\,d\sigma \tag{6.7}$$

and

$$\int_\Omega v(x)\triangle u(x)\,dx - \int_\Omega u(x)\triangle v(x)\,dx = \int_S (v\partial_\nu u - u\partial_\nu v)\,d\sigma \tag{6.8}$$

are useful. Here $\Omega \subset \mathbf{R}^n$ is a regular domain for the divergence theorem and u, v are two C^2 functions on $\bar\Omega$ and ν denotes the outer normal vector field on the positively oriented boundary S of Ω.

Note that (6.8) follows from (6.7), and (6.7) is obtained by taking $w = v\nabla u$ in the **divergence theorem**

$$\int_\Omega (\text{Div } w)(x)\,dx = \int_S w\cdot\nu\,d\sigma.$$

By the spherical symmetry of $\triangle$, we are led to try a radial function

$$E_n(x) = \psi(r), \quad r = |x|,$$

such that $\triangle E_n = 0$ on $r > 0$ as a possible fundamental function. In this case,

$$\triangle E_n(x) = \psi''(r) + \frac{n-1}{r}\psi'(r),$$

obtained from the chain rule of differentiation. Then, the radial solutions of $\triangle E_n = 0$ on $r > 0$ are the C^∞ functions given by

$$\psi(r) = \begin{cases} C\log r & (n = 2), \\ \frac{C}{2-n}r^{2-n} & (n > 2) \end{cases} \tag{6.9}$$

[8] In 1828, G. Green's included these identities in "An Essay on the Application of Mathematical Analysis to the Theories of Electricity and Magnetism". See footnote 17 in Chapter 2.

plus an additive constant.

Denote by ω_{n-1} the area of the unit sphere $S_{n-1} \subset \mathbf{R}^n$:

$$\omega_0 = 2, \quad \omega_1 = 2\pi, \quad \omega_2 = 4\pi, \quad \omega_{n-1} = \frac{2\pi^{n/2}}{\Gamma(n/2)}.$$

Theorem 6.26. *The function E_n defined on $\mathbf{R}^n \setminus \{0\}$ by*

$$E_n(x) = \begin{cases} \frac{1}{\omega_1} \log|x| & \text{if } n = 2, \\ -\frac{1}{(n-2)\omega_{n-1}} |x|^{2-n} & \text{if } n \neq 2 \end{cases} \tag{6.10}$$

is a locally integrable fundamental solution for the Laplacian, and it is of class C^∞ on $\mathbf{R}^n \setminus \{0\}$.

Proof. Polar integration shows that E_n is integrable on $B(0,1)$, and it is clearly a C^∞ function on $\mathbf{R}^n \setminus \{0\}$.

For $\varphi \in \mathcal{D}(\mathbf{R}^n)$, we require that

$$\int_{\mathbf{R}^n} (\Delta\varphi)(x) E_n(x)\, dx = \varphi(0). \tag{6.11}$$

Since E_n is singular at $x = 0$, we cut out from $\mathbf{R}^n$ a small ball $\bar{B}(0,r)$ and, since φ is supported by some ball $|x| < R$, we are led to integrate on $\Omega_r = \{x;\, r < |x| < R\}$ and to show that

$$\lim_{r\downarrow 0} \int_{\Omega_r} E_n(x)\Delta\varphi(x)\, dx = \varphi(0). \tag{6.12}$$

We apply Green's identity (6.8) with $v(x) = E_n(x) = \psi(r)$ given by (6.9) and $u = \varphi$. Since $\Delta v = 0$ on Ω_r and $\varphi = 0$ on a neighborhood of the outer boundary $|x| = R$ of Ω_r, we have

$$\int_{\Omega_r} E_n \Delta\varphi\, dx = \int_{\{|x|=r\}} (E_n \partial_\nu \varphi - \varphi \partial_\nu E_n)\, d\sigma.$$

On $|x| = r$ the exterior normal points towards 0 and $E(x) = \psi(r)$. Consequently, again using Green's identities with $v = 1$ and $u = \varphi$,

$$\begin{aligned} \int_{\Omega_r} E_n \Delta\varphi\, dx &= \psi(r) \int_{\{|x|=r\}} \partial_\nu \varphi\, d\sigma + \psi'(r) \int_{\{|x|=r\}} \varphi\, d\sigma \\ &= -\psi(r) \int_{B(0,r)} \Delta\varphi + C r^{1-n} \int_{\{|x|=r\}} \varphi\, d\sigma. \end{aligned}$$

Here $|B(0,r)| = Ar^n$ and $\lim_{r\to 0} r^n \psi(r) = 0$, so that, from the continuity of φ

$$\lim_{r\downarrow 0} \int_{\Omega_r} E_n(x)\Delta\varphi(x)\, dx = C\omega_n \varphi(0)$$

and we obtain (6.12) by taking $C = 1/\omega_n$. □

Remark 6.27. It will follow from Theorem 6.31 that $\triangle$ is hypoelliptic, a result known as **Weyl's lemma**.

The behavior of this fundamental solution will allow us to show that (6.6) gives a solution of the **Poisson equation**,[9] $\triangle u = f$, for any $f \in L^1$, if $n \geq 3$. In Exercise 6.37 we consider the case $n = 2$.

Theorem 6.28. *If $f \in L^1(\mathbf{R}^n)$ and $n > 2$, then $u := \triangle * f$ is a well-defined locally integrable function which is a distributional solution of*

$$\triangle u = f.$$

Proof. Define $E_0 = \chi_{B(0,1)} E_n \in L^1(\mathbf{R}^n)$ and $E_\infty = E_n - E_0 \in L^\infty(\mathbf{R}^n)$. Then $E_0 * f \in L^1(\mathbf{R}^n)$ and $E_\infty * f \in L^\infty(\mathbf{R}^n)$, so that $u = \triangle * f$ is well-defined and locally integrable.

Now consider

$$f^N = \chi_{B(0,N)} f \quad \text{and} \quad u^N = \triangle * f^N,$$

so that, by a standard application of the dominated convergence theorem, $u^N \to u$ in $\mathcal{D}'(\mathbf{R}^n)$ and also $\triangle u^N \to \triangle u$ in $\mathcal{D}'(\mathbf{R}^n)$.

But $\triangle u^N = f^N$, since

$$\begin{aligned} \langle \varphi, \triangle u^N \rangle &= \langle \triangle \varphi, E_n * f^N \rangle = \langle \widetilde{f^N} * \triangle \varphi, E_n \rangle \\ &= \langle \triangle(\widetilde{f^N} * \varphi), E_n \rangle = \langle \widetilde{f^N} * \varphi, L E_n \rangle \\ &= (\widetilde{f^N} * \varphi)(0) = \langle \varphi, f^N \rangle, \end{aligned}$$

and we are led to $\triangle u = \lim_N \triangle u^N = \lim_N f^N = f$. □

There are similar results for the **heat operator**[10]:

Theorem 6.29. *The function Γ defined on $\mathbf{R}^{1+n}$ by*

$$\Gamma(t,x) = \begin{cases} (4c\pi t)^{-n/2} e^{-|x|^2/(4ct)} & \text{if } t > 0, \\ 0 & \text{if } t \leq 0 \end{cases} \tag{6.13}$$

is a locally integrable fundamental solution, of class C^∞ on $\mathbf{R}^n \setminus \{0\}$, for the operator

$$L \equiv \partial_t - c\triangle = \partial_t - c\sum_{j=1}^{n} \partial_{x_j}^2.$$

Here $c > 0$.

[9]Named after Siméon-Denis Poisson, who in 1812 discovered that Laplace's equation $\triangle u = 0$ of potential theory is valid only outside a solid.

[10]In Chpter 7, the Fourier transform will show that $\Gamma(t,x)$ is a natural fundamental function for this operator.

Proof. Note that, if $t > 0$, after a change of variable, we obtain

$$\int_{\mathbf{R}^n} \Gamma(t,x)\,dx = \frac{1}{(2\pi)^{n/2}} \int_{\mathbf{R}^n} e^{-|u|^2/2}\,du = 1$$

and Γ is locally integrable on $\mathbf{R}^{1+n}$.

For every test function φ,

$$\langle \varphi, \partial_t \Gamma\rangle = -\lim_{\varepsilon\to 0} \int_\varepsilon^\infty \int_{\mathbf{R}^n} \partial_t\varphi(t,x)\Gamma(t,x)\,dt\,dx$$

and, by partial integration,

$$(6.14)\quad \langle \varphi, \partial_t \Gamma\rangle = \lim_{\varepsilon\to 0} \int_\varepsilon^\infty \int_{\mathbf{R}^n} \varphi(\varepsilon,x)\partial_t\Gamma(t,x)\,dt\,dx + \int_{\mathbf{R}^n} f(\varepsilon,x)\Gamma(t,x)\,dx.$$

Here $\Gamma(t,x) = t^{-n/2}\Gamma(1, t^{-1/2}x)$ and the substitution $x = u\sqrt{\varepsilon}$ gives

$$\int_{\mathbf{R}^n} \varphi(\varepsilon,x)\Gamma(t,x)\,dx = \int_{\mathbf{R}^n} \varphi(\varepsilon,u\sqrt{\varepsilon})\Gamma(1,u)\,du \to \varphi(0)\int_{\mathbf{R}^n} \Gamma(1,u)\,du = \varphi(0)$$

when $\varepsilon \downarrow 0$, by dominated convergence.

A direct computation shows that, on $t > 0$,

$$\partial_t\Gamma(t,x) = c\triangle\Gamma(t,x)$$

and integration by parts proves that

$$\int_{\{t>\varepsilon\}} \int_{\mathbf{R}^n} \varphi(t,x)\partial_t\Gamma(t,x)\,dt\,dx = \int_{\{t>\varepsilon\}} \int_{\mathbf{R}^n} c\triangle\varphi(t,x)\Gamma(t,x)\,dt\,dx$$

and (6.14) becomes

$$\langle \varphi, \partial_t\Gamma\rangle = \langle \varphi, c\triangle\Gamma\rangle + \varphi(0),$$

which means that $L\Gamma = \delta$.

It is also clear that Γ is C^∞ away from the origin, since the partial derivatives vanish as $t \downarrow 0$ when $x \neq 0$. □

This fundamental solution is zero for $t \le 0$ and, as for the Laplacian, we have

Theorem 6.30. *If $f \in L^1(\mathbf{R}^{1+n})$, then $u := \Gamma * f$ is a well-defined locally integrable function which is a distributional solution of the heat equation,*

$$Lu \equiv \partial_t u - c\triangle u = f.$$

Proof. Denote $\Gamma_t(x) = \Gamma(t,x)$ and $f_t(x) = f(t,x)$. To prove that

$$u(t,x) = \int_{-\infty}^t \int_{\mathbf{R}^n} \Gamma_{t-s}(x-y) f_s(y)\,dy\,ds$$

is well-defined as a distribution, we apply Young's inequality to the convolution $\Gamma_{t-s} * f_s$ to obtain

$$\sup_{t\in\mathbf{R}} \int_{\mathbf{R}^n} |u(t,x)|\,dx \leq \int_{-\infty}^{\infty} \int_{\mathbf{R}^n} |f(s,x)|\,ds\,dx = \|f\|_1$$

and we conclude that u is defined a.e. and locally integrable.

Now, as in Theorem 6.28, we define

$$f^N = \chi_{B(0,N)} f \quad \text{and} \quad u^N = \Gamma * f^N,$$

so that $u^N \to u$ in $\mathcal{D}'(\mathbf{R}^{1+n})$ and also $Lu^N \to Lu$ in $\mathcal{D}'(\mathbf{R}^{1+n})$.

But again $Lu^N = f^N$, since

$$\begin{aligned} \langle \varphi, Lu^N \rangle &= \langle L'\varphi, \Gamma * f^N \rangle = \langle \widetilde{f^N} * L'\varphi, \Gamma \rangle \\ &= \langle L'(\widetilde{f^N} * \varphi), \Gamma \rangle = \langle \widetilde{f^N} * \varphi, L\Gamma \rangle \\ &= (\widetilde{f^N} * \varphi)(0) = \langle \varphi, f^N \rangle. \end{aligned}$$

Thus $Lu = \lim_N Lu^N = \lim_N f^N = f$. □

The regularity of the solutions of both Poisson and heat equations appears as special cases of the following theorem about operators that have a fundamental function which is C^∞ away from zero.

We will use the fact that, if $f \in \mathcal{E}(\Omega)$ and $u \in \mathcal{D}'(\Omega)$, then

$$D^\alpha(fu) = fD^\alpha u + \sum_{0<|\beta|\leq|\alpha|} C_\beta D^\beta f D^{\alpha-\beta} u = fD^\alpha u + v_\alpha,$$

which follows from the Leibniz formula. Note that, if f is constant on an open set $G \subset \Omega$, then $v = 0$ on G, since $D^\beta f = 0$ on this open set for every $\beta \neq 0$.

Of course, by linearity, it follows that

(6.15) $\quad P(D)(fu) = fP(D)u + v$ and $v = 0$ on G if f is constant on G,

for every differential operator with constant coefficients $P(D)$.

Theorem 6.31. *If a differential operator with constant coefficients $P(D)$ has a fundamental solution which is of class C^∞ on $\mathbf{R}^n \setminus \{0\}$, then $P(D)$ is hypoelliptic.*

Proof. Suppose that $P(D)E = \delta$ with E of class C^∞ on $\{0\}^c$ and that $P(D)u = f$ on an open set $\Omega \subset \mathbf{R}^n$, with $f \in \mathcal{E}(\Omega)$ and $u \in \mathcal{D}'(\Omega)$. We start by showing that every compact set $K \subset \Omega$ has an open neighborhood $G \subset \Omega$ where u is C^∞.

We select a second compact set $K(2\delta) = K+\bar{B}(0,2\delta) \subset \Omega$ and we choose $G = K+B(0,\delta)$. Then let $K(2\delta) \prec \varrho \prec \Omega$ and $\bar{B}(0,\delta/2) \prec \gamma \prec B(0,\delta)$, and consider the compactly supported distributions ϱu and γE. As in (6.15),

$$P(D)(\varrho u) = \varrho P(D)u + v = \varrho f + v,$$

where $v \in \mathcal{D}'(\Omega)$ and $\operatorname{supp} v \cap K(2\delta) = \emptyset$, since v is zero on a neighborhood of $K(2\delta)$. Also

$$P(D)(\gamma E) = \gamma P(D)E + h = \delta + h$$

and $h \in \mathcal{D}(\mathbf{R}^n)$ since E is C^∞ on $\{0\}^c$ and $h = 0$ on $B(0,\delta/2)$.

Then

$$P(D)(\gamma E * \varrho u) = \gamma E * \varrho f + \gamma E * v$$

and

$$P(D)(\gamma E * \varrho u) = \varrho u + h * \varrho u,$$

so that

$$\varrho u = -h * \varrho u + \gamma E * \varrho f + \gamma E * v = \varphi + \gamma E * v, \quad \varphi \in \mathcal{D}(\mathbf{R}^n),$$

with $\operatorname{supp}(\gamma E * v) \subset \bar{B}(0,\delta) + \operatorname{supp} v \subset \Omega_K^c$; thus $u = \varphi$ on G.

By considering an exhaustive increasing sequence K_m of compact sets in Ω as in (3.2) and $K_m \subset G_m \subset \operatorname{Int} K_{m+1}$ so that u is a C^∞ function g_m on G_m, we have that $g_{m+1} = u = g_m$ on G_m, so that $u = g$ on Ω if g is the common extension to Ω of the functions g_m. □

Now the regularity of the solutions of the Poisson equation $\triangle u = f$ is an obvious corollary of the preceding results:

Theorem 6.32. *Let Ω be an open subset of $\mathbf{R}^n$. If $f \in \mathcal{E}(\Omega)$ and if $u \in \mathcal{D}'(\Omega)$ is a distributional solution of*

$$\triangle u = f,$$

then $u \in \mathcal{E}(\Omega)$.

Of course, there is the analogous result for the heat operator, but the situation is not the same for the **wave operator**, i.e.,

$$\square \equiv \partial_t^2 - \triangle = \partial_t^2 - \sum_{j=1}^{n} \partial_{x_j}^2$$

on $\mathbf{R}^{1+n}$.

This operator is no longer hypoelliptic and it is more involved. Here we describe only the elementary case $n = 1$, i.e.,

$$\square \equiv \partial_t^2 - \partial_x^2.$$

After the change of variables

$$s = t - x, \quad y = t + x$$

we are led to the operator

$$\partial_s \partial_y$$

since then $f(s, y) = f(t - x, t + x) = g(t, x)$ shows that

$$4\partial_s \partial_y f(s, y) = (\partial_t^2 - \partial_x^2) g(t, x).$$

We must find a locally integrable solution $F(s, y)$ of

$$4\partial_s \partial_y F(s, y) = \delta(s, y),$$

and the simplest one, obtained by separating variables, is

$$F(s, y) = \frac{1}{4} Y(s) Y(y) = \frac{1}{4} \chi_{\{s, y \geq 0\}}(s, y).$$

Our candidate is then $E(t, x) = cY(t - x)Y(t + x)$ with c such that

$$\begin{aligned}
\varphi(0,0) &= \int_{\mathbf{R}^2} E(t, x) \Box \varphi(t, x) \, dt \, dx \\
&= \frac{c}{2} \int_{\mathbf{R}^2} Y(y) Y(s) 4\partial_s \partial_y \varphi\left(\frac{s+y}{2}, \frac{s-y}{2}\right) ds \, dy \\
&= 2c\varphi(0, 0).
\end{aligned}$$

Thus

$$E(t, x) = \frac{1}{2} Y(t - x) Y(t + x);$$

that is, $E = 1/2$, constant in the sector $t + x \geq 0$, $t - x \geq 0$, and 0 elsewhere. This shows that the operator is not hypoelliptic.

6.5.3. Green's functions. For the boundary value problem

$$Lu \equiv (pu')' - qu = g \qquad (0 < p \in \mathcal{C}^1[a, b];\ q \in \mathcal{C}[a, b])$$

with

$$B_1(u) := A_1 u(a) + A_2 u'(a) = 0 \qquad (|A_1| + |A_2| \neq 0)$$

and

$$B_2(u) := B_1 u(b) + B_2 u'(b) = 0 \qquad (|B_1| + |B_2| \neq 0)$$

the Green's function $G(x, \xi)$ has been defined in (2.37) so that it is continuous and, for every $\xi \in (a, b)$, it satisfies the following properties:

(a) $G(\cdot, \xi) \in \mathcal{C}^2([a, \xi) \cup (\xi, b])$, $LG(\cdot, \xi) = 0$ on $[a, \xi) \cup (\xi, b]$, and $G(\cdot, \xi)$ satisfies the boundary conditions $B_1(G(\cdot, \xi)) = 0$, $B_2(G(\cdot, \xi)) = 0$.

(b) $G(\cdot, \xi) \in \mathcal{C}[a, b]$.

(c) The right side and left side derivatives of $G(\cdot, \xi)$ exist at $x = \xi$ and

$$\partial_x G(\xi+, \xi) - \partial_x G(\xi-, \xi) = \frac{1}{p(\xi)}.$$

Note that, as a distribution on (a,b), $LG(\cdot,\xi) = \delta_\xi$, which will also be represented by

$$L_x G(x,\xi) = \delta_\xi(x).$$

Indeed, for every $\varphi \in \mathcal{D}(a,b)$, and if we denote $G_\xi = G(\cdot,\xi)$, we can write

$$\langle \varphi, LG_\xi \rangle = \langle (p\varphi')', G_\xi \rangle + \langle \varphi, qG_\xi \rangle,$$

where, since $\operatorname{supp}\varphi \subset (a,b)$ and G_ξ is continuous, integration by parts produces

$$\langle (p\varphi')', G_\xi \rangle = -\Big\{ \int_a^{\xi-} - \int_{\xi+}^b \Big\} \varphi' p G'_\xi.$$

Finally we integrate by parts once again to obtain

$$\begin{aligned} \langle \varphi, LG_\xi \rangle &= \langle \varphi, qG_\xi \rangle + \Big\{ \int_a^{\xi-} - \int_{\xi+}^b \Big\} \varphi (pG_\xi)' - (\varphi p G'_\xi)(\xi-) + (\varphi p G'_\xi)(\xi-) \\ &= \Big\{ \int_a^{\xi-} - \int_{\xi+}^b \Big\} (\varphi L G_\xi) - (\varphi p)(\xi)(G'_\xi(\xi-) - G'_\xi(\xi+)) \\ &= (\varphi p)(\xi) \frac{1}{p(\xi)} = \varphi(\xi) = \langle \varphi, \delta_\xi \rangle. \end{aligned}$$

On the space

$$\{u \in \mathcal{C}^2[a,b];\ B_1(u) = 0,\ B_2(u) = 0\},$$

the operator $Tf(x) = \int_a^b G(x,\xi) f(\xi)\, d\xi = \langle f(\xi), G(x,\xi) \rangle$ solves $Lu = f$.

In a more general setting, we say that $G(x,\xi)$ is a **Green's function for a differential operator** L on an open set $\Omega \subset \mathbf{R}^n$ if $G(\cdot,\xi) \in \mathcal{D}'(\Omega)$ and

$$LG(\cdot,\xi) = \delta_\xi$$

for every $\xi \in \Omega$.

Formally, if $u(x) = \langle f(\xi), G(x,\xi) \rangle = \int_\Omega G(x,\xi) f(\xi)\, d\xi$, we obtain

$$Lu(x) = \int_\Omega LG(x,\xi) f(\xi)\, d\xi = \langle f(\xi), \delta_x(\xi) \rangle = f(x)$$

and $Tf(x) := \langle f(\xi), G(x,\xi) \rangle$ solves $Lu = f$.

Note that if E is a fundamental solution for a linear operator with constant coefficients $P(D)$, then $G(x,\xi) = E(x - \xi)$ is a Green's function for $P(D)$ on $\Omega = \mathbf{R}^n$, since

$$P(D)G(\cdot,\xi) = P(D)\tau_\xi E = \tau_\xi P(D) E = \tau_\xi \delta = \delta_\xi$$

and the formal solution is $u(x) = \int_\Omega G(x,\xi) f(\xi)\, d\xi = E * f$.

By requiring $G(\cdot,\xi)$ to satisfy some linear conditions that determine a subspace H of $\mathcal{D}'(\Omega)$, we can try to solve $Lu = f$ with $u \in H$.

This was the case in Theorem 2.53 for a boundary value problem.

As a very important example, let us describe a method to find a **Green's function for the Dirichlet problem**

$$\triangle u = f, \qquad u_{|\partial\Omega} = 0 \tag{6.16}$$

on a bounded open set $\Omega \subset \mathbf{R}^n$. For such a Green's function, G, the requirements are

$$\triangle G_\xi = \delta_\xi, \quad G_\xi \in \mathcal{C}(\bar{\Omega}), \quad G_\xi(y) = 0 \quad \forall y \in \partial\Omega \tag{6.17}$$

for every $\xi \in \Omega$, where we denote $G_\xi(x) = G(x, \xi)$.

The uniqueness of such a function will follow from Theorem 7.37, and the existence will be proved for instance if for every $\xi \in \Omega$ we can solve the Dirichlet problem

$$\triangle g_\xi = 0 \text{ on } \Omega, \quad g_\xi \in \mathcal{C}(\bar{\Omega}), \quad g_\xi(y) = E_n(y - \xi) \quad \forall y \in \partial\Omega,$$

with E_n the fundamental solution defined in (6.10), since by defining

$$G(x, \xi) = E_n(x - \xi) - g_\xi(x),$$

G_ξ satisfies (6.17).

From the condition $G_\xi = 0$ on $\partial\Omega$, the possible solution

$$u(x) = \int_\Omega G(x, \xi) f(\xi)\, d\xi$$

of $\triangle u = f$ will also satisfy $u(y) = \int_\Omega G(y, \xi) f(\xi)\, d\xi = 0$ when $y \in \partial\Omega$.

Similar heuristic arguments can be used to guess a method to solve the homogeneous Dirichlet problem on Ω with inhomogeneous boundary conditions

$$\triangle u = 0, \qquad u_{|S} = f. \tag{6.18}$$

It can be shown that, if the bounded open set Ω has a C^∞ boundary, the Green's function G exists, $G(x, y) = G(y, x)$, $G(x, \cdot)$ extends to $G(x, \cdot) \in \mathcal{C}^\infty(\bar{\Omega} \setminus \{x\})$ for every $x \in \Omega$, and

$$u(x) = \int_\Omega G(x, y) f(y)\, dy$$

is the solution of (6.16).

If u is this solution, since $\triangle u = 0$, a formal application of Green's identiy (6.8) leads to

$$\begin{aligned} u(x) &= \int_\Omega u(y)\delta(x - y)\, dy = \int_\Omega \Big(u(y) \triangle G_x(y) - G_x(y) \triangle u(y) \Big)\, dy \\ &= \int_S u(y) \partial_{\nu(y)} G_x(y) d\sigma(y) = \int_S f(y) \partial_{\nu(y)} G_x(y) d\sigma(y). \end{aligned}$$

The function

$$P(x,y) = \partial_{\nu(y)} G(x,y) \qquad ((x,y) \in \Omega \times S)$$

is called the **Poisson kernel** for Ω, and

$$u(x) = \int_S P(x,y) f(y)\, dy$$

is the **Poisson integral**, a candidate for the solution of the Dirichlet problem (6.18).

Usually it is hard to find Green's functions and Poisson kernels. We restrict ourselves here to the very special but important case of the ball $B = \{x \in \mathbf{R}^n;\ |x| < 1\}$ and we refer to Folland's "Introduction to Partial Differential Equations" [**14**] for further information on this subject.

6.5.4. Green's function of the Dirichlet problem in the ball. We write

$$E(x,y) = E(y,x) = E_n(x-y) = \begin{cases} \frac{1}{2\pi} \log|x-y| & \text{if } n = 2, \\ -\frac{1}{(n-2)\omega_{n-1}} |x-y|^{2-n} & \text{if } n \neq 2 \end{cases}$$

and note that, if $x \neq 0$ is given, the function

$$g_x := |x|^{2-n} E\Big(\frac{x}{|x|^2}, \cdot\Big)$$

is harmonic on $\mathbf{R}^n \setminus \{0, x/|x|^2\}$. We claim that $g_x(y) = E(x,y)$ if $y \in S = \partial B$.

Indeed, if $n > 2$ and $|y| = 1$,

$$E(x,y) - g_x(y) = -\frac{1}{(n-2)\omega_{n-1}} \Big(|x-y|^{2-n} - \big| |x|^{-1}x - |x| y \big|^{2-n} \Big)$$

and

$$\big| |x|^{-1}x - |x|y \big| = |x-y| \tag{6.19}$$

since

$$\begin{aligned} |x-y|^2 &= |x|^2 - 2x\cdot y + 1 \\ &= \big| |x|y \big|^2 - 2|x|^{-1}x \cdot (|x|y) + \big| |x|^{-1}x \big|^2 \\ &= \big| |x|y - |x|^{-1}x \big|^2. \end{aligned}$$

We define

$$G(x,y) := E(x,y) - g_x(y) = -\frac{1}{(n-2)\omega_{n-1}} \Big(|x-y|^{2-n} - \big| |x|^{-1}x - |x|y \big|^{2-n} \Big)$$

if $x \neq 0$ and

$$G(0,y) := -\frac{1}{(n-2)\omega_{n-1}} (|y|^{2-n} - 1).$$

Then G satisfies all the required properties.

If $n = 2$,

$$G(x, y) := \frac{1}{2\pi}\Big(\log|x - y| - \log\big||x|^{-1}x - |x|y\big|\Big) \tag{6.20}$$

when $x \neq 0$, and

$$G(0, y) := \frac{1}{2\pi}\log|y|.$$

We can compute the **Poisson kernel for the ball**,

$$P(x, y) = \partial_{\nu(y)}G(x, y) \qquad ((x, y) \in B \times S)$$

since $\nu(y) = y$ is the normal vector to the sphere S and the normal derivative on S is $\partial_{\nu(y)} = \sum_{j=1}^n y_j\partial_{y_j}$. Thus, when $n > 2$,

$$\begin{aligned} P(x, y) &= -\frac{1}{\omega_{n-1}}\sum_{j=1}^{n}\Big(\frac{y_j(x_j - y_j)}{|x - y|^n} + \frac{-|x|y_j(|x|^{-1}x_j - y_j|x|)}{\big||x|y - |x|^{-1}x\big|^n}\Big) \\ &= \frac{1}{\omega_{n-1}}\Big(\frac{|y|^2 - x\cdot y}{|x - y|^n} - \frac{|x|^2|y|^2 - x\cdot y}{\big||x|y - |x|^{-1}x\big|^n}\Big) \end{aligned}$$

and from (6.19) we obtain

$$P(x, y) = \frac{1}{\omega_{n-1}}\frac{1 - |x|^2}{|x - y|^n} \qquad ((x, y) \in B \times S) \tag{6.21}$$

when $n > 2$.

If $n = 2$, we recover the Poisson kernel of Theorem 5.8, since a direct computation of $\partial_{\nu(\xi)}G(z, \xi)$, with G defined as in (6.20), leads to

$$P(z, \xi) = \frac{1}{2\pi}\frac{1 - |z|^2}{|\xi - z|^2} = \frac{1}{2\pi}\Re\frac{\xi + z}{\xi - z} \qquad ((z, \xi) \in U \times \mathbf{T}).$$

Note that here we have included the normalizing factor $1/2\pi$.

Now the expected result for $n > 2$ can be proved:

Theorem 6.33. *If $f \in \mathcal{C}(S)$ and $P(x, y)$ is the Poisson kernel for the ball given by (6.21), then the function*

$$u(x) := \int_S P(x, y)f(y)\,dy \qquad (x \in B)$$

is harmonic on B and extends continuously to $\bar{B}$ and $u = f$ on S.

Proof. We will follow the argument used to prove Theorem 2.41, now based in the following facts:

(1) $\int_S P(x, y)\,d\sigma(y) = 1$.

(2) If $y_0 \in S$ and $V = B(y_0, \delta) \cap S$, then $\lim_{r\uparrow 1}\int_{S\setminus V} P(ry_0, y)\,d\sigma(y) = 0$.

To prove (1), we will apply the mean value theorem for harmonic functions[11] to $P_y = P(\cdot, y)$,

$$P_y(x) = \frac{1}{|S_r|}\int_{S_r(x)} P_y(z)\, d\sigma(z) = \frac{1}{\omega_{n-1}}\int_S P_y(x+rz)\, d\sigma(z),$$

at the point $x = 0$, if $0 < r < 1$, which gives

$$\frac{1}{\omega_{n-1}} = P(0, y) = \frac{1}{\omega_{n-1}}\int_S P(rz, y)\, d\sigma(z).$$

By (6.19), $P(rz, y) = P(ry, z)$ and

$$\int_S P(ry, z)\, d\sigma(z) = 1$$

is (1) if $x = ry$, with $r < 1$.

Fact (2) is almost obvious, since in

$$P(ry_0, y) = \frac{1}{\omega_{n-1}}\frac{1-|ry_0|^2}{|ry_0 - y|^n}$$

$\lim_{r\uparrow 1}(1-|ry_0|^2) = 0$ and $1/|ry_0 - y|^n$ is uniformly bounded for $1/2 < r < 1$ and $|y - y_0| \geq \delta$.

Proceeding to the proof of the theorem, if $\varepsilon > 0$ is given, choose $\delta > 0$ so that

$$|f(y) - f(z)| \leq \varepsilon \text{ if } |y - z| \leq \delta$$

and $V(y) = B(y, \delta) \cap S$. Then by (1)

$$f(y) - u(ry) = \left\{\int_{V(y)} + \int_{S\setminus V(y)}\right\}(f(y) - f(z))P(ry, z)\, d\sigma(z)$$

and we obtain

$$|f(y) - u(ry)| \leq \varepsilon + 2\|f\|_S \int_{S\setminus V(y)} P(ry, z)\, d\sigma(z) \leq 2\varepsilon$$

if r is close to 1, by (2).

This shows that $\lim_{r\uparrow 1} u(ry) = f(y)$ uniformly for $y \in S$, and u has a continuous extension to $\bar{B}$ given by $u(y) = f(y)$ if $y \in S$. □

In Exercise 6.38 we leave it to the reader to prove a similar result for the L^p convergence $u(ry) \to f(y)$ for every $f \in L^p(S)$.

---

[11] See Folland, "Introduction to Partial Differential Equations" [**14**, page 90].

6.6. Exercises

Exercise 6.1. Let $0 < r < 1$. From g as in Example 6.1, define a Urysohn function ϱ_r for $[-r, r] \subset \mathbf{R}$ supported by $[-1, 1]$ by choosing first

$$\psi(x) := \int_{-\infty}^{x} g(t)g(1 - r - t)\, dt$$

and then

$$\varrho_r(x) = \frac{1}{c^2}\psi(1 + x)\psi(1 - x).$$

On $\mathbf{R}^n$, if $0 < R < 1$, find r so that $H(x) = \varrho_r(|x|^2)$ defines a test function H such that $\bar{B}(0, R) \prec H \prec B(0, 1)$.

Exercise 6.2. If Ω is open in $\mathbf{R}^n$ and $1 \leq p < \infty$, prove that $\mathcal{D}(\Omega)$ is dense in $L^p(\Omega)$.

Exercise 6.3. Consider on $\mathcal{D}(\mathbf{R})$ the locally convex topology $\mathcal{T}$ defined by the sufficient sequence of norms

$$\|\varphi\|_N := \max_{m \leq N} \|\varphi^{(m)}\|_\infty.$$

Prove that this topology is metrizable but not complete and that its restriction to every $\mathcal{D}_K(\mathbf{R})$ is the topology that we have defined on this vector space. Show also that the convergence of sequences in $\mathcal{D}(\mathbf{R})$ for $\mathcal{T}$ is not the convergence that we are considering for sequences of test functions.

Exercise 6.4. Suppose $0 \neq \varphi \in \mathcal{D}(\mathbf{R})$ and $\varphi_n(t) = n^{-1}\varphi(n^{-1}t)$. Study the possible convergence of the sequence $\{\varphi_n\}_{n=1}^\infty$ in $\mathcal{D}(\mathbf{R})$ and in $\mathcal{E}(\mathbf{R})$.

Exercise 6.5. In $\mathcal{D}(\Omega)$, let $\mathcal{U}$ be the family of all convex balanced subsets U such that $U \cap \mathcal{D}_K(\Omega)$ is an open set in $\mathcal{D}_K(\Omega)$, for every $K \in \mathcal{K}(\Omega)$.

(a) Show that $\mathcal{U}$ is a local basis for a vector topology in $\mathcal{D}(\Omega)$.

(b) Show that the convergence for this topology is precisely the convergence we are considering in $\mathcal{D}(\Omega)$.

(c) Show that for this topology every Cauchy sequence is convergent.

(d) Show that for this topology the dual of $\mathcal{D}(\Omega)$ is $\mathcal{D}'(\Omega)$.

Exercise 6.6 (Borel). Let $\{c_n\}_{n=1}^\infty \subset \mathbf{C}$. Prove that there is a C^∞ function f on $\mathbf{R}$ such that $f^{(m)}(0) = c_m$ for every $m \in \mathbf{N}$ as follows:

(a) Consider $[-1, 1] \prec \varphi \prec (-R, R)$, define

$$f_n(t) := \frac{c_n}{n!} t^n \varphi(r_n t)$$

with $\{r_n\}_{n=1}^\infty \subset [1, \infty)$, and prove that, if $m < n$,

$$\|f_n^{(m)}\|_\mathbf{R} \leq |c_n| q_{n-1}(\varphi) r_n^{m-n} \sum_{j=0}^{m} \binom{m}{j} \frac{R^{n-j}}{(n-j)!}$$

with q_n as in (6.1) on the compact set $[-R, R]$.

(b) Show that it is possible to choose $\{r_n\}_{n=1}^{\infty} \subset [1, \infty)$ so that the functions f_n satisfy $q_{n-1}(f_n) \leq 1/2^n$.

(c) Prove that $\sum_{n=1}^{\infty} f_n(t) = f(t)$ for some $f \in \mathcal{E}(\mathbf{R})$ and $f^{(m)}(0) = c_m$.

Exercise 6.7. Prove that, for every function $f \in L^1_{\text{loc}}(\Omega)$ and every complex Borel measure μ on Ω, the distributions defined by

$$\langle \varphi, f \rangle = \int_{\Omega} \varphi(x) f(x)\, dx \quad \text{and} \quad \langle \varphi, \mu \rangle = \int_{\Omega} \varphi\, d\mu$$

are of order 0.

Exercise 6.8. Prove that the Dirac distribution δ_a is not a function.

Exercise 6.9 (δ' is not a measure). Show that the distribution δ' is not a complex measure.

Exercise 6.10. Find $\lim_{\lambda \to 0} K_\lambda$ in $\mathcal{D}'(\mathbf{R}^n)$ if $\{K_\lambda\}_{\lambda \in \Lambda}$ is a summability kernel on $\mathbf{R}^n$.

Exercise 6.11 (Dirac comb). Prove that $\text{Ш} = \sum_{k=-\infty}^{+\infty} \delta_k$ on $\mathbf{R}$ is well-defined as the sum of a convergent series in $\mathcal{D}'(\mathbf{R})$.

Exercise 6.12. Find the order of $\delta^{(m)} \in \mathcal{D}'(\mathbf{R})$.

Exercise 6.13. Prove that

$$u(\varphi) = \sum_{n=1}^{\infty} \varphi^{(n)}(1/n)$$

defines on $(0, \infty)$ a distribution of infinite order that is not the restriction of any distribution v on $\mathbf{R}$, meaning that $u \neq v_{|\mathcal{D}(0,\infty)}$.

Exercise 6.14. Every positive linear form $u : \mathcal{D}(\Omega) \to \mathbf{C}$ ($u(\varphi) \geq 0$ if $\varphi \geq 0$) is a distribution. The restrictions $u : \mathcal{D}_K(\Omega) \to \mathbf{C}$ satisfy $|u(\varphi)| \leq u(\varrho)\|\varphi\|_K$ if ϱ is a Urysohn function on $K \subset \Omega$, and they are continuous.

Exercise 6.15. Let $f \in L^1_{\text{loc}}(\mathbf{R}^n \setminus \{x_0\})$ and assume that x_0 is a nonintegrable singularity of f. We only know that $\int f\varphi$ exists for a test function φ if $x_0 \notin \operatorname{supp} \varphi$. A **regularization** of f is a distribution u_f on $\mathbf{R}^n$ which satisfies $u_f(\varphi) = \int f\varphi$ if $x_0 \notin \operatorname{supp} \varphi$.

For the function $f(t) = 1/t$, on $\mathbf{R}$ we obtain a regularization of f by taking $a, b > 0$ and defining

$$u_f(\varphi) := \int_{-\infty}^{-a} \frac{\varphi(t)}{t}\, dt + \int_{-a}^{b} \frac{\varphi(t) - \varphi(0)}{t}\, dt + \int_{b}^{+\infty} \frac{\varphi(t)}{t}\, dt.$$

There is a continuous function ψ_φ so that $u_2(\varphi) := \int_{-a}^{b} \frac{\varphi(t)-\varphi(0)}{t}\, dt = \int_{-a}^{b} \psi_\varphi(t)\, dt$ defines a distribution on $\mathbf{R}$, and u_f appears as the sum of three distributions. If $0 \notin \operatorname{supp} \varphi$, then $u_f(\varphi) = \int_{\mathbf{R}} \varphi(t)/t\, dt$.

Exercise 6.16. Prove the following statements concerning limits in $\mathcal{D}'(\mathbf{R})$.

(a) If $u_r(\varphi) := \int_{-r}^{r} \frac{\varphi(t)-\varphi(0)}{t}\, dt$ $(r > 0)$, then $u_r \in \mathcal{D}'(\mathbf{R})$ and $u_r \to 0$ as $r \downarrow 0$.

(b) $\mathrm{pv}\frac{1}{t} := \lim_{\varepsilon\downarrow 0} \chi_{[-\varepsilon,\varepsilon]^c}(t)\frac{1}{t}$ is a well-defined distribution. Moreover, if $\operatorname{supp}\varphi \subset [-r, r]$, then

$$\langle \varphi, \mathrm{pv}\frac{1}{t}\rangle = u_r(\varphi)$$

and we write

$$\langle \varphi, \mathrm{pv}\frac{1}{t}\rangle = \mathrm{pv}\int_{-\infty}^{+\infty} \frac{\varphi(t)}{t}\, dt.$$

It is a regularization of $f(t) = 1/t$.

Exercise 6.17. Let $f \in L^1_{\mathrm{loc}}(\mathbf{R}^n\setminus\{0\})$ so that $f(x)|x|^m$ is locally integrable for some positive integer m and, for a test function φ, denote by

$$T_m\varphi(x) := \sum_{k=0}^{m} \frac{1}{k!} d^k\varphi(0)(x, \dots, x)$$

the Taylor polynomial of degree m, so that $|\varphi(x) - T_m\varphi(x)| \le C_r|x|^m$ if $|x| \le r$. Show that

$$u_f(\varphi) := \int_{\mathbf{R}^n} f(x)\big(\varphi(x) - T_m\varphi(x)\chi_{(-\infty,1)}(|x|)\big)\, dx$$

defines a regularization of f.

Exercise 6.18. Find the distributional derivatives $Y^{(n)}$ of the Heaviside function $Y = \chi_{[0,\infty)}$.

Exercise 6.19. Show that $f(t) := \log|t|$ is locally integrable on $\mathbf{R}$ and prove that $f' = \mathrm{pv}\frac{1}{t}$. Note that $\lim_{\varepsilon\to 0} \log\varepsilon(\varphi(-\varepsilon) - \varphi(\varepsilon)) = 0$.

Exercise 6.20. Let $f(t) = \int_0^t g(x)\, dx$ $(t \in \mathbf{R})$ for some $g \in L^1(\mathbf{R})$; that is, f is an absolutely continuous function on $\mathbf{R}$, and $f' = g$ a.e.

Prove that $g = f'$ is the distributional derivative of f: $\langle\varphi, g\rangle = -\langle\varphi', f\rangle$ for every test function φ on $\mathbf{R}$.

Exercise 6.21. Prove that $D^\alpha : \mathcal{D}'(\Omega) \to \mathcal{D}'(\Omega)$ is a linear continuous operator that extends $D^\alpha : \mathcal{E}^m(\Omega) \to \mathcal{E}^{m-|\alpha|}(\Omega)$, if $|\alpha| \le m$.

Exercise 6.22. Suppose $P(D) = \sum_{|\alpha|\le m} c_\alpha D^\alpha$ is a differential operator with C^∞ coefficients c_α on Ω and denote $P(D)^\sharp u := \sum_{|\alpha|\le m} D^\alpha(c_\alpha u)$.

Prove that $P(D)^{\sharp\sharp} = P(D)$ and $(P(D)u)\varphi = u(P(D)^\sharp\varphi)$.

Exercise 6.23. Prove that, if at least one of $u, v \in \mathcal{D}'(\mathbf{R}^n)$ has compact support, then $\tau_a(u * v) = (\tau_a u) * v = u * (\tau_a v)$.

Exercise 6.24. Prove that $\operatorname{supp} D^\alpha u \subset \operatorname{supp} u$ for every distribution $u \in \mathcal{D}'(\Omega)$.

Exercise 6.25. Find the supports of Y and δ' on $\mathbf{R}$.

Exercise 6.26. Prove that the order of all distributions with compact support is finite.

Exercise 6.27. Prove that $1 * (\delta' * Y) \neq (1 * \delta') * Y$ and that this is not in contradiction to Theorem 6.21(b).

Exercise 6.28. If $u \in \mathcal{E}'(\mathbf{R}^n)$, prove that $u* : \mathcal{E}(\mathbf{R}^n) \to \mathcal{E}(\mathbf{R}^n)$ is a continuous linear operator which commutes with translations.

Exercise 6.29. We say that $\{d_m\}_{m=1}^\infty \subset L^1_{\mathrm{loc}}(\mathbf{R})$ is an **approximation of** δ if

$$\lim_{m\to\infty} \int_a^b d_m(t)\,dt = 0 \text{ if } 0 \notin [a,b] \text{ and } \lim_{m\to\infty} \int_a^b d_m(t)\,dt = 1 \text{ if } 0 \in (a,b).$$

(a) If $\{d_m\}_{m=1}^\infty$ is an approximation of δ and $Y_m(x) := \int_{-1}^x d_m(t)\,dt$, prove that $Y_m \to Y$ and $d_m \to \delta$ in $\mathcal{D}'(\mathbf{R})$.

(b) Find a concrete approximation of δ.

Exercise 6.30. Not all the solutions in $\mathcal{D}'(\mathbf{R})$ of the differential equation $tu'(t) = 0$ are classical solutions.

Exercise 6.31. Find the fundamental solutions of the differential operator $P(D)u = u''' - u'' + u' - u$.

Exercise 6.32. Find the fundamental solutions of $P(D)u = u''' + 3u'' + 3u' + u$.

Exercise 6.33. Here we assume that $\varphi \in \mathcal{C}_c(\mathbf{R}^2) \cap \mathcal{E}^2(\mathbf{R}^2)$ and that E_2 is the fundamental solution for $\triangle$ on $\mathbf{R}^2$. Prove that $u := \varphi * E_2$ is a classical solution of the Poisson equation $\triangle u = \varphi$; that is, $\triangle u(x)$ exists in the pointwise sense and coincides with $\varphi(x)$.

Exercise 6.34. On $\mathbf{C} = \mathbf{R}^2$, prove that

$$E(x,y) := \frac{1}{\pi z} \qquad (z = x + iy)$$

is a fundamental solution of the **Cauchy-Riemann operator**

$$\partial_{\bar z} := \frac{1}{2}(\partial_x + i\partial_y).$$

Exercise 6.35. (a) If $c \neq 0$ and $u \in \mathcal{D}'(\mathbf{R})$, show that we can define $u(x - cy) \in \mathcal{D}'(\mathbf{R}^2)$ by

$$\langle \varphi(x,y), u(x-cy)\rangle := \int_{\mathbf{R}} \langle \varphi(\cdot, y), \tau_{cy} u\rangle\,dy \qquad (\varphi \in \mathcal{D}(\mathbf{R}^2))$$

and that, if $u \in L^1_{\mathrm{loc}}(\mathbf{R})$, then this distribution is a locally integrable function on $\mathbf{R}^2$.

(b) Show that for any distribution $u \in \mathcal{D}'(\mathbf{R})$, $v = u(x - ct)$ is a distributional solution on $\mathbf{R}^2$ of the wave equation $\partial_t^2 v - c^2 \partial_x^2 v = 0$.

Exercise 6.36. Prove that the heat operator and the Cauchy-Riemann operator are hypoelliptic.

Exercise 6.37. Prove a version of Theorem 6.28 to find a solution of $\triangle u = f$ on $\mathbf{R}^2$ under the assumptions $f \in L^1(\mathbf{R}^2)$ and $f(x)\log^+(|x|) \in L^1(\mathbf{R}^2)$.

Exercise 6.38. Suppose $f \in L^p(S)$ $(1 \le p < \infty)$ and $P(x, y)$ is the Poisson kernel for the ball given by (6.21). Prove that

$$u(x) := \int_S P(x, y) f(y)\, d\sigma(y) \qquad (x \in B)$$

is a well-defined harmonic function on B which satisfies the boundary value condition

$$\lim_{r \uparrow 1} \int_S |f(y) - u(ry)|^p \, d\sigma(y) = 0.$$

References for further reading:

P. A. M. Dirac, *The Principles of Quantum Mechanics.*

R. E. Edwards, *Functional Analysis, Theory and Applications.*

G. B. Folland, *Introduction to Partial Differential Equations.*

I. M. Gelfand and G. E. Chilov, *Generalized Functions.*

L. Hörmander, *Linear Partial Differential Operators.*

J. Horvath, *Topological Vector Spaces and Distributions.*

E. H. Lieb and M. Loss, *Analysis.*

W. Rudin, *Functional Analysis.*

L. Schwartz, *Théorie des Distributions.*

K. Yosida, *Functional Analysis.*

Chapter 7

Fourier transform and Sobolev spaces

The Fourier transform is one of the most powerful operators in analysis. Its scope and applications have been extended to areas as different as harmonic analysis, partial differential equations, signal theory, probabilities, and algebraic number theory.

Its virtues depend on the use of the functions $e^{i\alpha t} = \cos(\alpha t) + i\sin(\alpha t)$, which are the homomorphisms of the additive group $\mathbf{R}$ to the multiplicative group $\mathbf{T}$, and on the translation-invariance of the Lebesgue measure.

These facts are intimately linked to the fundamental properties of converting convolution and linear differential operators into multiplication operators, changing convolution and partial differential equations into algebraic equations, and yielding explicit solutions in basic equations such as Laplace, heat, and wave equations.

With the extension to distributions, the scope of the Fourier transform increased substantially. By considering the Sobolev spaces of functions with distributional derivatives in L^2 up to a certain order, a control on the smoothness properties of these functions is obtained. The reason is that the Fourier transform, which changes differentiation into multiplication, is an L^2 isometry and L^2 is a Hilbert space.

With the fundamental properties of the Fourier transform of distributions, we present an introduction to the theory of Sobolev spaces.[1] For

[1] Around 1930 the Russian mathematician Sergei L'vovich Sobolev introduced his space $W^{1,2}(\Omega)$, or $H^1(\Omega)$, with the use of weak derivatives as the natural Hilbert space for solving the Laplace or Poisson equation $-\triangle u = f$ with boundary conditions. A little later, in France Jean Leray considered a similar method to find weak solutions for the Navier-Stokes equation.

completeness, we give the basic definitions in the L^p setting but, in fact, we only use the Hilbert space case, $p = 2$.

The estimates given by the Sobolev norms are a standard tool to prove the existence and regularity of solutions for partial differential equations. We illustrate this by means of an application to the Dirichlet problem and by studying the eigenfunctions of the Laplacian, and we include Rellich's compactness theorem, a result which is of great importance in the applications.

7.1. The Fourier integral

For each $\xi = (\xi_1, \ldots, \xi_n) \in \mathbf{R}^n$, e_ξ is the complex sinusoidal defined on $\mathbf{R}^n$ by

$$e_\xi(x) := \exp(2\pi i x \cdot \xi) = \exp\Big(2\pi i \sum_{k=1}^{n} x_k \xi_k\Big).$$

The **Fourier integral** $\widehat{f}$ of a function $f \in L^1(\mathbf{R}^n)$ is defined by

$$\widehat{f}(\xi) := \int_{\mathbf{R}^n} f(x) e^{-2\pi i x \cdot \xi}\, dx = \langle f, e_{-\xi} \rangle.$$

The **Fourier transform** $\mathcal{F} : L^1(\mathbf{R}^n) \to L^\infty(\mathbf{R}^n) \cap \mathcal{C}(\mathbf{R}^n)$, such that $\mathcal{F}f = \widehat{f}$, is a linear mapping and $\|\widehat{f}\|_\infty \leq \|f\|_1$ (i.e., $\|\mathcal{F}\| \leq 1$).

An application of Fubini's theorem, derivation under the integral, partial integration and elementary changes of variables show that the following useful properties of the Fourier transform hold on $L^1(\mathbf{R}^n)$:

(a) $\widehat{\tau_a f} = e_{-a}\widehat{f}$ and $\widehat{e_a f} = \tau_a \widehat{f}$.

(b) $\big[h^{-n} f(h^{-1}x)\big]\widehat{\ }(\xi) = \widehat{f}(h\xi)$ and $[f(hx)]\widehat{\ }(\xi) = h^{-n}\widehat{f}(h^{-1}\xi)$.

(c) $\widehat{f * g} = \widehat{f} \cdot \widehat{g}$ and $\int_{\mathbf{R}^n} f(y)\widehat{g}(y)\, dy = \int_{\mathbf{R}^n} \widehat{f}(y) g(y)\, dy$.

(d) $\partial_1 \widehat{f}(\xi) = (-2\pi i)[\widehat{x_1 f(x)}](\xi)$, if $x_1 f(x)$ is also integrable.

(e) $\widehat{\partial_1 f}(\xi) = 2\pi i \xi_1 \widehat{f}(\xi)$, if f is of class C^1 and $\partial_1 f$ is also integrable.

Note that to check (e) we can suppose $n = 1$. Then $f(t) = \int_0^t f'(x)\, dx + f(0)$ and $\lim_{t \to \pm\infty} f(t) = \int_0^t f'(x)\, dx + f(0)$ exists and is finite, since f' is assumed to be integrable, and this limit has to be 0, since f is also integrable. Integration by parts gives

$$\int_{-\infty}^{+\infty} f'(t) e^{-2\pi i \xi t}\, dt = f(t) e^{-2\pi i \xi t}\Big]_{t=-\infty}^{t=+\infty} + 2\pi i \xi \int_{-\infty}^{+\infty} f(t) e^{-2\pi i \xi t}\, dt$$

with $f(t) e^{-2\pi i \xi t}]_{t=-\infty}^{t=+\infty} = 0$.

The **inverse Fourier transform** is the mapping $\widetilde{\mathcal{F}}$ such that

$$(\widetilde{\mathcal{F}}f)(\xi) := \int_{\mathbf{R}^n} f(x)e^{2\pi i x\cdot\xi}\,dx = \langle f, e_\xi\rangle = \widehat{f}(-\xi).$$

Example 7.1. The Fourier integral of **the square wave**, $\chi_{(-1/2,1/2)}$, is the function sinc defined as

$$\operatorname{sinc}(\xi) := \frac{\sin(\pi\xi)}{\pi\xi}$$

since

$$\int_{-1/2}^{1/2} e^{-2\pi i\xi x}\,dx = \int_{-1/2}^{1/2} \cos(2\pi\xi x)\,dx - i\int_{-1/2}^{1/2} \sin(2\pi\xi x)\,dx = \operatorname{sinc}(\xi).$$

This function plays an important role in signal analysis and will appear again in Theorem 7.18.

Example 7.2. If W is the function defined on $\mathbf{R}^n$ by $W(x) = e^{-\pi|x|^2}$, then $\widehat{W} = W$ and

$$\int_{\mathbf{R}^n} e^{-\pi a|x|^2} e^{-2\pi i x\cdot\xi}\,dx = \frac{1}{a^{n/2}} e^{-\pi|\xi|^2/a} \tag{7.1}$$

for every $a > 0$.

If $n = 1$, from $W'(t) = -2\pi t W(t)$, the Fourier transform gives

$$2\pi\xi\widehat{W}(\xi) + (\widehat{W})'(\xi) = 0$$

and then

$$(e^{\pi\xi^2}\widehat{W}(\xi))' = 2\pi\xi\widehat{W}(\xi)e^{\pi\xi^2} + (\widehat{W})'(\xi)e^{\pi\xi^2} = 0.$$

Thus $e^{\pi\xi^2}\widehat{W}(\xi) = K$, a constant. The value of this constant is obtained from the Euler-Gauss integral $\int_{-\infty}^{\infty} e^{-x^2}\,dx = \sqrt{\pi}$ with the substitution $x = \sqrt{\pi}\,t$, which gives

$$K = \widehat{W}(0) = \int_{-\infty}^{\infty} e^{-\pi t^2}\,dt = 1,$$

so that $\widehat{W}(\xi) = W(\xi)$.

For n variables, $W(x) := e^{-\pi|x|^2} = W(x_1)\cdots W(x_n)$, and also $\widehat{W} = W$.

Another simple change of variables or an application of property (b) of the Fourier transform yields (7.1).

The operators $\mathcal{F}$ and $\widetilde{\mathcal{F}}$ will be extended to certain distributions, known as temperate distributions, and their extensions will essentially keep properties (a)–(e).

To show why $\mathcal{F}$ is a valuable tool for solving some partial differential equations, consider the example of an initial value problem for the heat equation on $[0, \infty) \times \mathbf{R}^n$,

$$\partial_t u(t,x) - \triangle u(t,x) = 0, \qquad u(0,x) = f(x), \tag{7.2}$$

with $\triangle = \sum_{j=1}^n \partial_{x_j}^2$, and formally apply $\mathcal{F}$ with respect to the variable $x \in \mathbf{R}^n$. By property (d),

$$\partial_t \widehat{u}(t,\xi) + 4\pi^2|\xi|^2\widehat{u}(t,\xi) = 0, \qquad \widehat{u}(0,\xi) = \widehat{f}(\xi),$$

and, taking ξ as a parameter, we note that this is a very simple initial value problem for an ordinary linear differential equation whose solution is

$$\widehat{u}(t,\xi) = \widehat{f}(\xi)e^{-4\pi^2|\xi|^2 t}.$$

According to Example 7.2, if $a = 1/(4\pi t)$,

$$e^{-4\pi^2|\xi|^2 t} = (4\pi t)^{-n/2}\int_{\mathbf{R}^n} e^{-|x|^2/4t}e^{-2\pi i x\cdot\xi}\,dx$$

so that, by property (c),

$$\widehat{u}(t,\xi) = \widehat{f}(\xi)\widehat{W_t}(\xi) = \widehat{f * W_t}(\xi)$$

if $W_t(x) = (4\pi t)^{-n/2}e^{-|x|^2/4t}$. Hence, the function

$$u(t,x) = (f * W_t)(x) = \frac{1}{(4\pi t)^{n/2}}\int_{\mathbf{R}^n} e^{-|y|^2/4t} f(x-y)\,dy \tag{7.3}$$

is a candidate for a solution of (7.2).

Note that

$$W_t(x) = \frac{1}{(\sqrt{4\pi t})^n}W\Big(\frac{x}{\sqrt{4\pi t}}\Big)$$

is a summability kernel, known as the **Gauss-Weierstrass kernel**, associated to the positive integrable function W, so that the initial value condition $\lim_{t\downarrow 0} f * W_t = f$ will hold.[2]

Moreover, equation (7.3) suggests $\Gamma(t,x) = W_t(x)Y(t)$ as a possible fundamental solution of the heat operator, which is the case as we have seen in Theorem 6.29.[3]

The following **Poisson theorem** relates Fourier integrals with Fourier series:

---

[2]See the details in Exercise 7.2.

[3]This was how Poisson constructed solutions for the heat equation in the work contained in his "Théorie mathématique de la chaleur" (1835).

Theorem 7.3. *Suppose that $f \in L^1(\mathbf{R})$ satisfies the condition*

$$\sum_{k=-\infty}^{\infty} \left|\widehat{f}\left(\frac{k}{T}\right)\right| < \infty. \tag{7.4}$$

Then there exists a continuous T-periodic function f_T on $\mathbf{R}$ such that

$$f_T(t) = \sum_{k=-\infty}^{\infty} f(t - kT) \text{ a.e.}$$

and

$$f_T(t) = \sum_{k=-\infty}^{\infty} \frac{1}{T}\widehat{f}\left(\frac{k}{T}\right)e^{2\pi ikt/T},$$

a series which is uniformly convergent on $\mathbf{R}$, that is, in $\mathcal{C}_T(\mathbf{R})$.

Proof. Denote $L = T/2$. On $[-L, L)$ (and on every $[kT - L, kT + L)$), we can define a.e. the periodic function

$$f_T(t) := \sum_{k=-\infty}^{\infty} f(t - kT),$$

with convergence in $L^1(-L, L)$, since

$$\int_{-L}^{L} \sum_{k=-\infty}^{\infty} |f(t - kT)|\, dt = \|f\|_1 < \infty$$

and then $\sum_{k=-\infty}^{\infty} |f(t - kT)| < \infty$ a.e.

The Fourier coefficients of $f_T \in L^1_T(\mathbf{R})$ are

$$c_k(f_T) = \frac{1}{T}\widehat{f}\left(\frac{k}{T}\right),$$

since

$$\begin{aligned} c_k(f_T) &= \frac{1}{T}\sum_{k=-\infty}^{\infty} \int_{-L}^{L} f(t - kT)e^{-2k\pi it/T}\, dt \\ &= \frac{1}{T}\sum_{k=-\infty}^{\infty} \int_{-L-kT}^{L-kT} f(t)e^{-2k\pi it/T}\, dt \\ &= \frac{1}{T}\int_{\mathbf{R}} f(t)e^{-2k\pi it/T}\, dt. \end{aligned}$$

From condition (7.4) and by the M-test of Weierstrass, it follows that the Fourier series is absolutely and uniformly convergent to a continuous function which coincides with f_T a.e. □

If the support of $\widehat{f}$ is compact, or if there are two constants $A, \delta > 0$ such that

$$|\widehat{f}(\xi)| \le A(1+|\xi|)^{-1-\delta},$$

then the integrable function f satisfies condition (7.4).

The function f_T is called the periodized extension of f. If $\operatorname{supp} f \subset [-T/2, T/2]$, f_T is the T-periodic extension of the restriction of f to the interval $[-T/2, T/2]$.

7.2. The Schwartz class $\mathcal{S}$

To define the Fourier transform of distributions, instead of $\mathcal{D}(\mathbf{R}^n)$ we need to consider a class of C^∞ functions that is invariant under the Fourier transform. Properties (c) and (d) of the Fourier transform suggest that we consider the complex vector space

$$\mathcal{S}(\mathbf{R}^n) := \{\varphi \in \mathcal{E}(\mathbf{R}^n);\ q_N(\varphi) < \infty \ \text{ for } N = 0, 1, 2, \dots\},$$

where

$$q_N(\varphi) := \sup_{x\in\mathbf{R}^n;\, |\alpha|\le N} (1+|x|^2)^N |D^\alpha \varphi(x)|.$$

Note that $|x^\alpha| \le (1+|x|^2)^N$ if $|\alpha| \le N$, so that $\varphi \in \mathcal{E}(\mathbf{R}^n)$ is in $\mathcal{S}(\mathbf{R}^n)$ if and only if, for every couple P, Q of polynomials, the function $P(x)Q(D)\varphi(x)$ is bounded.

The topology of $\mathcal{S}(\mathbf{R}^n)$ is defined by the sequence $q_0 \le q_1 \le q_2 \le \cdots$ of norms, so that the convergence $\varphi_k \to \varphi$ in $\mathcal{S}(\mathbf{R}^n)$ is the uniform convergence on $\mathbf{R}^n$

$$x^\beta D^\alpha \varphi_k(x) \to x^\beta D^\alpha \varphi(x)$$

for all $\alpha, \beta \in \mathbf{N}^n$, which is equivalent to the uniform convergence

$$P \cdot Q(D)\varphi_k \to P \cdot Q(D)\varphi$$

for every couple P, Q of polynomials.

The following theorem collects some basic properties of this new space.

Theorem 7.4. *$\mathcal{S}(\mathbf{R}^n)$, with the topology defined by the increasing sequence of norms $\{q_N\}_{N=0}^\infty$, is a Fréchet space.*

The inclusions $\mathcal{D}(\mathbf{R}^n) \subset \mathcal{S}(\mathbf{R}^n) \subset L^1(\mathbf{R}^n)$ are continuous (that is, for every compact set $K \subset \mathbf{R}^n$, the mappings $\mathcal{D}_K(\mathbf{R}^n) \hookrightarrow \mathcal{S}(\mathbf{R}^n) \hookrightarrow L^1(\mathbf{R}^n)$ are continuous), and $\mathcal{D}(\mathbf{R}^n)$ is dense in $\mathcal{S}(\mathbf{R}^n)$.

The differential operators $P(D)$, the multiplication by polynomials, translations, dilations, the symmetry, and every modulation or multiplication by a complex sinusoidal e_a are also continuous linear mappings of $\mathcal{S}(\mathbf{R}^n)$ into $\mathcal{S}(\mathbf{R}^n)$.

Proof. If $\{\varphi_k\} \subset \mathcal{S}(\mathbf{R}^n)$ is such that every $\{(1+|x|^2)^N D^\alpha \varphi_k(x)\}_{k=1}^\infty$ is a uniformly Cauchy sequence, then $(1+|x|^2)^N D^\alpha \varphi_k(x) \to \varphi_{N,\alpha}(x)$ uniformly as $k \to \infty$, so that $\varphi = \varphi_{0,0} \in \mathcal{E}(\mathbf{R}^n)$, since $D^\alpha \varphi_k(x) \to \varphi_{0,\alpha}(x)$ uniformly and then $\varphi \in \mathcal{E}(\mathbf{R}^n)$ and $\varphi_{0,\alpha}(x) = D^\alpha \varphi(x)$.

Hence $(1+|x|^2)^N D^\alpha \varphi_k(x) \to (1+|x|^2)^N D^\alpha \varphi(x)$ uniformly, which means that $\varphi_k \to \varphi$ in $\mathcal{S}(\mathbf{R}^n)$.

Note that

$$\int_{\mathbf{R}^n} \frac{1}{(1+|x|^2)^N}\, dx = \omega_n \int_0^\infty \frac{r^{n-1}}{(1+r^2)^N}\, dr < \infty$$

if $n-1-2N < -1$.

Then, if $N > n/2$ and $\varphi \in \mathcal{S}(\mathbf{R}^n)$,

$$\|\varphi\|_1 \le \int_{\mathbf{R}^n} q_N(\varphi)(1+|x|^2)^{-N}\, dx = C q_N(\varphi)$$

and $\mathcal{S}(\mathbf{R}^n) \hookrightarrow L^1(\mathbf{R}^n)$ is continuous.

Recall that the topology of $\mathcal{D}_K(\mathbf{R}^n)$ is defined by the increasing sequence of norms

$$p_N(\varphi) := \sup_{|\alpha|\le N} \|D^\alpha \varphi\|_K = \sup_{|\alpha|\le N} \|D^\alpha \varphi\|_{\mathbf{R}^n},$$

so that, if $\varphi \in \mathcal{D}_K(\mathbf{R}^n)$,

$$q_N(\varphi) \le \sup_{x\in K} (1+|x|^2)^N p_N(\varphi),$$

and $\mathcal{D}_K(\mathbf{R}^n) \hookrightarrow \mathcal{S}(\mathbf{R}^n)$ is continuous.

To prove that $\mathcal{D}(\mathbf{R}^n)$ is dense in $\mathcal{S}(\mathbf{R}^n)$, let $\varphi \in \mathcal{S}(\mathbf{R}^n)$. To find functions $\varphi_N \in \mathcal{D}(\mathbf{R}^n)$ such that $\varphi_N \to \varphi$ in $\mathcal{S}(\mathbf{R}^n)$, choose $\bar{B}(0,1) \prec \varrho \prec \mathbf{R}^n$ and define $\varrho_N(x) = \varrho(N^{-1}x)$ $(N \in \mathbf{N})$, so that $\bar{B}(0,N) \prec \varrho_N \prec \mathbf{R}^n$.

Then, if $\varphi_N = \varrho_N \varphi$,

$$\begin{aligned} |D^\alpha(\varphi - \varphi_N)(x)| &\le \sum_{\beta\le\alpha} \binom{\alpha}{\beta} |D^{\alpha-\beta}\varphi(x)||D^\beta(1-\varrho_N)(x)| \\ &= \sum_{\beta\le\alpha} \binom{\alpha}{\beta} |D^{\alpha-\beta}\varphi(x)| N^{-|\beta|} \sup_{x\in\mathbf{R}^n} |D^\beta(1-\varrho_N)(x/N)|, \end{aligned}$$

where we can select N so that $\sup_{|x|\ge N} |D^{\alpha-\beta}\varphi(x)| \le \varepsilon$ for all $\beta \le \alpha$, since $D^{\alpha-\beta}\varphi \in \mathcal{S}(\mathbf{R}^n)$. Note that $D^\beta(1-\varrho_N)(x) = 0$ if $|x| \le N$, and we obtain

$$\sup_{x\in\mathbf{R}^n} |D^\alpha(\varphi - \varphi_N)(x)| \le \sum_{\beta\le\alpha} \binom{\alpha}{\beta} \varepsilon N^{-|\beta|} \sup_{|x|\ge N} |D^\beta(1-\varrho_N)(x/N)| \le C_{N,\alpha}\varepsilon.$$

A similar estimate holds for every $\sup_{x\in\mathbf{R}^n}(1+|x|^2)^m|D^\alpha(\varphi-\varphi_N)(x)|$, since also

$$\sup_{|x|\geq N}(1+|x|^2)^m|D^{\alpha-\beta}\varphi(x)|\leq\varepsilon.$$

Thus, $q_m(\varphi-\varphi_N)\to 0$ if $N\to\infty$.

We leave the remaining part of the proof as an easy exercise. □

Theorem 7.5 (Inversion Theorem). *The Fourier transform is a continuous bijective linear operator* $\mathcal{F}:\mathcal{S}(\mathbf{R}^n)\to\mathcal{S}(\mathbf{R}^n)$ *and* $\mathcal{F}^{-1}=\widetilde{\mathcal{F}}$.

Proof. For every $\varphi\in\mathcal{S}(\mathbf{R}^n)$, the functions $D^\beta x^\alpha\varphi(x)$ are also in $\mathcal{S}(\mathbf{R}^n)$ and it follows from the properties of $\mathcal{F}$ that $\xi^\beta D^\alpha\widehat{\varphi}(\xi)$ are bounded and $\widehat{\varphi}\in\mathcal{S}(\mathbf{R}^n)$.

To prove that $\mathcal{F}:\mathcal{S}(\mathbf{R}^n)\to\mathcal{S}(\mathbf{R}^n)$ is continuous, note that

$$|\widehat{\varphi}(\xi)|\leq\int_{\mathbf{R}^n}|\varphi(x)|(1+|x|^2)^N(1+|x|^2)^{-N}\,dx\leq Cq_N(\varphi)\qquad(N>n/2),$$

so that, if $\varphi_k\to 0$ and $\widehat{\varphi}_k\to\psi$ in $\mathcal{S}(\mathbf{R}^n)$, then $\widehat{\varphi}_k(\xi)\to 0$ and $\psi(\xi)=0$, and the continuity now follows from the closed graph theorem.

To show that $\widetilde{\mathcal{F}}\mathcal{F}=I$, if we try a direct calculation of $\widetilde{\mathcal{F}}(\mathcal{F}\varphi)$, the integral

$$\int_{\mathbf{R}^n}\int_{\mathbf{R}^n}\widehat{f}(x)e^{2\pi i(z-x)\cdot\xi}\,d\xi\,dx$$

is not absolutely convergent on $\mathbf{R}^{2n}$.

We avoid this problem by using properties (b) and (c) of the Fourier transform to obtain

$$\int_{\mathbf{R}^n}f(y)\widehat{g}(hy)\,dy=\int_{\mathbf{R}^n}f(y)[h^{-n}g(h^{-1}x)]^\frown(y)\,dy=\int_{\mathbf{R}^n}\widehat{f}(y)h^{-n}g(h^{-1}y)\,dy$$

which, with the substitution $hy=x$, becomes

$$\frac{1}{h^n}\int_{\mathbf{R}^n}f(h^{-1}x)\widehat{g}(x)\,dx=\int_{\mathbf{R}^n}\widehat{f}(y)\frac{1}{h^n}g(h^{-1}y)\,dy;$$

that is, $\int_{\mathbf{R}^n}f(h^{-1}x)\widehat{g}(x)\,dx=\int_{\mathbf{R}^n}\widehat{f}(y)g(h^{-1}y)\,dy$. By letting $h\to\infty$, the dominated convergence theorem gives

$$\int_{\mathbf{R}^n}f(0)\widehat{g}(x)\,dx=\int_{\mathbf{R}^n}\widehat{f}(y)g(0)\,dy.$$

If we choose $g=W$, since $\int_{\mathbf{R}^n}\widehat{W}(x)\,dx=\int_{\mathbf{R}^n}W(x)\,dx=1$ and $W(0)=1$,

$$f(0)=\int_{\mathbf{R}^n}\widehat{f}(y)\,dy.$$

An application of this identity to $f(x) = (\tau_{-x}f)(0)$ combined with property (a) gives

$$f(x) = \int_{\mathbf{R}^n} \widehat{\tau_{-x}f}(y)\,dy = \int_{\mathbf{R}^n} e_x(y)\widehat{f}(y)\,dy,$$

which is $f = \widetilde{\mathcal{F}}\mathcal{F}f$. The identity $f = \mathcal{F}\widetilde{\mathcal{F}}f$ is similar. □

As an application, let us present a new proof of the Riemann-Lebesgue lemma.

Corollary 7.6 (Riemann-Lebesgue). If $f \in L^1(\mathbf{R}^n)$, then $\widehat{f} \in \mathcal{C}_0(\mathbf{R}^n)$; that is,

$$\lim_{|\xi|\to\infty} \widehat{f}(\xi) = 0.$$

Proof. We know that $\mathcal{D}(\mathbf{R}^n)$ is dense in $L^1(\mathbf{R}^n)$ (see Exercise 6.2) and, if $\varphi_k \to f$ in $L^1(\mathbf{R}^n)$, then $\widehat{\varphi}_k \to \widehat{f}$ uniformly with $\widehat{\varphi}_k \in \mathcal{S}(\mathbf{R}^n)$ so that they are null at infinity. □

Not only multiplication by a polynomial and by e_a is continuous on $\mathcal{S}(\mathbf{R}^n)$:

Theorem 7.7. *If $\psi \in \mathcal{S}(\mathbf{R}^n)$ and $\omega_s(x) := (1 + |x|^2)^{s/2}$ $(s \in \mathbf{R})$, the pointwise multiplications $\psi\cdot$ and $\omega_s\cdot$ and the convolution $\psi *$ are continuous linear operators of $\mathcal{S}(\mathbf{R}^n)$.*

Proof. Let $\varphi \in \mathcal{S}(\mathbf{R}^n)$ and note that, if $|\alpha| \le N$,

$$(1+|x|^2)^N|D^\alpha(\psi\varphi)| \le \sum_{\beta\le\alpha}\binom{\alpha}{\beta}(1+|x|^2)^N|D^{\alpha-\beta}\psi(x)||D^\beta\varphi(x)| \le Cq_N(\varphi),$$

since every $|D^{\alpha-\beta}\psi(x)|$ is bounded. The case of $\omega_s\cdot$ is similar, since $|\omega_s\varphi| \le (1+|x|^2)^N|\varphi(x)|$ and $|\partial_j\omega_s(x)| \le |sx_j\omega_{s-2}(t)|$.

Finally, $\mathcal{F}(\psi * \varphi) = \mathcal{F}(\psi)\mathcal{F}(\varphi)$, so that $\psi * \varphi = \widetilde{\mathcal{F}}(\mathcal{F}(\psi)\mathcal{F}(\varphi))$ and $\psi *$ is the product of continuous operators. □

7.3. Tempered distributions

If $\mathcal{S}'(\mathbf{R}^n)$ is the topological dual space of $\mathcal{S}(\mathbf{R}^n)$, it follows from Theorem 7.4 that the mapping $u \in \mathcal{S}'(\mathbf{R}^n) \mapsto v = u_{|\mathcal{D}(\mathbf{R}^n)} \in \mathcal{D}'(\mathbf{R}^n)$ is linear and one-to-one.

A distribution $v \in \mathcal{D}'(\mathbf{R}^n)$ is the restriction of an element $u \in \mathcal{S}'(\mathbf{R}^n)$ if and only if there exist $N \in \mathbf{N}$ and a constant $C_N > 0$ such that

$$|u(\varphi)| \le C_N q_N(\varphi) = C_N \sup_{x\in\mathbf{R}^n;\,|\alpha|\le N}(1+|x|^2)^N|D^\alpha\varphi(x)| \qquad (\varphi \in \mathcal{D}(\mathbf{R}^n)).$$

The elements of $\mathcal{S}'(\mathbf{R}^n)$, or their restrictions to $\mathcal{D}(\mathbf{R}^n)$, are called **tempered distributions**. On $\mathcal{S}'(\mathbf{R}^n)$ we consider the topology w^*, so that $u_k \to u$ in $\mathcal{S}'(\mathbf{R}^n)$ means that $\langle \varphi, u_k \rangle \to \langle \varphi, u \rangle$ for every $\varphi \in \mathcal{S}(\mathbf{R}^n)$.

It is customary to identify every $u \in \mathcal{S}'(\mathbf{R}^n)$ with its restriction $v \in \mathcal{D}'(\mathbf{R}^n)$, so that

$$\mathcal{S}'(\mathbf{R}^n) \subset \mathcal{D}'(\mathbf{R}^n).$$

Example 7.8. Suppose $1 \le p < \infty$ and $N \in \mathbf{N}$. If $(1+|x|^2)^{-N} f(x)$ is in $L^p(\mathbf{R}^n)$, then, as a distribution, $f \in \mathcal{S}'(\mathbf{R}^n)$. In particular,

$$L^p(\mathbf{R}^n) \hookrightarrow \mathcal{S}'(\mathbf{R}^n) \qquad (1 \le p \le \infty),$$

and this inclusion is continuous.

If $p = 1$,

$$\begin{aligned} |\langle \varphi, f \rangle| &= \int_{\mathbf{R}^n} (1+|x|^2)^{-N} f(x)(1+|x|^2)^N \varphi(x)\, dx \\ &\le \int_{\mathbf{R}^n} (1+|x|^2)^{-N} f(x) q_N(\varphi)\, dx \\ &= C_N q_N(\varphi). \end{aligned}$$

If $p > 1$, using Hölder's inequality, we obtain

$$|\langle \varphi, f \rangle| \le C_N^{1/p} \Big(\int_{\mathbf{R}^n} |(1+|x|^2)^N \varphi(x)|^{p'}\, dx \Big)^{1/p'}$$

and, if M is such that $\int_{\mathbf{R}^n} (1+|x|^2)^{N-M}\, dx = C < \infty$, a division and multiplication by $(1+|x|^2)^M$ give

$$|\langle \varphi, f \rangle| \le C_N^{1/p} C^{1/p'} \sup_{x \in \mathbf{R}^n} (1+|x|^2)^M |\varphi(x)| \le K q_M(\varphi).$$

Example 7.9. Every distribution with compact support, $u \in \mathcal{E}'(\mathbf{R}^n)$, is a tempered distribution.

The inclusion $\mathcal{S}(\mathbf{R}^n) \hookrightarrow \mathcal{E}(\mathbf{R}^n)$ is continuous, since the topology of $\mathcal{E}(\mathbf{R}^n)$ is defined by the seminorms

$$p_{K,N}(\varphi) = \sup_{|\alpha| \le N} \|D^\alpha \varphi\|_K$$

and $p_{K,N}(\varphi) \le q_N(\varphi)$ if $\varphi \in \mathcal{S}(\mathbf{R}^n)$.

Moreover $\mathcal{S}(\mathbf{R}^n)$ is dense in $\mathcal{E}(\mathbf{R}^n)$, since $\mathcal{D}(\mathbf{R}^n)$ is dense, and the restriction of $u \in \mathcal{E}'(\mathbf{R}^n)$ to $\mathcal{S}(\mathbf{R}^n)$ is a tempered distribution.

Let P be a polynomial with constant coefficients, $\psi \in \mathcal{S}(\mathbf{R}^n)$, and $u \in \mathcal{S}'(\mathbf{R}^n)$. As in the case of general distributions, we define Pu, $P(D)u$, and ψu by

$$\langle \varphi, P(D)u \rangle = \langle P(-D)\varphi, u \rangle, \quad \langle \varphi, Pu \rangle = \langle P\varphi, u \rangle, \quad \langle \varphi, \psi u \rangle = \langle \psi\varphi, u \rangle,$$

where $P(-D) = \sum_{|\alpha|\le N} c_\alpha(-1)^{|\alpha|}D^\alpha$ if $P(x) = \sum_{|\alpha|\le N} c_\alpha x^\alpha$ (the substitution of every $x_j^{\alpha_j}$ by $(-\partial_j)^{\alpha_j}$, and $D^\alpha = \partial_1^{\alpha_1}\cdots\partial_n^{\alpha_n}$).

They belong to $\mathcal{S}'(\mathbf{R}^n)$, since they are the composition of continuous linear mappings. For instance, $P(D)u : \varphi \mapsto P(-D)\varphi \mapsto u(P(-D)\varphi)$, where $P(-D)$, $P\cdot$, and $\psi\cdot$ are continuous on $\mathcal{S}(\mathbf{R}^n)$, their transposes are $P(D)$, $P\cdot$, and $\psi\cdot$, and they are continuous linear operators of $\mathcal{S}'(\mathbf{R}^n)$.

Also translations, modulations, dilations, and the symmetry defined, respectively, by

$$\langle\varphi, \tau_a u\rangle = \langle\tau_{-a}\varphi, u\rangle, \quad \langle\varphi, e_a u\rangle = \langle e_a\varphi, u\rangle,$$
$$\langle\varphi(x), u(hx)\rangle = \langle h^{-n}\varphi(h^{-1}x), u(x)\rangle, \quad \langle\varphi, \tilde{u}\rangle = \langle\tilde{\varphi}, u\rangle,$$

are continuous linear mappings of $\mathcal{S}'(\mathbf{R}^n)$ into $\mathcal{S}'(\mathbf{R}^n)$.

7.3.1. Fourier transform of tempered distributions. Property (c) of the Fourier integral on $L^1(\mathbf{R}^n)$ suggests that we may also define the Fourier transform $\widehat{u}$ of any tempered distribution u by

$$\langle\varphi, \widehat{u}\rangle := \langle\widehat{\varphi}, u\rangle \qquad (\varphi \in \mathcal{S}(\mathbf{R}^n)).$$

Since $\mathcal{F}$ is continuous on $\mathcal{S}(\mathbf{R}^n)$, $\widehat{u} = u\circ\mathcal{F} \in \mathcal{S}'(\mathbf{R}^n)$ for any $u \in \mathcal{S}'(\mathbf{R}^n)$. We still write $\mathcal{F}u = \widehat{u}$.

Similarly, $\widetilde{\mathcal{F}}$ is defined on $\mathcal{S}'(\mathbf{R}^n)$ by $\langle\varphi, \widetilde{\mathcal{F}}u\rangle := \langle\widetilde{\mathcal{F}}\varphi, u\rangle$.

Theorem 7.10. *The Fourier transform $\mathcal{F} : \mathcal{S}'(\mathbf{R}^n) \to \mathcal{S}'(\mathbf{R}^n)$ is a bijective continuous linear extension of $\mathcal{F} : L^1(\mathbf{R}^n) \to L^\infty(\mathbf{R}^n)$. The inverse of $\mathcal{F}$ on $\mathcal{S}'(\mathbf{R}^n)$ is $\widetilde{\mathcal{F}}$.*

The behavior of $\mathcal{F}$ on $L^1(\mathbf{R}^n)$ and on $\mathcal{S}(\mathbf{R}^n)$ with respect to derivatives, translations, modulations, dilations, and symmetry extends to the Fourier transform on $\mathcal{S}'(\mathbf{R}^n)$, and $\widetilde{\mathcal{F}}u = \widetilde{\mathcal{F}u}$.

Proof. If $f \in L^1(\mathbf{R}^n)$ and $u_f = \langle\cdot, f\rangle$, then $\mathcal{F}u_f = \langle\cdot, \widehat{f}\rangle$.

As the transpose of $\mathcal{F} : \mathcal{S}(\mathbf{R}^n) \to \mathcal{S}(\mathbf{R}^n)$, $\mathcal{F} : \mathcal{S}'(\mathbf{R}^n) \to \mathcal{S}'(\mathbf{R}^n)$ is weakly continuous. By property (c) of the Fourier integral, if $f \in L^1(\mathbf{R}^n)$, then $\mathcal{F}u_f = u_{\widehat{f}}$ and $\mathcal{F} : L^1(\mathbf{R}^n) \to L^\infty(\mathbf{R}^n)$ is the restriction of $\mathcal{F} : \mathcal{S}'(\mathbf{R}^n) \to \mathcal{S}'(\mathbf{R}^n)$.

Also, $\widetilde{\mathcal{F}}\mathcal{F} = \mathrm{Id}$ and $\mathcal{F}\widetilde{\mathcal{F}} = \mathrm{Id}$, since

$$\langle\varphi, \widetilde{\mathcal{F}}\mathcal{F}u\rangle = \langle\mathcal{F}\widetilde{\mathcal{F}}\varphi, u\rangle = \langle\varphi, u\rangle.$$

Let us consider the behavior of $\mathcal{F}$ with respect to dilations:

$$\begin{aligned}\langle\varphi(y), \mathcal{F}[u(hx)](y)\rangle &= \langle(\mathcal{F}\varphi)(x), u(hx)\rangle = \langle h^{-n}(\mathcal{F}\varphi)(h^{-1}x), u(x)\rangle \\ &= \langle\mathcal{F}[\varphi(hy)](x), u(x)\rangle = \langle\varphi(hy), (\mathcal{F}u)(y)\rangle \\ &= \langle\varphi(y), h^{-n}(\mathcal{F}u)(h^{-1}y)\rangle,\end{aligned}$$

and $\widehat{u(hx)}$ is the distribution $h^{-n}\widehat{u}(h^{-1}y)$. We leave it to the reader to check the remaining statements.

For differential operators, $P(D) = \sum_{|\alpha|\le m} c_\alpha D^\alpha$ ($c_\alpha \in \mathbf{C}$),

$$\langle \varphi, \mathcal{F}P(D)u\rangle = \langle P(-D)\mathcal{F}\varphi, u\rangle = \langle P(2\pi i\xi)\mathcal{F}\varphi(\xi), u(\xi)\rangle$$

and $\mathcal{F}P(D)u$ is the distribution $P(2\pi i\xi)\mathcal{F}u(\xi)$. □

Example 7.11. $\widehat{1} = \delta$, since $\langle \widehat{\varphi}, 1\rangle = \int_{\mathbf{R}^n} \widehat{\varphi}(\xi)\, d\xi = \varphi(0) = \langle \varphi, \delta\rangle$. As a modulation of 1, the Fourier transform in $\mathcal{S}'(\mathbf{R}^n)$ of the function $e^{2k\pi i a\cdot x}$ is $\tau_a\delta = \delta_a$.

Similarly, $\widehat{\delta} = 1$, and $\widehat{\delta}_a(\xi) = e^{2k\pi i a\cdot\xi}$.

Example 7.12. Every $f \in L^2_T(\mathbf{R})$ is a tempered distribution such that, in $\mathcal{S}'(\mathbf{R})$,

$$f = \sum_{k=-\infty}^{\infty} c_k(f) e^{2k\pi i t/T} \quad \text{and} \quad \widehat{f} = \sum_{k=-\infty}^{+\infty} c_k(f)\delta_{k/T}.$$

Here

$$c_k(f) = \frac{1}{T}\int_a^{a+T} f(t)e^{-2\pi i k t/T}\, dt \qquad (k \in \mathbf{Z})$$

are the Fourier coefficients of f.

Since

$$\int_{\mathbf{R}} \frac{|f(t)|^2}{1+t^2}\, dt \le \sum_{k=-\infty}^{+\infty} \int_{kT}^{(k+1)T} \frac{|f(t)|^2}{1+k^2}\, dt = C\|f\|_2^2,$$

it follows from Example 7.8 that $L^2_T(\mathbf{R}) \hookrightarrow \mathcal{S}'(\mathbf{R})$, continuously. Then

$$f = \sum_{k=-\infty}^{\infty} c_k(f)e^{2k\pi i t/T}$$

in $\mathcal{S}'(\mathbf{R})$, since this is true in $L^2_T(\mathbf{R})$.

But the Fourier transform is linear and continuous in $\mathcal{S}'(\mathbf{R})$, and maps $e^{2k\pi i t/T}$ into $\delta_{k/T}$, so that $f = \sum_{k=-\infty}^{\infty} c_k(f)e^{2k\pi i t/T}$ in $\mathcal{S}'(\mathbf{R})$.[4]

7.3.2. Plancherel Theorem. We also have $L^p(\mathbf{R}^n) \subset \mathcal{S}'(\mathbf{R}^n)$, and the action of $\mathcal{F}$ on $L^2(\mathbf{R}^n)$ is especially important.

Theorem 7.13 (Plancherel[5]). *The restrictions of $\widetilde{\mathcal{F}}$ and $\mathcal{F}$ to $L^2(\mathbf{R}^n)$ are linear bijective isometries such that $\widetilde{\mathcal{F}} = \mathcal{F}^{-1}$.*

---

[4]These results are also true for $f \in L^p_T(\mathbf{R})$ ($p > 1$), since $f = \sum_{k=-\infty}^{\infty} c_k(f)e^{2k\pi i t/T}$ in $L^p_T(\mathbf{R})$.

[5]Named after the Swiss mathematician Michel Plancherel, who in 1910 established conditions under which the theorem holds. It was first used in 1889 by Lord Rayleigh (John William Strutt) in the investigation of blackbody radiation.

Proof. Since $\mathcal{D}(\mathbf{R}^n)$ is dense in $L^2(\mathbf{R}^n)$ ($\langle\varphi, f\rangle = 0$ for every φ implies $f = 0$, so that $\mathcal{D}(\mathbf{R}^n)^\perp = \{0\}$ in $L^2(\mathbf{R}^n)$), the larger subspace $\mathcal{S}(\mathbf{R}^n)$ is also dense.

The identity $\int_{\mathbf{R}^n} \varphi(y)\widehat{g}(y)\,dy = \int_{\mathbf{R}^n} \widehat{\varphi}(y)g(y)\,dy$ holds for all $\varphi, g \in \mathcal{S}(\mathbf{R}^n)$. If $g = \overline{\widehat{\psi}}$, then $\widehat{g} = \bar{\psi}$ and $\int_{\mathbf{R}^n} \varphi(y)\overline{\psi(y)}\,dy = \int_{\mathbf{R}^n} \widehat{\varphi}(y)\overline{\widehat{\psi}(y)}\,dy$, i.e., $(\varphi, \psi)_2 = (\widehat{\varphi}, \widehat{\psi})_2$. This shows that $\mathcal{F}$ is an L^2-isometry on $\mathcal{S}(\mathbf{R}^n)$ that by continuity extends to a unique isometry $\mathcal{F}_2$ of $L^2(\mathbf{R}^n)$.

The restriction of $\mathcal{F}$ on $\mathcal{S}'(\mathbf{R}^n)$ to $\mathcal{F} : L^2(\mathbf{R}^n) \to \mathcal{S}'(\mathbf{R}^n)$ is continuous, since the inclusion $L^2(\mathbf{R}^n) \hookrightarrow \mathcal{S}'(\mathbf{R}^n)$ is continuous, and it coincides with $\mathcal{F}_2$ on the dense subspace $\mathcal{S}(\mathbf{R}^n)$; hence $\mathcal{F}_2 = \mathcal{F}$ on $L^2(\mathbf{R}^n)$.

The operator $\widetilde{\mathcal{F}}$, such that $\widetilde{\mathcal{F}}u = \widetilde{\mathcal{F}u}$, has a similar behavior. To show that $\widetilde{\mathcal{F}} = \mathcal{F}^{-1}$ on $L^2(\mathbf{R}^n)$, note that $\widetilde{\mathcal{F}}\mathcal{F} = \mathrm{Id} = \mathcal{F}\widetilde{\mathcal{F}}$ on $\mathcal{S}'(\mathbf{R}^n)$, so that also $\widetilde{\mathcal{F}}\mathcal{F} = \mathrm{Id} = \mathcal{F}\widetilde{\mathcal{F}}$ on $L^2(\mathbf{R}^n)$. □

Remark 7.14. If $f, g \in L^2(\mathbf{R}^n)$, $\int_{\mathbf{R}^n} f(y)\overline{g(y)}\,dy = \int_{\mathbf{R}^n} \widehat{f}(y)\overline{\widehat{g}(y)}\,dy$, but on $L^2(\mathbf{R}^n)$ the Fourier transform can be seen as an improper integral for the convergence in $L^2(\mathbf{R}^n)$,

$$\widehat{f}(\xi) = \lim_{R\uparrow\infty} \int_{B(0,R)} f(x)e^{-2\pi i x\xi}\,dx.$$

Note that $\chi_{B(0,R)}f \in L^1(\mathbf{R}^n) \cap L^2(\mathbf{R}^n)$ and

$$f = \lim_{R\uparrow\infty} \chi_{B(0,R)}f$$

in $L^2(\mathbf{R}^n)$, so that

$$\widehat{f} = \lim_{R\uparrow\infty} \mathcal{F}(\chi_{B(0,R)}f), \quad \text{with } \mathcal{F}(\chi_{B(0,R)}f)(\xi) = \int_{B(0,R)} f(x)e^{-2\pi i x\xi}\,dx.$$

Obviously, instead of $\chi_{B(0,R)}$ we can use more regular functions, such as $\varrho(R^{-1}x)$ with $\bar{B}(0,1) \prec \varrho \prec \mathbf{R}^n$.[6]

Example 7.15. If $\Omega > 0$ and $h \in \mathbf{R}$, then $\operatorname{sinc}(2\Omega t + h)$ is in $L^2(\mathbf{R})$, and

$$2\Omega[\operatorname{sinc}(2\Omega t + h)]\widehat{\ }(\xi) = e^{\pi i h\xi/\Omega}\chi_{[-\Omega,\Omega]}(\xi).$$

Indeed, since sinc is the Fourier transform of the square wave of Example 7.1, it belongs to $L^2(\mathbf{R})$, and the same happens with $\operatorname{sinc}(2\Omega t + h) = \operatorname{sinc}(2\Omega(t + h/2\Omega))$. By the Plancherel theorem $\widehat{\operatorname{sinc}} = \chi_{[-1/2,1/2]}$ and, using the properties of the Fourier transform,

$$\Big[\operatorname{sinc}\Big(2\Omega(t + \frac{h}{2\Omega})\Big)\Big]\widehat{\ }(\xi) = e^{2\pi i \frac{h}{2\Omega}\xi}[\operatorname{sinc}(2\Omega t)]\widehat{\ }(\xi) = \frac{1}{2\Omega}e^{\pi i\frac{h}{\Omega}\xi}\widehat{\operatorname{sinc}}(\xi/2\Omega)$$

[6] If $n = 1$, $\widehat{f}(\xi) = \lim_{M\to\infty}\int_{-M}^{M} f(x)e^{-2\pi i x\xi}\,dx$ for almost all $\xi \in \mathbf{R}$ holds if $f \in L^2$. This result is equivalent to the Carleson theorem on the almost everywhere convergence of Fourier series, one of the most celebrated theorems in Fourier analysis.

with $\widehat{\text{sinc}}\,(\xi/2\Omega) = \chi_{[-1/2,1/2]}(\xi/2\Omega) = \chi_{[-\Omega,\Omega]}(\xi)$.

We also know from Example 7.9 that the distributions with compact support are tempered, so that we can consider the restriction of $\mathcal{F}$ to the space $\mathcal{E}'(\mathbf{R}^n)$. Let us denote

$$e_z(\xi) := e^{2\pi i z\cdot\xi} \qquad (\xi \in \mathbf{R}^n,\ z \in \mathbf{C}^n).$$

Theorem 7.16. *If $u \in \mathcal{E}'(\mathbf{R}^n)$, then $\widehat{u}$ is the restriction to $\mathbf{R}^n$ of the entire function*

$$F(z) := \langle e_{-z}, u\rangle.$$

For every $\alpha \in \mathbf{N}^n$ there is an integer N such that the function

$$(1+|x|^2)^{-N/2} D^\alpha \widehat{u}(x)$$

is bounded.

Proof. The function F is continuous on $\mathbf{C}$, since $e_z \to e_{z_0}$ in $\mathcal{E}(\mathbf{R}^n)$ as $z \to z_0$. Indeed, $D^\alpha e_z \to D^\alpha e_{z_0}$ uniformly on every compact subset of $\mathbf{R}$, since for one variable

$$e^{2\pi i t z} - e^{2\pi i t z_0} = \int_{[z_0,z]} 2\pi i t e^{2\pi i t\zeta}\, d\zeta.$$

Let us show that F is holomorphic in every variable z_j by an application of Morera's theorem; that is, $\int_\gamma F(z)\,dz_j = 0$ if γ is the oriented boundary of a rectangle in $\mathbf{C}$. By writing the integral as a limit of Riemann sums,

$$\int_\gamma F(z)\,dz_j = \int_\gamma u(e_{-z})\,dz_j = \Big\langle \int_\gamma e_{-z}(t)\,dz_j, u(t)\Big\rangle = 0,$$

since $\int_\gamma e^{-2\pi i z_j t_j}\,dz_j = 0$ for every $t_j \in \mathbf{R}$.

For every $\varphi \in \mathcal{D}_{[a,b]^n}(\mathbf{R}^n)$,

$$\langle \varphi(t), u(e_{-t})\rangle = \int_{[a,b]^n} \langle \varphi(t)e_{-t}, u\rangle\,dt = \Big\langle \int_{[a,b]^n} \varphi(t)e_{-t}(x)\,dt, u(x)\Big\rangle = \langle \widehat{\varphi}, u\rangle$$

and $\widehat{u} = F$ on $\mathbf{R}^n$.

Note that by the continuity of u on $\mathcal{E}(\mathbf{R}^n)$,

$$|D^\alpha \widehat{u}(\xi)| = |(-2\pi i)^{|\alpha|}\langle x^\alpha e_{-\xi}(x), u(x)\rangle| \le C \sup_{|\beta|\le N,\,|x|\le N} |D^\beta(x^\alpha e_{-\xi}(x))|$$

and it follows that $|D^\alpha \widehat{u}(\xi)| \le C'(1+|\xi|^2)^{N/2}$. □

7.4. Fourier transform and signal theory

A first main topic in the digital processing of signals is the analog-to-digital conversion by means of sampling, which changes a continuous time signal $f(t)$ into a discrete time signal $x = \{x[k]\}_{k=-\infty}^{\infty} \subset \mathbf{C}$, $x[k] := f(kT)$.

The band of an analog signal f is the smallest interval $[-\Omega, \Omega]$ which supports its Fourier transform $\widehat{f}$ and we shall see that for a band-limited signal, that is, with $\Omega < \infty$, sampling can be done in an efficient manner. It is worth observing that in this case f is analytic:

We know from Theorem 7.16 that, if $u \in \mathcal{S}'(\mathbf{R})$ has a Fourier transform with compact support $\widehat{u} \in \mathcal{E}'(\mathbf{R})$, then u is the restriction to $\mathbf{R}$ of the entire function,

$$F(z) := \langle e^{2\pi i \xi z}, \widehat{u}(\xi) \rangle.$$

This shows that a signal cannot be simultaneously band-limited and time-limited.[7] Usually, analog signals are of finite time, so they are not of limited band, but they are almost band-limited in the sense that $\widehat{u} \simeq 0$ outside of some finite interval $[-\Omega, \Omega]$. Sometimes, filtering of the analogical signal is convenient in order to reduce it to a band-limited signal.

We will suppose that $f \in L^2(\mathbf{R})$ and $\operatorname{supp} \widehat{f} \subset [-\Omega, \Omega]$, so f is analytic. The minimal value Ω_N of Ω is called the **Nyquist frequency**[8] of f.

Let us consider the T-periodized extension of $\widehat{f}$ with $T = 2\Omega$,

$$\widehat{f}_T(\xi) = \sum_{k=-\infty}^{\infty} \widehat{f}(\xi - kT) \qquad (P = 2\Omega),$$

which is in $L^2_T(\mathbf{R})$. Then

$$c_k(\widehat{f}_T) = \frac{1}{2\Omega} \widehat{\widehat{f}} \left(\frac{k}{2\Omega} \right) = \frac{1}{2\Omega} f \left(\frac{-k}{2\Omega} \right) \tag{7.5}$$

and

$$\widehat{f}_T(\xi) = \sum_{k=-\infty}^{\infty} \frac{1}{2\Omega} f \left(\frac{-k}{2\Omega} \right) e^{\pi i k \xi / \Omega},$$

with $L^2(-\Omega, \Omega)$-convergence of this Fourier series.

According to the inversion theorem, $f(t) = \int_{-\Omega}^{\Omega} \widehat{f}(\xi) e^{2\pi i t \xi}\, d\xi$ for every $x \in \mathbf{R}$, and the scalar product by $e^{-2\pi i t \xi / \Omega}$ on $[-\Omega, \Omega]$ gives the pointwise identity

$$f(t) = \sum_{k=-\infty}^{\infty} \frac{1}{2\Omega} f \left(\frac{-k}{2\Omega} \right) \int_{-\Omega}^{\Omega} e^{2\pi i (t + k/2\Omega)\xi}\, d\xi$$

---

[7] This is a version of the uncertainty principle

[8] Named after the Swedish engineer Harry Nyquist, who in 1927 determined that the number of independent pulses that could be put through a telegraph channel per unit of time is limited to twice the bandwidth of the channel.

and, by Example 7.15,

$$\frac{1}{2\Omega}\int_{-\Omega}^{\Omega} e^{2\pi i(t+k/2\Omega)\xi}\, d\xi = \operatorname{sinc}(2\Omega t + k). \tag{7.6}$$

This shows that we obtain a complete reconstruction of $f(t)$,

$$f(t) = \sum_{k=-\infty}^{\infty} f\Big(\frac{k}{2\Omega}\Big) \operatorname{sinc}(2\Omega t - k), \tag{7.7}$$

with pointwise convergence, by sampling with the sampling period $T_m = 1/2\Omega$.

Our aim is to prove that in fact we have uniform and L^2 convergence.

Lemma 7.17. *The functions*

$$\sqrt{2\Omega}\operatorname{sinc}(2\Omega t - k) \qquad (k \in \mathbf{Z}) \tag{7.8}$$

form an orthonormal system in $L^2(\mathbf{R})$.

Proof. If

$$\varphi_k := [\operatorname{sinc}(2\Omega t - k)]\widehat{\ } = [\operatorname{sinc}(2\Omega(t - k/2\Omega))]\widehat{\ },$$

according to Example 7.15

$$\varphi_k(\xi) = e^{\pi i k\xi/\Omega}[\operatorname{sinc}(2\Omega t)]\widehat{\ }(\xi) = e^{\pi i k\xi/\Omega}\frac{1}{2\Omega}\chi_{[-\Omega,\Omega]}(\xi).$$

From the Plancherel theorem,

$$([\operatorname{sinc}(2\Omega t - m)], [\operatorname{sinc}(2\Omega t - n)])_2 = (\varphi_m, \varphi_n)_2,$$

and then

$$(2\Omega)^2(\varphi_m, \varphi_n) = \int_{-\Omega}^{\Omega} e^{\pi i(m-n)\xi/\Omega}\, d\xi = 0$$

if $m \neq n$, and $(2\Omega)^2\|\varphi_m\|_2^2 = 2\Omega$, so that

$$\|[\operatorname{sinc}(2\Omega t - m)]\|_2 = 1/\sqrt{2\Omega}.$$

□

The family of functions (7.8) is called the **Shannon system**.

Theorem 7.18 (Shannon[9]). *Suppose* $f \in L^2(\mathbf{R})$ *and* $\operatorname{supp}\widehat{f} \subset [-\Omega, \Omega]$, *so that we can assume that* f *is continuous. Then*

$$f(t) = \sum_{k=-\infty}^{\infty} f\Big(\frac{k}{2\Omega}\Big) \operatorname{sinc}(2\Omega t - k)$$

in $L^2(\mathbf{R})$ *and uniformly.*

[9] Named after the electrical engineer and mathematician Claude Elwood Shannon, the founder of information theory in 1947.

Proof. We have seen in (7.7) that the sequence

$$\begin{aligned} s_N(f,t) &= \sum_{k=-N}^{N} f\Big(\frac{k}{2\Omega}\Big) \operatorname{sinc}(2\Omega t - k) \\ &= \sum_{k=-N}^{N} f\Big(\frac{k}{2\Omega}\Big) \frac{1}{2\Omega} \int_{-\Omega}^{\Omega} e^{2\pi i t\xi} e^{-2\pi i k\xi/2\Omega} \end{aligned}$$

is pointwise convergent to $f(t)$ and that

$$f(t) = \int_{-\Omega}^{\Omega} \widehat{f}(\xi) e^{2\pi i \xi t}\, d\xi.$$

Hence,

$$|f(t) - s_N(f,x)| = \Big| \int_{-\Omega}^{\Omega} \Big\{ \widehat{f}(\xi) - \sum_{k=-N}^{N} \frac{1}{2\Omega} f\Big(\frac{-k}{2\Omega}\Big) e^{2\pi i k\xi/2\Omega} \Big\} e^{2\pi i \xi t} d\xi \Big|$$

and, by Schwarz inequality,

$$|f(t) - s_N(f,x)| \le \Big(\int_{-\Omega}^{\Omega} \Big| \widehat{f}(\xi) - \sum_{k=-N}^{N} \frac{1}{2\Omega} f\Big(\frac{-k}{2\Omega}\Big) e^{2\pi i k\xi/2\Omega} \Big|^2 d\xi \Big)^{1/2} (2\Omega)^{1/2}.$$

We know from (7.5) that the Fourier coefficients of $\widehat{f}_T$ with respect to the trigonometric system are

$$c_k(\widehat{f}_T) = \frac{1}{2\Omega} f\Big(\frac{-k}{2\Omega}\Big).$$

Hence,

$$\sum_{k=-N}^{N} \frac{1}{2\Omega} f\Big(\frac{-k}{2\Omega}\Big) e^{2\pi i k\xi/2\Omega} = S_N(\widehat{f}_T, t),$$

and $S_N(\widehat{f}_T) \to \widehat{f}_T$ in $L^2_T(\mathbf{R})$.

Then,

$$\sup_{t\in\mathbf{R}} |f(t) - s_N(f,x)| \le (2\Omega)^{1/2} \|\widehat{f} - S_N(\widehat{f}_T)\|_{L^2(-\Omega,\Omega)},$$

which yields the uniform convergence.

Since $c(\widehat{f}_T) = \{c_k(\widehat{f}_T)\} \in \ell^2$ and the Shannon system is orthonormal,

$$g(t) = \sum_{k=-\infty}^{\infty} \frac{1}{2\Omega} f\Big(\frac{k}{2\Omega}\Big) \sqrt{2\Omega}\, \operatorname{sinc}(2\Omega t - k) \tag{7.9}$$

in $L^2(\mathbf{R})$. But some subsequence of the partial sums is a.e. convergent to g and uniformly convergent to f, so that $g = f$ as elements of $L^2(\mathbf{R})$. □

The Shannon theorem shows that the sampling rate of $\Omega_s = 2\Omega_N$ samples/seg is optimal, and it is called the **Nyquist rate.** If $\Omega_s > 2\Omega_N$, we are considering $\operatorname{supp} \widehat{f} \subset [-\Omega_s/2, \Omega_s/2]$, wider than the symmetric interval which supports $\widehat{f}$ and an unnecessary oversampling if Ω_s is much greater than $2\Omega_N$. If $\Omega_s > 2\Omega_N$, the function g obtained in the sum (7.9) differs from the original signal f and is called an **alias** of f.

In digital processing, the discrete time signals $x = \{x[k]\}_{k=-\infty}^{\infty}$ obtained by sampling from analogical signals are usually of finite time, so that $x[k] = 0$ if $|k| > N$ for some N, but for technical reasons it is convenient to consider more general signals. We say that x is **slowly increasing**, and write $x \in \ell$, if there exist two constants N and C, such that

$$|x[k]| \leq C|k|^N \qquad (k \neq 0).$$

The class ℓ is a vector space with the usual operations and slowly increasing sequences can be considered tempered distributions:

Theorem 7.19. *If $x \in \ell$, then*

$$u_x := \sum_{k=-\infty}^{+\infty} x[k]\delta_k$$

defines a tempered distribution, and the correspondence $x \in \ell \mapsto u_x \in \mathcal{S}'(\mathbf{R})$ is an injective linear mapping which shows that we can consider $\ell \subset \mathcal{S}'(\mathbf{R})$.

The Fourier transform of x as a tempered distribution is

$$\widehat{x} = \sum_{k=-\infty}^{+\infty} x[k]e^{-2\pi i k\xi}.$$

Proof. If $u = \sum_{k=-\infty}^{+\infty} x[k]\delta_k = 0$, for every $n \in \mathbf{Z}$ we can choose $\varphi \in \mathcal{S}$ with support in $(n-1, n+1)$ such that $u(\varphi) = x[n]$. Since

$$\sum_{k=-\infty}^{+\infty} |x[k]\varphi(k)| \leq C \sum_{k\neq 0} |k|^{-2N}|k|^{4N}|\varphi(k)| \leq K q_{2N}(\varphi), \tag{7.10}$$

$u : \varphi \mapsto \sum_{k=-\infty}^{+\infty} x[k]\varphi(k)$ is linear and continuous on $\mathcal{S}(\mathbf{R})$. It is the limit in $\mathcal{S}'$ of the partial sums $u_N = \sum_{k=-N}^{N} x[k]\delta_k$, $u_N(\varphi) \to u(\varphi)$ if $\varphi \in \mathcal{S}$.

Moreover, $\widehat{u} = \sum_{k=-\infty}^{+\infty} x[k]\widehat{\delta_k} = \sum_{k=-\infty}^{+\infty} x[k]e^{-2\pi i k\xi}$. □

This Fourier transform (7.10) is called the **spectrum** of x.

If $\{x[k]\} \in \ell^2$, we have convergence in $L^2_1(\mathbf{R})$ and $x \in \ell^2 \mapsto \widehat{x} \in L^2_1(\mathbf{R})$ is a bijective isometry, such that $x[-k] = c_k(\widehat{x})$ $(k \in \mathbf{Z})$.

If $\{x[k]\} \in \ell^1$, then $\widehat{x} \in \mathcal{C}_1(\mathbf{R})$. Of course, $\ell^1 \subset \ell^2 \subset \ell$ (see Exercise 7.13).

Example 7.20. The Fourier transform of $x[j] = c\,\mathrm{sinc}\,(cj)$ $(0 < c < 1/2)$ is the 1-periodic square wave such that $\chi_{[-c/2,c/2]}(\xi)$ on $[-1/2, 1/2]$.

In this example, $x \in \ell^2$, since $\widehat{x} \in L^2_1(\mathbf{R})$, but $\widehat{x}$ is not continuous, so that $x \notin \ell^1$.

The Fourier transform of a signal and that of its samples are related as follows:

Theorem 7.21. *Let $f \in L^2(\mathbf{R})$ be a band-limited signal and let $x[j] := f(j/2\Omega)$, with $\Omega \geq \Omega_N$. Then*

$$\widehat{f}(\xi) = \frac{1}{2\Omega}\widehat{x}\Big(\frac{\xi}{2\Omega}\Big) \qquad (|\xi| \leq \Omega).$$

Proof. According to the inversion theorem,

$$x[j] = \int_{-1/2}^{1/2} \widehat{x}(\xi)e^{2\pi i\xi j}\,d\xi = \frac{1}{2\Omega}\int_{-\Omega}^{\Omega} \widehat{x}\Big(\frac{\xi}{2\Omega}\Big)e^{2\pi i\xi j/2\Omega}\,d\xi$$

so, by the 2Ω-periodicity of $e^{2\pi i\xi j/2\Omega}$,

$$\begin{aligned} x[j] &= f\Big(\frac{j}{2\Omega}\Big) = \int_{-\infty}^{\infty} \widehat{f}(\xi)e^{2\pi i\xi j/2\Omega}\,d\xi = \sum_{k=-\infty}^{\infty}\int_{-\Omega}^{\Omega} \widehat{f}(\xi - 2\Omega k)e^{2\pi i\xi j/2\Omega}\,d\xi \\ &= \int_{-\Omega}^{\Omega}\Big(\sum_{k=-\infty}^{\infty} \widehat{f}(\xi - 2\Omega k)\Big)e^{2\pi i\xi j/2\Omega}\,d\xi \end{aligned}$$

and, from the uniqueness property of the Fourier coefficients,

$$\frac{1}{2\Omega}\widehat{x}\Big(\frac{\xi}{2\Omega}\Big) = \sum_{k=-\infty}^{\infty} \widehat{f}(\xi - 2\Omega k) = \widehat{f}_{2\Omega}.$$

This relation means that

$$\widehat{x}(\xi) = 2\Omega\sum_{k=-\infty}^{\infty} \widehat{f}(2\Omega(\xi - k)),$$

where the right side is $\widehat{f}$ scaled by the factor 2Ω. Since $\Omega \geq \Omega_N$, we have $\widehat{f}_{2\Omega}(\xi) = \widehat{f}(\xi)$ if $|\xi| \leq 1/2$ and then $\widehat{x}(\xi) = 2\Omega\widehat{f}(2\Omega\xi)$. □

7.5. The Dirichlet problem in the half-space

In Theorem 6.33 we have obtained the solution $u(x) = \int_S P(x, y)f(y)\,dy$ of the homogeneous Dirichlet problem[10] for the ball with the inhomogeneous boundary condition $u_{|S} = f$ by means of its Poisson kernel P.

Here, as in (7.2) for the heat equation, we will use the Fourier transform as a tool to solve the homogeneous Dirichlet problem in the half-space $\mathbf{R}^{1+n}_+ := \{(t, x) \in \mathbf{R} \times \mathbf{R}^n;\ t > 0\}$,

$$\partial_t^2 u + \triangle u = 0, \qquad u(0, x) = f(x), \tag{7.11}$$

where $\triangle = \sum_{j=1}^n \partial_{x_j}^2$.

We will be looking for bounded solutions,[11] that is, for bounded harmonic functions u on the half-space $t > 0$ such that, in some sense, $u(t, x) \to f(x)$ as $t \downarrow 0$.

7.5.1. The Poisson integral in the half-space.

The Fourier transform changes a linear differential equation with constant coefficients

$$P(D)u = f,$$

with $P(D) = \sum_{|\alpha|\le m} c_\alpha D^\alpha$, into the algebraic equation

$$P(2\pi i\xi)\widehat{u}(\xi) = \widehat{f}(\xi),$$

where $P(x) = \sum_{|\alpha|\le m} c_\alpha x^\alpha = \sum_{|\alpha|\le m} c_\alpha x_1^{\alpha_1} \cdots x_n^{\alpha_n}$.

For this reason, to find an integral kernel for the Dirichlet problem (7.11) similar to the Poisson kernel for the ball, we apply the Fourier transform in x to convert the partial differential equation in (7.11) into an ordinary differential equation.

Assuming for the moment that $f \in \mathcal{S}(\mathbf{R}^n)$, for every $t > 0$ we obtain

$$\partial_t^2 \widehat{u}(t, \xi) - 4\pi^2|\xi|^2\widehat{u}(t, \xi) = 0, \qquad (\widehat{u}(0, \xi) = \widehat{f}(\xi)),$$

and we are led to solve an ordinary differential equation in t for every $\xi \in \mathbf{R}^n$.

The general solution of this equation is

$$\widehat{u}(t, \xi) = A(\xi)e^{2\pi|\xi|t} + B(\xi)e^{-2\pi|\xi|t}, \quad A(\xi) + B(\xi) = \widehat{f}(\xi).$$

If we want to apply the inverse Fourier transform, and also because of the boundedness condition, we must have $A(\xi) = 0$. Then $B(\xi) = \widehat{f}(\xi)$ and

[10] The work of Johann Peter Gustav Lejeune Dirichlet included potential theory, integration of hydrodynamic equations, convergence of trigonometric series and Fourier series, and the foundation of analytic number theory and algebraic number theory. In 1837 he proposed what is today the modern definition of a function. After Gauss's death, Dirichlet took over his post in Göttingen.

[11] This boundedness requirement is imposed to obtain uniqueness; see Exercise 7.12.

with this election

$$u(t,x) = (P_t * f)(x), \qquad \widehat{P_t}(\xi) = e^{-2\pi|\xi|t},$$

where $P(t,x) := P_t(x)$ will be the **Poisson kernel for the half-space**.

If $n = 1$, an easy computation will show that the Fourier transform of $g(x) = e^{-2\pi|x|}$ on $\mathbf{R}$ is

$$\widehat{g}(\xi) = \frac{1}{\pi}\frac{1}{1+\xi^2}. \tag{7.12}$$

Note that the family of functions

$$P_t(x) = \frac{1}{\pi}\frac{x}{t^2+x^2} \qquad (t > 0)$$

is the Poisson summability kernel of Example 2.42, obtained from P_1 by letting $P_t(x) = (1/t)P_1(x/t)$.

To check (7.12), a double partial integration in

$$\int_{\mathbf{R}} e^{-2\pi|x|}e^{-2\pi i\xi x}\,dx = 2\int_0^\infty e^{-2\pi x}\cos(2\pi\xi x)\,dx$$

shows that

$$\begin{aligned}
\widehat{g}(\xi) &= -\frac{1}{\pi}\Big[e^{-2\pi x}\cos(2\pi\xi x)\Big]_0^\infty - 2\pi\int_0^\infty e^{-2\pi x}\sin(2\pi\xi x)\,dx \\
&= \frac{1}{\pi} - 2\pi\int_0^\infty e^{-2\pi x}\sin(2\pi\xi x)\,dx = \frac{1}{\pi} - \xi^2\widehat{g}(\xi).
\end{aligned}$$

Thus

$$\widehat{g}(\xi) = \frac{1}{\pi}\frac{1}{1+\xi^2}.$$

Obviously $P_1 = \widehat{g} > 0$ and $\int_{\mathbf{R}} P_1(x)\,dx = g(0) = 1$.

This result is extended for $n > 1$, but the calculation is somewhat more involved and it will be obtained from (7.1) and from

$$\int_0^\infty \frac{e^{-t}}{\sqrt{t}}e^{-a^2/4t}\,dt = \sqrt{\pi}e^{-a} \tag{7.13}$$

for $a > 0$.

To prove (7.13), we will use the obvious identity

$$\int_0^\infty e^{-(1+s^2)t}\,dt = 1/(1+s^2)$$

and also

$$\int_{\mathbf{R}} \frac{e^{iat}}{1+t^2}\,dt = \pi e^{-a},$$

which follows from (7.12) by a change of variables.

Indeed,

$$\begin{aligned} e^{-a} &= \frac{1}{\pi}\int_{\mathbf{R}} e^{iat}\int_0^\infty e^{-(1+t^2)s}\,ds\,dt = \frac{1}{\pi}\int_0^\infty e^{-s}\int_{\mathbf{R}} e^{iat}e^{-st^2}\,dt\,ds \\ &= 2\int_0^\infty e^{-s}\int_{\mathbf{R}} e^{2\pi iax}e^{-4\pi^2 sx^2}\,dx\,ds = \int_0^\infty e^{-s}e^{-a^2/4s}\frac{1}{\sqrt{s\pi}}\,ds \end{aligned}$$

and (7.13) follows.

Now we are ready to prove that

$$P(t,x) = c_n\frac{t}{(t^2+|x|^2)^{(n+1)/2}} \qquad \left(c_n = \frac{\Gamma\left(\frac{n+1}{2}\right)}{\pi^{(n+1)/2}}\right); \tag{7.14}$$

that is,

$$\int_{\mathbf{R}^n} e^{-2\pi|\xi|t}e^{-2\pi ix\cdot\xi}\,d\xi = c_n\frac{t}{(t^2+|x|^2)^{(n+1)/2}}.$$

A change of variables allows us to suppose $t=1$ and, using (7.1) and (7.13),

$$\begin{aligned} P(1,x) &= \int_{\mathbf{R}^n} e^{-2\pi|\xi|}e^{-2\pi ix\cdot\xi}\,d\xi \\ &= \int_{\mathbf{R}^n}\left(\frac{1}{\sqrt{\pi}}\int_0^\infty \frac{e^{-t}}{\sqrt{t}}e^{-4\pi^2|\xi|^2/4t}\,dt\right)e^{-2\pi ix\cdot\xi}\,d\xi \\ &= \int_0^\infty \frac{e^{-t}}{\sqrt{t\pi}}\int_{\mathbf{R}^n} e^{-\pi^2|\xi|^2/t}e^{-2\pi ix\cdot\xi}\,d\xi\,dt \\ &= \pi^{-(n+1)/2}(1+|x|^2)^{-(n+1)/2}\int_0^\infty e^{-x}x^{(n+1)/2}\,dx \\ &= \frac{c_n}{(1+|x|^2)^{(n+1)/2}}. \end{aligned}$$

According to Theorem 2.41, since $P_t = P(t,\cdot)$ is a summability kernel, if $f\in\mathcal{C}(\mathbf{R}^n)$ tends to 0 at infinity, then $u(t,x) = (P_t * f)\to f(x)$ uniformly as $t\downarrow 0$. If $f\in L^p(\mathbf{R}^n)$ $(1\le p<\infty)$, then $u(t,\cdot)\to f$ in L^p when $t\downarrow 0$. Thus, in both cases, f can be considered as the boundary value of u, defined on $t>0$.

Moreover, a direct calculation shows that $(\partial_t^2+\triangle)P(t,x)=0$, and then u is harmonic on the half-space $t>0$, since

$$(\partial_t^2+\triangle)\int_{\mathbf{R}^n} P(t,x-y)f(y)\,dy = 0.$$

Note also that if $|f|\le C$, then $|u(t,x)|\le C\int_{-\infty}^{+\infty} P(t,x-y)\,dt = C$.

7.5.2. The Hilbert transform. In the two-variables case we write

$$P(x,y) = \frac{1}{\pi}\frac{y}{x^2+y^2}.$$

We have seen that it is a harmonic function on the half-plane $y > 0$. This also follows from the fact that it is the real part of the holomorphic function

$$\frac{i}{\pi z} = \frac{1}{\pi}\frac{y+ix}{x^2+y^2} = P_y(x) + iQ_y(x).$$

The imaginary part,

$$Q_y(x) = Q(x,y) = \frac{1}{\pi}\frac{x}{x^2+y^2},$$

which is the conjugate function of $P(x,y)$, is the **conjugate Poisson kernel**, which for every $x \neq 0$ satisfies

$$\lim_{y\downarrow 0} Q_y(x) = \frac{1}{\pi}\frac{1}{x}.$$

This limit is not locally integrable and does not appear directly as a distribution, but we can consider its principal value as a regularization of $1/x$ as in Exercise 6.16. Let us describe it as a limit in $\mathcal{S}'(\mathbf{R})$ of Q_y:

By definition, for every $\varphi \in \mathcal{S}(\mathbf{R})$,

$$\langle \varphi(x), \mathrm{pv}\frac{1}{x}\rangle := \mathrm{pv}\int_{-\infty}^{+\infty} \frac{\varphi(x)}{x}\,dx = \lim_{\varepsilon\downarrow 0}\int_{\mathbf{R}} h_\varepsilon(x)\varphi(x)\,dx$$

if $h_\varepsilon(x) = x^{-1}\chi_{\{|x|>\varepsilon\}}(x)$. This limit exists since $\int_{\varepsilon<|x|<1} x^{-1}\varphi(0)\,dx = 0$ and then

$$\langle \varphi(x), \mathrm{pv}\frac{1}{x}\rangle = \int_{|x|<1} \frac{\varphi(x)-\varphi(0)}{x}\,dx + \int_{|x|>1}\frac{\varphi(x)}{x}\,dx.$$

Theorem 7.22. *In $\mathcal{S}'(\mathbf{R})$,*

$$\frac{1}{\pi}\,\mathrm{pv}\frac{1}{x} = \lim_{y\downarrow 0} Q_y$$

and

$$\mathcal{F}(\frac{1}{\pi}\,\mathrm{pv}\frac{1}{x}) = -i\,\mathrm{sgn}\,.$$

Proof. For the first equality, we only need to show that $F_\varepsilon := \pi Q_\varepsilon - h_\varepsilon \to 0$ in $\mathcal{S}'(\mathbf{R})$ as $\varepsilon \to 0$. But we note that, for every $\varphi \in \mathcal{S}(\mathbf{R})$,

$$\begin{aligned}\langle \varphi, F_\varepsilon\rangle &= \int_{\{|x|<\varepsilon\}} \frac{x\varphi(x)}{\varepsilon^2+x^2}\,dx + \int_{\{|x|>\varepsilon\}}\left(\frac{x\varphi(x)}{\varepsilon^2+x^2} - \frac{\varphi(x)}{x}\right)dx \\ &= \int_{\{|x|<1\}} \frac{x\varphi(\varepsilon x)}{1+x^2}\,dx - \int_{\{|x|>1\}}\frac{\varphi(\varepsilon x)}{x(1+x^2)}\,dx\end{aligned}$$

and both integrals tend to 0 as $\varepsilon \to 0$, by dominated convergence.

For the second formula, a direct computation of the Fourier transform of the function $-i(\operatorname{sgn}\xi)e^{-2\pi y|\xi|}$ shows that

$$\widehat{Q}_y(\xi) = -i(\operatorname{sgn}\xi)e^{-2\pi y|\xi|}$$

and then

$$\mathcal{F}(\frac{1}{\pi}\operatorname{pv}\frac{1}{x})(\xi) = \lim_{y\downarrow 0} -i(\operatorname{sgn}\xi)e^{-2\pi y|\xi|} = -i(\operatorname{sgn}\xi)$$

in $\mathcal{S}'(\mathbf{R})$, since

$$\lim_{y\downarrow 0}\int_{\mathbf{R}} -i(\operatorname{sgn}\xi)e^{-2\pi y|\xi|}\varphi(\xi)\,d\xi = \int_{\mathbf{R}} -i(\operatorname{sgn}\xi)\varphi(\xi)\,d\xi$$

for every $\varphi \in \mathcal{S}(\mathbf{R})$. □

The **Hilbert transform**[12] is the fundamental map of harmonic analysis and signal theory $H : L^2(\mathbf{R}) \to L^2(\mathbf{R})$ defined by

$$\widehat{Hf}(\xi) = -i\operatorname{sgn}(\xi)\widehat{f}(\xi) \qquad (f \in L^2(\mathbf{R})).$$

Theorem 7.23. *The Hilbert transform is a bijective linear isometry such that*

$$H^2 = -I \text{ and } H^* = -H.$$

For every $\varphi \in \mathcal{S}(\mathbf{R})$,

$$H\varphi = \frac{1}{\pi}\operatorname{pv}\frac{1}{x} * \varphi = \lim_{y\downarrow 0}(Q_y * \varphi) \text{ in } \mathcal{S}'(\mathbf{R}).$$

Proof. If $m(\xi) := -i\operatorname{sgn}\xi$, it is clear that $M : f \mapsto mf$ is a bijective linear isometry of $L^2(\mathbf{R})$ and, by the Plancherel theorem, $H = \widetilde{\mathcal{F}}M\mathcal{F}$ is also a bijective linear isometry. Moreover $H^2 = -I$, since $m^2 = -1$, and

$$(Hf, g)_2 = (\widetilde{\mathcal{F}}M\mathcal{F}f, g)_2 = (M\mathcal{F}f, \mathcal{F}g)_2 = (\mathcal{F}f, -M\mathcal{F}g)_2 = (f, -Hg)_2.$$

With Theorem 7.22 in hand and from the properties of the convolution,

$$\mathcal{F}(H\varphi) = \lim_{y\downarrow 0}\widehat{Q}_y\widehat{\varphi} = \lim_{y\downarrow 0}\mathcal{F}(Q_y * \varphi),$$

so that $H\varphi = \lim_{y\downarrow 0}(Q_y * \varphi)$ in $\mathcal{S}'(\mathbf{R})$.

Also $H\varphi = \widetilde{\mathcal{F}}(M\mathcal{F}\varphi) = (\widetilde{\mathcal{F}}m) * \varphi = \frac{1}{\pi}\operatorname{pv}\frac{1}{x} * \varphi$. □

[12]The name was coined by the English mathematician G. H. Hardy after David Hilbert, who was the first to observe the conjugate functions in 1912. He also showed that the function $\sin(\omega t)$ is the Hilbert transform of $\cos(\omega t)$ and this gives the $\pm\pi/2$ phase-shift operator, which is a basic property of the Hilbert transform in signal theory.

If $\varphi \in \mathcal{S}(\mathbf{R})$, its bounded harmonic extension to the half-plane $y > 0$,

$$u(x, y) = (P_y * \varphi)(x)$$

has

$$v(x, y) = (Q_y * \varphi)(x)$$

as the conjugate function, so that

$$\begin{aligned} F(z) &= u(x, y) + iv(x, y) \\ &= \frac{1}{\pi} \int_{\mathbf{R}} \frac{y}{(x-t)^2 + y^2} \varphi(t)\, dt + \frac{i}{\pi} \int_{\mathbf{R}} \frac{x-t}{(x-t)^2 + y^2} \varphi(t)\, dt \\ &= \frac{1}{\pi i} \int_{\mathbf{R}} \frac{t - \bar{z}}{|t - z|^2} \varphi(t)\, dt \end{aligned}$$

is holomorphic on $\Im z = y > 0$, and it is continuous on $y \geq 0$ with

$$\lim_{y \downarrow 0} F(z) = \varphi(x) + i(H\varphi)(x).$$

Since $F^2(z) = u^2(x, y) - v^2(x, y) + i2u(x, y)v(x, y)$ is also holomorphic, $2uv$ is the conjugate function of $u^2 - v^2$, and $H(\varphi^2 - (H\varphi)^2) = 2\varphi H\varphi$, where $H^{-1} = -H$. Thus

$$(H\varphi)^2 = \varphi^2 + 2H(\varphi H\varphi). \tag{7.15}$$

We can write

$$H\varphi(x) = \frac{1}{\pi} \int_{\mathbf{R}} \frac{\varphi(x - y)}{y}\, dy = \frac{1}{\pi} \int_{\mathbf{R}} \frac{\varphi(y)}{x - y}\, dy$$

in the sense of the principal value, and the integrals are called singular integrals. The kernel

$$K(x, y) = \frac{1}{x - y}$$

is far from satisfying the conditions of the Young inequalities (2.20), but H will still be an operator of type (p, p) if $1 < p < \infty$:

Theorem 7.24 (M. Riesz). *For every $1 < p < \infty$, H is a bounded operator of $L^p(\mathbf{R})$.*

Proof. We claim that if $\|H\varphi\|_p \leq C_p\|\varphi\|_p$, then $\|H\varphi\|_{2p} \leq (2C_p + 1)\|\varphi\|_{2p}$.

Indeed, either $\|H\varphi\|_{2p} \leq \|\varphi\|_{2p}$ and there is nothing to prove, or $\|\varphi\|_{2p} \leq \|Hf\|_{2p}$. In this last case, by (7.15),

$$\begin{aligned} \|H\varphi\|_{2p}^2 &= \|(H\varphi)^2\|_p \leq \|\varphi^2\|_p + 2\|H(\varphi H\varphi)\|_p \\ &\leq \|\varphi\|_{2p}^2 + 2C_p\|\varphi H\varphi\|_p \\ &\leq \|\varphi\|_{2p}^2 + 2C_p\|\varphi\|_{2p}\|H\varphi\|_{2p} \\ &\leq \|\varphi\|_{2p}(1 + 2C_p)\|H\varphi\|_{2p} \end{aligned}$$

as claimed.

From this claim, starting from $\|H\varphi\|_2 = \|\varphi\|_2$, we obtain by induction

$$\|H\varphi\|_{2^n} \leq (2^n - 1)\|\varphi\|_{2^n} \qquad (n = 1, 2, \ldots)$$

and an application of the Riesz-Thorin interpolation theorem[13] gives

$$\|H\varphi\|_p \leq C_p\|\varphi\|_p \qquad (\varphi \in \mathcal{S}(\mathbf{R}))$$

for every $2 \leq p < \infty$, so that $H(L^p(\mathbf{R})) \subset L^p(\mathbf{R})$ and H is of type (p, p) for these values of p.

Suppose now that $1 < p < 2$, so that $p' > 2$ and H is a bounded linear operator $L^{p'}(\mathbf{R}) \to L^{p'}(\mathbf{R})$.

Then

$$\begin{aligned} \|Hf\|_p &= \sup\left\{\left|\int_{\mathbf{R}} g(t)H\varphi(t)\,dt\right|;\ \|g\|_{p'} \leq 1\right\} \\ &= \sup\left\{\left|\int_{\mathbf{R}} \varphi(t)Hg(t)\,dt\right|;\ \|g\|_{p'} \leq 1\right\} \leq \|\varphi\|_p C_{p'}. \end{aligned}$$

□

7.6. Sobolev spaces

7.6.1. The spaces $W^{m,p}$. Let Ω be a nonempty open subset of $\mathbf{R}^n$, $1 \leq p \leq \infty$ and $m \in \mathbf{N}$.

The **Sobolev space of order** $m \in \mathbf{N}$ on Ω is defined by

$$W^{m,p}(\Omega) := \{u \in L^p(\Omega);\ D^\alpha u \in L^p(\Omega),\ |\alpha| \leq m\},$$

where the $D^\alpha u$ represent the distributional derivatives of u. We endow $W^{m,p}(\Omega)$ with the topology of the norm

$$\|u\|_{(m,p)} := \sum_{|\alpha| \leq m} \|D^\alpha u\|_p.$$

Note that the linear maps $W^{m,p}(\Omega) \hookrightarrow L^p(\Omega)$, $W^{m+1,p}(\Omega) \hookrightarrow W^{m,p}(\Omega)$, and $D^\alpha : W^{m,p}(\Omega) \to W^{m-|\alpha|,p}(\Omega)$ are continuous, if $|\alpha| \leq m$.

In $\mathbf{R}^N$ all the norms are equivalent, so that $\|\cdot\|_{(m,p)}$ is equivalent to the norm

$$\|u\| := \max_{|\alpha| \leq m} \|D^\alpha u\|_p.$$

It is easy to show that $W^{m,p}(\Omega)$ is a Banach space by describing it as the subspace of $\prod_{|\alpha| \leq m} L^p(\Omega)$ of all elements of the type $\{D^\alpha u\}_{|\alpha| \leq m}$. It is closed, since, if $\{D^\alpha u_k\}_{|\alpha| \leq m} \to \{u^{(\alpha)}\}_{|\alpha| \leq m}$ in the product space, then $D^\alpha u_k \to u^{(\alpha)}$ $(|\alpha| \leq m)$ in $L^p(\Omega)$.

---

[13] Marcel Riesz proved his convexity theorem, the Riesz-Thorin Theorem 2.45 for $p(\vartheta) \leq q(\vartheta)$, to use it in the proof of this fact and related results of harmonic analysis. See footnote 14 in Chapter 2.

Note that if $f_k \to f$ in $L^p(\Omega)$, then also $f_k \to f$ in $\mathcal{D}'(\Omega)$ since

$$|\langle \varphi, f - f_k \rangle \leq \|\varphi\|_{p'} \|f - f_k\|_p.$$

Hence $D^\alpha u_k \to D^\alpha u = u^{(\alpha)}$ in $\mathcal{D}'(\Omega)$, so that $\{D^\alpha u_k\}_{|\alpha|\leq m} \to \{D^\alpha u\}_{|\alpha|\leq m}$.

In the case $p = 2$, we can renorm the space with the equivalent norm

$$\|u\|_{m,2} = \Big(\sum_{|\alpha|\leq m} \|D^\alpha u\|_2^2 \Big)^{1/2},$$

so that $W^{m,2}(\Omega)$ becomes a Hilbert space with the scalar product

$$(u, v)_{m,2} = \sum_{|\alpha|\leq m} (D^\alpha u, D^\alpha v)_2.$$

Every $u \in W^{m,p}(\Omega) \subset L^p(\Omega)$ is a class of functions, and it is said that it is a C^m function if it has a C^m representative.

In the case of one variable, we can consider $W^{m,p}(a,b) \subset \mathcal{C}[a,b]$ for every $m \geq 1$, since, if $u = v$ a.e. and both of them are continuous, then $u = v$. Moreover, the following regularity result holds:

Theorem 7.25. *If $u \in W^{1,p}(a,b)$ and $v(t) := \int_c^t u'(s)\, ds$, then $u(t) = v(t) + C$ a.e., so that u coincides a.e. with a continuous function, which we still denote by u, such that*

$$u(x) - u(y) = \int_y^x u'(s)\, ds \qquad (x, y \in (a,b)).$$

The distributional derivative u' is the a.e. derivative of u, and the inclusion $W^{1,p}(a,b) \hookrightarrow \mathcal{C}[a,b]$ is continuous.

Proof. Function v, as a primitive of $u' \in L^1_{\text{loc}}(a,b)$, is absolutely continuous on $[a,b]$, and, by the Lebesgue differentiation theorem, u' is its a.e. derivative. The distributional derivatives v' and u' are the same, since, by partial integration and from $\varphi(a) = \varphi(b) = 0$ when $\varphi \in \mathcal{D}(a,b)$, we obtain

$$\langle -\varphi', v \rangle = -\int_a^b \varphi'(t) \int_c^t u'(s)\, ds\, dt = \int_a^b u'(t)\varphi(t)\, dt = \langle \varphi, u' \rangle.$$

But $(u - v)' = 0$ implies $u - v = C$, and u is continuous on $[a,b]$.

It follows from $u = v + C$ that $u(x) - u(y) = v(x) - v(y) = \int_y^x u'(s)\, ds$.

If $u_k \to u$ in $W^{1,p}(a,b)$ and $u_k \to v$ in $\mathcal{C}[a,b]$, then $v = u$, since there exists a subsequence of $\{u_k\}$ which is a.e. convergent to u. Hence, $W^{1,p}(a,b) \hookrightarrow \mathcal{C}[a,b]$ has a closed graph. □

Remark 7.26. It can be shown that, if $m > n/p$ and $1 \leq p < \infty$, $W^{m,p}(\Omega) \subset \mathcal{E}^k(\Omega)$ whenever $k < m - (n/p)$.

We will prove this result in the Hilbert space case $p = 2$ (see Theorem 7.29).

7.6.2. The spaces $H^s(\mathbf{R}^n)$. There is a Fourier characterization of the space $W^{m,2}(\mathbf{R}^n)$ which will allow us to define the Sobolev spaces of fractional order $s \in \mathbf{R}$ on $\mathbf{R}^n$.

In Theorem 7.7 we saw that the pointwise multiplication by the function

$$\omega_s(\xi) := (1 + |\xi|^2)^{s/2}$$

is a continuous linear operator of $\mathcal{S}(\mathbf{R}^n)$, and it can be extended to $\mathcal{S}'(\mathbf{R}^n)$, with $\omega_s u \in \mathcal{S}'(\mathbf{R}^n)$ for every $u \in \mathcal{S}'(\mathbf{R}^n)$, defined as usual by

$$\langle \varphi, \omega_s u \rangle = \langle \omega_s \varphi, u \rangle.$$

We define the operator Λ^s on $\mathcal{S}'(\mathbf{R}^n)$ in terms of the Fourier transform

$$\Lambda^s u = \mathcal{F}^{-1}(\omega_s \widehat{u}).$$

It is a bijective continuous linear operator of $\mathcal{S}(\mathbf{R}^n)$ and of $\mathcal{S}'(\mathbf{R}^n)$, with $(\Lambda^s)^{-1} = \Lambda^{-s}$ and such that

$$\widehat{\Lambda^s u} = \omega_s \widehat{u}.$$

It is called the **Fourier multiplier with symbol** ω_s, since it is the result of the multiplication by ω_s "at the other side of the Fourier transform".

Since $\Lambda^{2m} = (\operatorname{Id} - (4\pi)^{-1}\triangle)^m$, we can also write

$$\Lambda^s = (\operatorname{Id} - (4\pi)^{-1}\triangle)^{s/2}.$$

We define the **Sobolev space of order** $s \in \mathbf{R}$,

$$H^s(\mathbf{R}^n) = \{u \in \mathcal{S}'(\mathbf{R}^n);\ \Lambda^s u \in L^2(\mathbf{R}^n)\},$$

i.e., $H^s(\mathbf{R}^n) = \Lambda^{-s}(L^2(\mathbf{R}^n))$, and we provide it with the norm

$$\|u\|_{(s)} = \|\Lambda^s u\|_2 = \|\widehat{\Lambda^s u}\|_2 = \|\omega_s \widehat{u}\|_2,$$

associated with the scalar product $(u, v)_{(s)} = (\Lambda^s u, \Lambda^s v)_2$.

It is a Hilbert space, since Λ^s is a linear bijective isometry between $H^s(\mathbf{R}^n)$ and $L^2(\mathbf{R}^n) = H^0(\mathbf{R}^n)$ and also from $H^r(\mathbf{R}^n)$ onto $H^{r-s}(\mathbf{R}^n)$, which corresponds to $\widehat{u} \mapsto \omega_s \widehat{u}$:

$$\|\Lambda^s u\|_{(r-s)} = \|\omega_{r-s}\widehat{\Lambda^s u}\|_2 = \|\omega_r \widehat{u}\|_2 = \|u\|_{(r)}.$$

If $t < s$, $H^s(\mathbf{R}^n) \hookrightarrow H^t(\mathbf{R}^n)$, and $\mathcal{S}(\mathbf{R}^n)$ is a dense subspace of every $H^s(\mathbf{R}^n)$, since it is dense in $L^2(\mathbf{R}^n)$ and $\Lambda^{-s} : L^2(\mathbf{R}^n) \to H^s(\mathbf{R}^n)$ and $\Lambda^{-s}(\mathcal{S}(\mathbf{R}^n)) = \mathcal{S}(\mathbf{R}^n)$.

Theorem 7.27. *If $k \in \mathbf{N}$ and $s \in \mathbf{R}$,*

$$H^s(\mathbf{R}^n) = \{u \in \mathcal{S}'(\mathbf{R}^n);\ D^\alpha u \in H^{s-k}(\mathbf{R}^n)\ \forall |\alpha| \le k\},$$

and $\|\cdot\|_{(s)}$ and $u \mapsto \sum_{|\alpha|\le k} \|D^\alpha u\|_{(s-k)}$ are two equivalent norms.

In particular, if $s = m$, then $H^s(\mathbf{R}^n) = W^{m,2}(\mathbf{R}^n)$ with equivalent norms.

Proof. If $u \in H^s(\mathbf{R}^n)$, note that every

$$\widehat{D^\alpha u}(\xi) = (2\pi i\xi)^\alpha \widehat{u}(\xi) = (2\pi i \xi)^\alpha \omega_{-s}(\xi)\widehat{\Lambda^s u}(\xi)$$

is a locally integrable function, since $\widehat{\Lambda^s u} \in L^2(\mathbf{R}^n)$ and $(2\pi i\xi)^\alpha\omega_{-s}(\xi)$ is continuous.

If $|\alpha| \le k$, then

$$|\xi^\alpha| = |\xi_1^{\alpha_1}\cdots\xi_n^{\alpha_n}| \le |\xi|^{|\alpha|} = \Big(\sum_{j=1}^n \xi_j^2\Big)^{|\alpha|/2}$$

since, for every j, $|\xi_j|^{\alpha_j} \le (\sum_{i=1}^n \xi_i^2)^{\alpha_j/2} = |\xi|^{\alpha_j}$. So $|\xi^\alpha| \le (1+|\xi|^2)^{k/2} = \omega_k(\xi)$.

Moreover $(1+|\xi|^2)^{k/2} \le C\sum_{|\alpha|\le k}|\xi^\alpha|$, since $\sum_{|\alpha|\le k}|\xi^\alpha| > 0$ everywhere and

$$F(\xi) := \frac{(1+|\xi|^2)^{k/2}}{\sum_{|\alpha|\le k}|\xi^\alpha|}$$

is a continuous function on $\mathbf{R}^n$ such that, if $|\xi| \ge 1$,

$$F(\xi) \le \frac{(1+|\xi|^2)^{k/2}}{\sum_{j=1}^n |\xi_j|^k} \le 2^{k/2}\frac{|\xi|^k}{\sum_{j=1}^n |\xi_j|^k} \le C.$$

Thus,

$$(1+|\xi|^2)^{k/2} \simeq \sum_{|\alpha|\le k} (2\pi)^{|\alpha|}|\xi^\alpha| = \sum_{|\alpha|\le k} |(2\pi i\xi)^\alpha|.$$

From these estimates we obtain

$$\|u\|_{(s)} = \|\omega_{s-k}\omega_k\widehat{u}\|_2 \le C\sum_{|\alpha|\le k}\|\omega_{s-k}|(2\pi i\xi)^\alpha\widehat{u}|\|_2 = C\sum_{|\alpha|\le k}\|D^\alpha u(\xi)\|_{(s-k)}$$

and also, if $|\alpha| \le k$,

$$\|D^\alpha u\|_{(s-k)} = \|\omega_{s-k}|(2\pi i\xi)^\alpha\widehat{u}(\xi)|\|_2 \le C\|\omega_{s-k}\omega_k\widehat{u}\|_2 = C\|u\|_{(s)}.$$

□

Theorem 7.28 (Sobolev). *If $s - k > n/2$, $H^s(\mathbf{R}^n) \subset \mathcal{E}^k(\mathbf{R}^n)$.*

Proof. By polar integration, $\int_{\mathbf{R}^n}(1+|\xi|^2)^{k-s}\,d\xi<\infty$ if and only if

$$\int_1^\infty (1+r^2)^{k-s}r^{n-1}\,dr \simeq \int_1^\infty r^{2k-2s}r^{n-1}\,dr<\infty,$$

i.e., when $2k-2s+n-1<-1$, which is equivalent to condition $s-k>n/2$.

Then, if $\varphi\in\mathcal{S}(\mathbf{R}^n)$, multiplication and division in

$$D^\alpha\varphi(x)=\int_{\mathbf{R}^n}\widehat{D^\alpha\varphi}(\xi)e^{2\pi i x\cdot\xi}\,d\xi=\int_{\mathbf{R}^n}(2\pi i\xi)^\alpha\widehat{\varphi}(\xi)e^{2\pi i x\cdot\xi}\,d\xi$$

by $\omega_{s-k}(\xi)=(1+|\xi|^2)^{(s-k)/2}$ followed by an application of the Schwarz inequality yields

$$\begin{aligned}|D^\alpha\varphi(x)| &\le \int_{\mathbf{R}^n}|(2\pi\xi)^\alpha|\omega_{s-k}(\xi)|\widehat{\varphi}(\xi)|\omega_{k-s}(\xi)\,d\xi\\ &\le \|D^\alpha\varphi\|_{(s-k)}\Big(\int_{\mathbf{R}^n}(1+|\xi|^2)^{k-s}\,d\xi\Big)^{1/2}\end{aligned}$$

and

$$\|D^\alpha\varphi\|_\infty\le C\|D^\alpha\varphi\|_{(s-k)}. \tag{7.16}$$

When $u\in H^s(\mathbf{R}^n)$, we can consider $\varphi_m\to u$ in $H^s(\mathbf{R}^n)$ with $\varphi_k\in\mathcal{S}(\mathbf{R}^n)$, the estimate (7.16) ensures that $\{D^\alpha\varphi_m\}_{m=1}^\infty\subset\mathcal{S}(\mathbf{R}^n)$ is uniformly Cauchy, and we obtain that $D^\alpha\varphi_m\to D^\alpha u$ uniformly, if $|\alpha|\le k$. This proves that $u\in\mathcal{E}^k(\mathbf{R}^n)$. □

7.6.3. The spaces $H^m(\Omega)$. If $m\in\mathbf{N}$ and for any open set $\Omega\subset\mathbf{R}^n$, we will use the notation

$$H^m(\Omega)=W^{m,2}(\Omega)$$

suggested by Theorem 7.27.

In this case, as a consequence of the Sobolev Theorem 7.28 we obtain

Theorem 7.29. *If $m-k>n/2$, $H^m(\Omega)\subset\mathcal{E}^k(\Omega)$.*

Proof. It is sufficient to prove that $u\in H^m(\Omega)\subset L^2(\Omega)$ is a C^k function on a neighborhood of every point. By multiplying u by a test function if necessary, we can suppose that as a distribution its support is a compact subset K of Ω.

If $K\prec\eta\prec\Omega$ and if $\bar{u}$ is the extension of $u\in L^2(\Omega)$ by zero to $\bar{u}\in L^2(\mathbf{R}^n)$, then $\bar{u}\in H^m(\mathbf{R}^n)$, since for every $|\alpha|\le m$ we can apply the Leibniz rule to

$$\langle\varphi,D^\alpha\bar{u}\rangle=\langle\varphi,D^\alpha(\eta\bar{u})\rangle$$

to show that $\langle\cdot,D^\alpha\bar{u}\rangle$ is L^2-continuous on test functions, so $D^\alpha\bar{u}\in L^2(\mathbf{R}^n)$ by the Riesz representation theorem, and $\bar{u}\in H^m(\mathbf{R}^n)$. By Theorem 7.28 we know that $\bar{u}\in\mathcal{E}^k(\mathbf{R}^n)$ and then $u\in\mathcal{E}^k(\Omega)$. □

Remark 7.30. It is useful to approximate functions in $H^m(\Omega)$ by C^∞ functions. The space $H^m(\Omega)$ can be defined as the completion of $\mathcal{E}(\Omega) \cap H^m(\Omega)$ under the norm $\|\cdot\|_{m,2}$. In fact it can be proved that $\mathcal{E}(\Omega) \cap W^{m,p}(\Omega)$ is dense in $W^{m,p}(\Omega)$ for any $1 \le p < \infty$ and $m \ge 1$ an integer.[14]

We content ourselves with the following easier approximation result known as the Friedrichs theorem.

Theorem 7.31. *If $u \in H^1(\Omega)$, then there exists a sequence $\{\varphi_m\} \subset \mathcal{D}(\mathbf{R}^n)$ such that $\lim_m \|u - \varphi_m\|_{L^p(\Omega)} = 0$ and $\lim_m \|\partial_j u - \partial_j \varphi_m\|_{L^p(\omega)} = 0$ for every $1 \le j \le n$ and for every open set ω such that $\bar{\omega}$ is a compact subset of Ω.*

Proof. Denote by u^o the extension of u by zero on $\mathbf{R}^n$ and choose a mollifier ϱ_ε. Then $\varrho_\varepsilon * u^o \to u^o$ in $L^2(\mathbf{R}^n)$ and so $\lim_{\varepsilon\to 0} \|u - \varrho_\varepsilon * u^o\|_{L^p(\Omega)} = 0$. Allow $0 < \varepsilon < d(\bar{\omega}, \Omega^c)$, so that $(\varrho_\varepsilon * u^o)(x) = (\varrho_\varepsilon * u)(x)$ and $\partial_j(\varrho_\varepsilon * u^o)(x) = (\varrho_\varepsilon * (\partial_j u)^o)(x)$ for every $x \in \omega$. So $\lim_{\varepsilon\to 0} \|\partial_j u - \partial_j(\varrho_\varepsilon * u^o)\|_{L^2(\omega)} = 0$, since $(\partial_j u)^o \in L^2(\mathbf{R}^n)$.

If $\varepsilon_m \downarrow 0$, let us multiply the functions $f_m = \varrho_{\varepsilon_m} * u^o \in \mathcal{E}(\mathbf{R}^n)$ by the cut-off functions χ_m such that $\bar{B}(0,m) \prec \chi_m \prec \mathbf{R}^n$. By the dominated convergence theorem, if $\varphi_m = \chi_m f_m$, then

$$\|\varphi_m - u^o\|_2 \le \|\chi_m(f_m - u^o)\|_2 + \|\chi_m u^o - u^o\|_2 \to 0 \text{ as } m \to \infty$$

and $\bar{\omega} \subset B(0,m)$ for large m. It follows that the test functions φ_m satisfy all the requirements. □

As an application, we can prove the following **chain rule** for functions $v \in H^1(\Omega)$ and any $\varrho \in \mathcal{E}(\mathbf{R})$ with bounded derivative and such that $\varrho(0) = 0$:

$$\partial_j(\varrho \circ v) = (\varrho' \circ v)\partial_j v \qquad (1 \le j \le n). \tag{7.17}$$

Indeed, since $|\varrho'| \le M$ and $\varrho(0) = 0$, by the mean value theorem $|\varrho(t)| \le M|t|$. Thus $|\varrho \circ v| \le M|v|$ and $\varrho \circ v \in L^2(\Omega)$. It is also clear that $(\varrho' \circ v)\partial_j v \in L^2(\Omega)$.

Pick $\varphi_m \in \mathcal{D}(\mathbf{R}^n)$ as in Theorem 7.31. Then, for every $\varphi \in \mathcal{D}(\Omega)$ with $\operatorname{supp}\varphi \subset \omega$, from the usual chain rule

$$\int_\Omega (\varrho \circ \varphi_m)(x)\partial_j\varphi(x)\,dx = \int_\Omega (\varrho' \circ \varphi_m)(x)\partial_j\varphi_m(x)\varphi(x)\,dx.$$

Here $\varrho \circ \varphi_m \to \varrho \circ v$ in $L^2(\Omega)$ and $(\varrho' \circ \varphi_m)\partial_j\varphi_m \to (\varrho' \circ v)\partial_j v$ in $L^2(\omega)$ by dominated convergence, and (7.17) follows.

---

14. This is the Meyer-Serrin theorem and a proof can be found in [1] or [**17**].

7.6.4. The spaces $H_0^m(\Omega)$. When looking for distributional solutions $u \in H^m(\Omega)$ in boundary value problems such as the Dirichlet problem with a homogeneous boundary condition, it does not make sense to consider the pointwise values $u(x)$ of u.

Vanishing on the boundary in the distributional sense is defined by considering u as an element of a convenient subspace of $H^m(\Omega)$.

The Sobolev space $H_0^m(\Omega)$ is defined as the closure of

$$\mathcal{D}^m(\Omega) = \bigcup_{K\in\mathcal{K}(\Omega)} \mathcal{D}_K^m(\Omega)$$

in $H^m(\Omega)$, and it is endowed with the restriction of the norm of $H^m(\Omega)$.

In this definition, $\mathcal{D}^m(\Omega)$ can be replaced by $\mathcal{D}(\Omega)$:

Theorem 7.32. *For every $m \in \mathbf{N}$, $\mathcal{D}(\Omega)$ is dense in $H_0^m(\Omega)$.*

Proof. Let $\psi \in \mathcal{D}^m(\Omega) \subset \mathcal{D}^m(\mathbf{R}^n)$. If $\varrho \geq 0$ is a test function supported by $\bar{B}(0,1)$ such that $\|\varrho\|_1 = 1$, then $\varrho_k(x) := k^n\varrho(kx)$ is another test function such that $\|\varrho\|_1 = 1$, now supported by $\bar{B}(0,1/k)$. Moreover $\varrho_k * \psi \in \mathcal{D}(\mathbf{R}^n)$ and

$$D^\alpha(\varrho_k * \psi) = \varrho_k * D^\alpha\psi \to D^\alpha\psi$$

in $L^2(\mathbf{R}^n)$ for every $|\alpha| \leq m$ with

$$\operatorname{supp}\varrho_k * \psi \subset \operatorname{supp}\varrho_k + \operatorname{supp}\psi \subset \Omega$$

if k is large enough. Hence, $\mathcal{D}(\mathbf{R}^n) \ni \varrho_k * \psi \to \psi$ in $H^m(\Omega)$ and $\psi \in \overline{\mathcal{D}(\Omega)}$, closure in $H_0^m(\Omega)$. □

The class $\mathcal{S}(\mathbf{R}^n)$ is dense in $H^m(\mathbf{R}^n)$, $\mathcal{D}(\mathbf{R}^n)$ is also dense in $\mathcal{S}(\mathbf{R}^n)$, and the inclusion $\mathcal{S}(\mathbf{R}^n) \hookrightarrow H^m(\mathbf{R}^n)$ is continuous, so that $\mathcal{D}(\mathbf{R}^n)$ is dense in $H^m(\mathbf{R}^n)$ and

$$H_0^m(\mathbf{R}^n) = H^m(\mathbf{R}^n).$$

The fact that the elements in $H_0^m(\Omega)$ can be considered as distributions that vanish on the boundary $\partial\Omega$ of Ω is explained by the following results, where for simplicity we restrict ourselves to the special and important case $m = 1$.

Theorem 7.33. *If $u \in H^1(\Omega)$ is compactly supported, then $u \in H_0^1(\Omega)$, and its extension $\bar{u}$ by zero on $\mathbf{R}^n$ belongs to $H^1(\mathbf{R}^n)$.*

Proof. If $u \in H^1(\Omega)$ has a compact support $K \subset \Omega$, it is shown as in Theorem 7.29 that $\bar{u} \in L^2(\mathbf{R}^n)$ belongs to $H^1(\mathbf{R}^n)$ by using η such that $K \prec \eta \prec \Omega$.

But $\mathcal{D}(\mathbf{R}^n)$ is dense in $H^1(\mathbf{R}^n)$ and it follows from $\varphi_m \to \bar{u}$ in $H^1(\mathbf{R}^n)$ that $\varphi_m\eta \to u$ in $H^1(\Omega)$ with $\varphi_m\eta \in \mathcal{D}(\Omega)$, so that $u \in H_0^1(\Omega)$. □

Theorem 7.34. *Let $\Omega \subset \mathbf{R}^n$ be a bounded open set. If $u \in H^1(\Omega) \cap \mathcal{C}(\bar{\Omega})$ and $u_{|\partial\Omega} = 0$, then $u \in H_0^1(\Omega)$.*

Proof. Assume that u is real and let $\varrho \in \mathcal{E}^1(\mathbf{R})$ be such that $|\varrho(t)| \leq |t|$ on $\mathbf{R}$, $\varrho = 0$ on $[-1, 1]$, and $\varrho(t) = t$ on $(-2, 2)^c$. If $[-1, 1] \prec \varphi \prec (-2, 2)$, just take $\varrho(t) = t(1 - \varphi(t))$.

If $v \in H^1(\Omega)$, then $\varrho \circ v \in L^2(\Omega)$ and by the chain rule (7.17) also $\partial_j(\varrho \circ v) = (\varrho' \circ v) \cdot \partial_j v \in L^2(\Omega)$, so $\varrho \circ v \in H^1(\Omega)$.

Hence $u_m := m^{-1}\varrho(mu) \in H_0^1(\Omega)$ since $\operatorname{supp} u_m \subset \{|u| \geq 1/m\}$, which is a compact subset of the bounded open set Ω. By dominated convergence, $u_m \to u$ in $H^1(\Omega)$ and it follows that $u \in H_0^1(\Omega)$. □

It can be shown that this result is also true for unbounded open sets and that the converse holds when Ω is of class C^1:

$$u \in \mathcal{C}(\bar{\Omega}) \cap H_0^1(\Omega) \Rightarrow u_{|\partial\Omega} = 0. \tag{7.18}$$

We only include here the proof in the easy case $n = 1$:

Theorem 7.35. *If $n = 1$ and $\Omega = (a, b)$, then $H_0^1(a, b)$ is the class of all functions $u \in H^1(a, b) \subset \mathcal{C}[a, b]$ such that $u(a) = u(b) = 0$.*

Proof. By Theorem 7.34, we only need to show that $u(a) = u(b) = 0$ for every $u \in H_0^1(a, b) \subset \mathcal{C}[a, b]$. But, if $\mathcal{D}(a, b) \ni \varphi_k \to u$ in $H_0^1(a, b)$, also $\varphi_k \to u$ uniformly, since $H_0^1(a, b) \hookrightarrow \mathcal{C}[a, b]$ is continuous by Theorem 7.25, and then $\varphi_k(a) = \varphi_k(b) = 0$. □

7.7. Applications

To show how Sobolev spaces provide a good framework for the study of differential equations, let us start with a one-dimensional problem.

7.7.1. The Sturm-Liouville problem. We consider here the problem of solving

$$-(pu')' + qu = f, \;\; u(a) = u(b) = 0 \tag{7.19}$$

when $q \in \mathcal{C}[a, b]$, $p \in \mathcal{C}^1[a, b]$, and $p(t) \geq \delta > 0$.

If $f \in \mathcal{C}[a, b]$, a **classical solution** is a function $u \in \mathcal{C}^2[a, b]$ that satisfies (7.19) at every point.

If $f \in L^2(a, b)$, a **weak solution** is a function $u \in H_0^1(a, b)$ whose distributional derivatives satisfy $-(pu')' + qu = f$, i.e.,

$$\int_a^b p(t)u'(t)\varphi'(t)\,dt + \int_a^b q(t)u(t)\varphi(t)\,dt = \int_a^b f(t)\varphi(t)\,dt \qquad (\varphi \in \mathcal{D}(a, b)).$$

If $v \in H_0^1(a,b)$, by taking $\varphi_k \to v$ ($\varphi_k \in \mathcal{D}(a,b)$), the identity

$$\int_a^b p(t)u'(t)v'(t)\,dt + \int_a^b q(t)u(t)v(t)\,dt = \int_a^b f(t)v(t)\,dt$$

also holds.

To prove the existence and uniqueness of a weak solution for this Sturm-Liouville problem, with $f \in L^2(a,b)$, we define

$$B(u,v) := \int_a^b p(t)u'(t)\overline{v'(t)}\,dt + \int_a^b q(t)u(t)\overline{v(t)}\,dt.$$

Then we obtain a sesquilinear continuous form on $H_0^1(a,b) \times H_0^1(a,b)$ and $(\cdot, f)_2 \in H_0^1(a,b)'$, since

$$|B(u,v)| \le \|p\|_\infty \|u'\|_2 \|v'\|_2 + \|q\|_\infty \|u\|_2 \|v\|_2 \le c\|u\|_{(1,2)}\|v\|_{(1,2)}$$

and $|(u,f)_2| \le \|f\|_2 \|u\|_{(1,2)}$.

If B is coercive, we can apply the Lax-Milgram theorem and, for a given $f \in L^2(a,b)$, there exists a unique $u \in H_0^1(a,b)$ such that $B(v,u) = (v,f)_2$, which means that u is the uniquely determined weak solution of problem (7.19).

For instance, if also $q(t) \ge \delta > 0$, then

$$B(u,u) = \int_a^b (p(t)|u'(t)|^2 + q(t)|u(t)|^2)\,dt \ge \delta \|u\|^2_{H_0^1(a,b)}$$

and B is coercive.

Finally, if $f \in \mathcal{C}[a,b]$, the weak solution u is a C^2 function, and then it is a classical solution. Indeed, $pu' \in L^2(a,b)$ satisfies $(pu')' = qu - f$, which is continuous; then $g := pu'$ and $u' = g/p$ are C^1 functions on $[a,b]$, so that $u \in \mathcal{C}^2[a,b]$, and $u(a) = u(b) = 0$ by Theorem 7.35.

7.7.2. The Dirichlet problem. Now let Ω be a nonempty bounded open domain in $\mathbf{R}^n$ with $n > 1$, and consider the Dirichlet problem

$$-\Delta u = f, \qquad u = 0 \text{ on } \partial\Omega \qquad (f \in L^2(\Omega)). \tag{7.20}$$

If f is continuous on $\bar\Omega$ and u is a classical solution, then $u \in \mathcal{C}^2(\Omega) \cap \mathcal{C}(\bar\Omega)$ and (7.20) holds in the pointwise sense.

Let us write $(\nabla u, \nabla v)_2 := \sum_{j=1}^n (\partial_j u, \partial_j v)_2$. When trying to obtain existence and uniqueness of such a solution, we again start by looking for solutions in a weak sense. After multiplying by test functions $\varphi \in \mathcal{D}(\Omega)$, by integration we are led to consider functions $u \in \mathcal{C}^2(\Omega) \cap \mathcal{C}(\bar\Omega)$ such that $u = 0$ on $\partial\Omega$ and

$$(\varphi, -\Delta u)_2 = (\varphi, f)_2 \qquad (\varphi \in \mathcal{D}(\Omega)),$$

or, equivalently, such that

$$(\nabla\varphi, \nabla u)_2 = (\varphi, f)_2 \qquad (\varphi \in \mathcal{D}(\Omega)).$$

Then it follows from Theorem 7.34 that $u \in H_0^1(\Omega)$, and

$$(\nabla v, \nabla u)_2 = (v, f)_2 \qquad (v \in H_0^1(\Omega)) \tag{7.21}$$

since for every $v \in H_0^1(\Omega)$ we can take $\varphi_k \to v$ in $H^1(\Omega)$, so that $\varphi_k \to v$ and $\partial_j\varphi_k \to \partial_j u$ in $L^2(\Omega)$.

A **weak solution** of the Dirichlet problem is a function $u \in H_0^1(\Omega)$ such that $-\triangle u = f$ in the distributional sense or, equivalently, such that property (7.21) is satisfied.

Every classical solution is a weak solution, and we can look for weak solutions even for $f \in L^2(\Omega)$.

To prove the existence and uniqueness of such a weak solution, we will use the **Dirichlet norm** $\|\cdot\|_D$ on $H_0^1(\Omega)$, defined by

$$\|u\|_D^2 = \int_\Omega |\nabla u|^2 = \int_\Omega \sum_{j=1}^n |\partial_j u(x)|^2\, dx.$$

It is a true norm, associated to the scalar product

$$(u, v)_D := (\nabla u, \nabla v)_2,$$

and it is equivalent to the original one:

Lemma 7.36 (Poincaré). *There is a constant C depending on the bounded domain Ω such that*

$$\|u\|_2 \le C\|u\|_D \qquad (u \in H_0^1(\Omega)), \tag{7.22}$$

and on $H_0^1(\Omega)$ the Dirichlet norm $\|\cdot\|_D$ and the Sobolev norm $\|\cdot\|_{(1,2)}$ are equivalent.

Proof. Since $\mathcal{D}(\Omega)$ is dense in $H_0^1(\Omega)$, we only need to prove (7.22) for test functions $\varphi \in \mathcal{D}(\Omega) \subset \mathcal{D}(\mathbf{R}^n)$.

If $\Omega \subset [a, b]^n$, let us consider any $x = (x_1, x') \in \Omega$ and write

$$\varphi(x) = \int_a^{x_1} \partial_1\varphi(t, x')\, dt.$$

By the Schwarz inequality,

$$|\varphi(x)| \le (b-a)^{1/2}\Big(\int_a^b |\partial_1\varphi(t, x')|^2\, dt\Big)^{1/2}$$

and then, by Fubini's theorem,

$$\|\varphi\|_2^2 \le (b-a)\int_{[a,b]^n} dx \int_a^b |\partial_1\varphi(t, x')|^2\, dt = (b-a)^2\|\partial_1\varphi\|_2^2$$

with $|\partial_1\varphi| \le |\nabla\varphi|$, and (7.22) follows.

From this estimate,

$$\|u\|^2_{(1,2)} = \|u\|_2^2 + \||\nabla u|\|_2^2 \le (C^2+1)\|u\|_D^2$$

and obviously also $\|u\|_D \le \|u\|_{(1,2)}$. □

Theorem 7.37. *The Dirichlet problem (7.20) on the bounded domain Ω has a uniquely determined weak solution $u \in H_0^1(\Omega)$ for every $f \in L^2(\Omega)$, and the operator*

$$\Delta^{-1} : L^2(\Omega) \to H_0^1(\Omega)$$

is continuous.

Proof. If C is the constant that appears in the Poincaré lemma, then, by the Schwarz inequality,

$$|(v,f)_2| \le \|f\|_2\|v\|_2 \le C\|f\|_2\|v\|_D$$

and $(\cdot, f)_2 \in H_0^1(\Omega)'$ with $\|(\cdot,f)_2\| \le C\|f\|_2$. By the Riesz representation theorem, there is a uniquely determined function $u \in H_0^1(\Omega)$ such that $(v,f)_2 = (v,u)_D$ for all $v \in H_0^1(\Omega)$, which is property (7.21).

The estimate $\|u\|_D = \|(\cdot,f)_2\|_{H_0^1(\Omega)'} \le C\|f\|_2$ shows that $\|\Delta^{-1}\| \le C$. □

An application of (7.18) shows that a weak solution of class C^2 is also a classical solution if Ω is C^1:

Theorem 7.38. *Let $u \in \mathcal{C}^2(\Omega)\cap\mathcal{C}(\bar\Omega)$ and $f \in \mathcal{C}(\Omega)$. If u is a weak solution of the Dirichlet problem (7.20) and Ω is a C^1 domain, then u is a classical solution; that is, $-\Delta u(x) = f(x)$ for every $x \in \Omega$ and $u(x) = 0$ for every $x \in \partial\Omega$.*

Proof. By (7.18), $u_{|\partial\Omega} = 0$. Since $u \in \mathcal{C}^2(\Omega)$, the distribution Δu is the function $\Delta u(x)$ on Ω, and the distributional relation $-\Delta u = f$ is an identity of functions. □

We have not proved (7.18) if $n > 1$, and the proof of the regularity of the weak solutions is more delicate. For instance, if $f \in \mathcal{C}^\infty(\bar\Omega)$ and the boundary $\partial\Omega$ is C^∞, it can be shown that every weak solution u is also in $\mathcal{E}(\bar\Omega)$, so that it is also a classical solution. More precisely, the following result holds:

Theorem 7.39. *Let Ω be a bounded open set of $\mathbf{R}^n$ of class C^{m+2} with $m > n/2$ (or $\mathbf{R}^n$ or $\mathbf{R}^n_+ = \{x : x_n > 0\}$) and let $f \in H^m(\Omega)$. Then every weak solution u of the Dirichlet problem (7.20) belongs to $\mathcal{C}^2(\bar\Omega)$, and it is a classical solution.*

7.7.3. Eigenvalues and eigenfunctions of the Laplacian. We are going to apply the spectral theory for compact operators. The following result will be helpful:

Theorem 7.40. *Suppose that $\Phi \subset L^2(\mathbf{R}^n)$ satisfies the following three conditions:*

(a) *Φ is bounded in $L^2(\mathbf{R}^n)$,*

(b) *$\lim_{R\to\infty} \int_{|x|>R} |f(x)|^2\,dx = 0$ uniformly on $f \in \Phi$, and*

(c) *$\lim_{h\to 0} \|f - \tau_h f\|_2 = 0$ uniformly on $f \in \Phi$.*

Then the closure $\bar{\Phi}$ of Φ is compact in $L^2(\mathbf{R}^n)$.

Proof. Let $\varepsilon > 0$. By (b), we can choose $R > 0$ so that

$$\int_{|x|>R} |f(x)|^2\,dx \le \varepsilon^2 \qquad (f \in \Phi).$$

Choose $0 \le \varphi \in \mathcal{D}(B(0,1))$ with $\int \varphi = 1$, so that $\varphi_k(x) = k^n\varphi(kx)$ is a summability kernel on $\mathbf{R}^n$ such that $\operatorname{supp}\varphi_k \subset \bar{B}(0,1/k)$, and we know that $\lim_{k\to\infty} \|f * \varphi_k - f\|_2 = 0$ if $f \in L^2(\mathbf{R}^n)$. In fact, since $\varphi_k = 0$ on $|y| \ge 1/k$, it follows from the proof of Theorem 2.41 that

$$|(f * \varphi_k)(x) - f(x)| = \Big|\int_{|y|<1/k} [f(x-y) - f(y)]\varphi_k(y)\,dy\Big|$$

and then $\|f * \varphi_k - f\|_2 \le \sup_{|h|\le 1/k} \|\tau_h f - f\|_2$. Thus, by (c), we can choose N so that

$$\|f - f * \varphi_N\|_2 \le \varepsilon \qquad (f \in \Phi).$$

Moreover it follows very easily from the Schwarz inequality that

$$|(f * \varphi_N)(x) - (f * \varphi_N)(y)| \le \|\tau_{x-y} f - f\|_2 \|\varphi_N\|_2$$

and also

$$|(f * \varphi_N)(x)| \le \|f\|_2 \|\varphi_N\|_2.$$

These estimates, with conditions (a) and (c), allow us to apply the Ascoli-Arzelà theorem on $\bar{B}(0,R) \subset \mathbf{R}^n$ to the restrictions of the functions $f * \varphi_N$ with $f \in \Phi$, which can be covered by a finite family of balls in $\mathcal{C}(\bar{B}(0,R))$ with the centers in Φ,

$$B_{\mathcal{C}(\bar{B}(0,R))}(f_1,\delta), \ldots, B_{\mathcal{C}(\bar{B}(0,R))}(f_m,\delta),$$

for every $\delta > 0$.

Note that at every point $x \in \mathbf{R}^n$

$$\begin{aligned} |f(x) - f_j(x)| \le\ & \chi_{\{|x|>R\}}(x)|f(x)| + \chi_{\{|x|>R\}}(x)|f_j(x)| \\ & + |f(x) - (f * \varphi_n)(x)| + |f_j(x) - (f_j * \varphi_N)(x)| \\ & + \chi_{\{|x|\le R\}}(x)|(f * \varphi_N)(x) - (f_j * \varphi_N)(x)| \end{aligned}$$

and the previous estimates yield

$$\|f - f_j\|_2 \leq 4\varepsilon + |B(0,R)|^{1/2} \sup_{|x|\leq R} |(f * \varphi_N)(x) - (f_j * \varphi_N)(x)|,$$

so that, by choosing $\delta = \varepsilon/|B(0,R)|^{1/2}$, it follows that $\|f - f_j\|_2 \leq 5\varepsilon$ and, by Theorem 1.1, the closure of Φ in $L^2(\mathbf{R}^n)$ is compact. □

Remark 7.41. Obvious changes in the proof, such as applying Hölder's inequality instead of the Schwarz inequality, shows that the above theorem has an evident extension to $L^p(\mathbf{R}^n)$ if $1 \leq p < \infty$.

Theorem 7.42 (Rellich[15]). *If Ω is a bounded open set of $\mathbf{R}^n$, the natural inclusion $H_0^1(\Omega) \hookrightarrow L^2(\Omega)$ is compact.*

Proof. The extension by zero mapping $L^2(\Omega) \hookrightarrow L^2(\mathbf{R}^n)$ is isometric, so that it is sufficient to prove the compactness of the extension by zero map of Theorem 7.33, i.e.,

$$f \in H_0^1(\Omega) \mapsto \tilde{f} \in H^1(\mathbf{R}^n) = H_0^1(\mathbf{R}^n) \subset L^2(\mathbf{R}^n) \qquad (\tilde{f}(x) = 0 \text{ if } x \in \Omega^c).$$

This follows as an application of Theorem 7.40 when Φ is $\tilde{B} = \{\tilde{f};\, f \in B\}$, if B is the closed unit ball in $H_0^1(\Omega)$.

Indeed, $\tilde{B}$ is contained in the unit ball of $L^2(\mathbf{R}^n)$ and, if $\Omega \subset B(0,R)$, then $\int_{|x|>R} |\tilde{f}|^2 = 0$, so that conditions (a) and (b) of Theorem 7.40 are satisfied. To prove (c), note that

$$\|\tau_h u - u\|_2 \leq |h| \| |\nabla u| \|_2 \qquad (u \in H^1(\mathbf{R}^n)) \tag{7.23}$$

since we can consider $\varphi_k \in \mathcal{D}(\mathbf{R}^n)$ so that $\varphi_k \to u$ in $H^1(\mathbf{R}^n)$ as $k \to \infty$ and for every test function φ we have

$$|\tau_h\varphi(x) - \varphi(x)|^2 = \left| \int_0^1 h \cdot \nabla\varphi(x - th)\, dt \right|^2 \leq |h|^2 \int_0^1 |\nabla\varphi(x - th)|^2\, dt,$$

by the Schwarz inequality, and (7.23) follows for φ by integration.

Then $\|\tau_h\tilde{f} - \tilde{f}\|_2 \leq |h|$ for every $f \in B$, which is property (c) for $\tilde{B}$. □

Theorem 7.43. *If Ω is a bounded open set of $\mathbf{R}^n$, then $(-\triangle)^{-1}$ is a compact and injective self-adjoint operator on $L^2(\Omega)$ and on $H_0^1(\Omega)$.*

Proof. The compactness of $(-\triangle)^{-1} : L^2(\Omega) \to L^2(\Omega)$ follows by considering the decomposition

$$(-\triangle)^{-1} : L^2(\Omega) \to H_0^1(\Omega) \hookrightarrow L^2(\Omega),$$

[15]The theorem, in this case $p = 2$, is attributed to the South-Tyrolian mathematician Franz Rellich (1930 in Göttingen) and to Vladimir Kondrachov (1945) for the more general case stating that $W_0^{1,p}(\Omega)$ is compactly embedded in $L^q(\Omega)$ for any $q < np/(n-p)$ if $p < n$, and in $\mathcal{C}(\bar{\Omega})$ if $p > n$. For a proof we refer the reader to Gilbarg and Trudinger **[17]** and Brezis **[5]**.

where $(-\triangle)^{-1} : L^2(\Omega) \to H_0^1(\Omega)$ is continuous by Theorem 7.37 and, by the Rellich Theorem 7.42, $H_0^1(\Omega) \hookrightarrow L^2(\Omega)$ is compact.

Similarly,

$$(-\triangle)^{-1} : H_0^1(\Omega) \hookrightarrow L^2(\Omega) \overset{(-\triangle)^{-1}}{\longrightarrow} H_0^1(\Omega)$$

is also compact.

Note that $(-\triangle)^{-1} : L^2(\Omega) \to H_0^1(\Omega)$ is bijective, by Theorem 7.37.

If $u = (-\triangle)^{-1}\varphi$ and $v = (-\triangle)^{-1}\psi$, with $\varphi, \psi \in \mathcal{D}(\Omega)$, then

$$(u, v)_D = (\nabla u, \nabla v)_2 = (-\triangle u, v)_2 = (\varphi, v)_2,$$

so that

$$((-\triangle)^{-1}\varphi, \psi)_D = (\varphi, (-\triangle)^{-1}\psi)_D$$

and

$$((-\triangle)^{-1}u, v)_D = (u, (-\triangle)^{-1}v)_D$$

for all $u, v \in H_0^1(\Omega)$ by continuity. Also

$$((-\triangle)^{-1}\varphi, \psi)_2 = (\nabla\varphi, \nabla\psi)_2 = (\varphi, \psi)_D = (\varphi, (-\triangle)^{-1}\psi)_2.$$

This shows that $(-\triangle)^{-1}$ is self-adjoint on $H_0^1(\Omega)$ and on $L^2(\Omega)$. □

Note that

$$((-\triangle)^{-1}u, v)_D = (u, v)_2 \qquad (u, v \in H_0^1(\Omega)), \tag{7.24}$$

from the density of $\mathcal{D}(\Omega)$ in $H_0^1(\Omega)$. Thus $(-\triangle)^{-1}$ is a positive operator on $H_0^1(\Omega)$ in the sense that

$$((-\triangle)^{-1}u, u)_D > 0 \tag{7.25}$$

if $0 \neq u \in H_0^1(\Omega)$.

An eigenfunction for the Laplacian on $H_0^1(\Omega)$ is an element $u \in H_0^1(\Omega)$ such that $\triangle u = \lambda u$ for some λ, which is said to be an eigenvalue of $\triangle$ if there exists some nonzero eigenfunction u such that $\triangle u = \lambda u$.

Hence 0 is not an eigenvalue, since $\triangle : H_0^1(\Omega) \to L^2(\Omega)$ is injective, and $u \in H_0^1(\Omega)$ is an eigenfunction for the eigenvalue λ of $\triangle$ if and only if

$$(-\triangle)^{-1}u = -\frac{1}{\lambda}u.$$

The solutions of this equation form the eigenspace for this eigenvalue λ. Note that $\lambda < 0$, as a consequence of the positivity property (7.25) of $(-\triangle)^{-1}$.

From the spectral theory of compact self-adjoint operators,

$$(-\triangle)^{-1} : H_0^1(\Omega) \to H_0^1(\Omega)$$

has a spectral representation

$$(7.26)\qquad (-\triangle)^{-1}v = \sum_{k=0}^{\infty} \mu_k (v, u_k)_D u_k \qquad (v \in H_0^1(\Omega)),$$

a convergent series in $H_0^1(\Omega)$, where $\mu_k = -1/\lambda_k \downarrow 0$ is the sequence of the eigenvalues of $(-\triangle)^{-1}$ and $\{u_k\}_{k=0}^{\infty}$ is an orthonormal system in $H_0^1(\Omega)$ with respect to $(\cdot,\cdot)_D$ such that $(-\triangle)^{-1}u_k = \mu_k u_k$. Since $(-\triangle)^{-1}$ is injective, $\{u_k\}_{k=0}^{\infty}$ is a basis in $H_0^1(\Omega)$. Moreover $(u_k, u_j)_2 = 0$ if $k \neq j$, by (7.24).

Theorem 7.44. *Suppose $f \in L^2(\Omega)$. The weak solution of the Dirichlet problem*

$$-\triangle u = f \quad (u \in H_0^1(\Omega))$$

is given by the sum

$$u = -\sum_{k=0}^{\infty} (f, u_k)_2 u_k$$

in $H_0^1(\Omega)$. The sequence of eigenfunctions $\sqrt{-\lambda_k}u_k$ is an orthonormal basis of $L^2(\Omega)$.

Proof. It follows from (7.24) applied to the elements $u_k \in H_0^1(\Omega)$ that

$$\|u_k\|_2^2 = \mu_k = -1/\lambda_k, \qquad (u_k, u_m)_2 = 0 \ \text{ if } m \neq k.$$

Moreover, since $H_0^1(\Omega)$ contains $\mathcal{D}(\Omega)$, it is densely and continuously included in $L^2(\Omega)$, $\{u_k\}_{k=0}^{\infty}$ is total in $L^2(\Omega)$, and the orthonormal system $\{\sqrt{-\lambda_k}u_k\}_{k=0}^{\infty}$ is complete in $L^2(\Omega)$.

For every $f \in L^2(\Omega)$,

$$(7.27)\qquad f = \sum_{k=0}^{\infty} (f, \sqrt{-\lambda_k}u_k)_2 \sqrt{-\lambda_k}u_k = -\sum_{k=0}^{\infty} \lambda_k (f, u_k)_2 u_k$$

in $L^2(\Omega)$ and $\{\sqrt{-\lambda_k}(f, u_k)_2\}_{k=0}^{\infty} \in \ell^2$. Also $\{(f, u_k)_2\}_{k=0}^{\infty} \in \ell^2$, since $\sqrt{-\lambda_k} \to \infty$.

We can define

$$u = \sum_{k=0}^{\infty} (f, u_k)_2 u_k$$

since the series converges in $H_0^1(\Omega)$. Then

$$\triangle u = \sum_{k=0}^{\infty} (f, u_k)_2 \triangle u_k = \sum_{k=0}^{\infty} \lambda_k (f, u_k)_2 u_k$$

in $\mathcal{D}'(\Omega)$. In (7.27) we have a sum in $\mathcal{D}'(\Omega)$, so that $-\triangle u = f$. □

To complete our discussion, we want to show that the eigenfunctions u_k are in $\mathcal{E}(\Omega)$, so that they are classical solutions of $-\Delta u_k = \lambda_k u_k$. The method we are going to use is easily extended to any elliptic linear differential operator L with constant coefficients.

Theorem 7.45. *Suppose $L = \Delta + \lambda$ and $u \in \mathcal{D}'(\Omega)$, where Ω is a nonempty open set in $\mathbf{R}^n$. If $Lu \in \mathcal{E}(\Omega)$, then $u \in \mathcal{E}(\Omega)$.*

Proof. If $Lu \in \mathcal{E}(\Omega)$, then

$$\varphi Lu \in H^s(\mathbf{R}^n) \quad \forall \varphi \in \mathcal{D}(\Omega) \tag{7.28}$$

for every $s \in \mathbf{R}$. We claim that it follows from (7.28) that

$$\varphi u \in H^{s+2}(\mathbf{R}^n) \quad \forall \varphi \in \mathcal{D}(\Omega).$$

Then an application of Theorem 7.29 shows that $\varphi u \in \mathcal{E}(\Omega)$ for every $\varphi \in \mathcal{D}(\Omega)$ and then $u \in \mathcal{E}(\Omega)$.

To prove this claim, let $\operatorname{supp} \varphi \subset U$, U an open set with compact closure $\bar{U}$ in Ω, and choose $\bar{U} \prec \psi \prec \Omega$. Note that $\psi u \in \mathcal{E}'(\mathbf{R}^n)$ and it follows from Theorem 7.16 that $\psi u \in H^t(\mathbf{R}^n)$. By decreasing t if necessary, we can suppose that $s + 2 - t = k \in \mathbf{N}$.

Let $\psi_0 = \psi$, $\psi_k = \varphi$ and define $\psi_1, \dots, \psi_{k-1}$ by recurrence so that

$$\operatorname{supp} \psi_{j+1} \prec \psi_j \prec U_j \subset \{\psi_{j-1} = 1\}.$$

It is sufficient to show that $\psi_j u \in H^{t+j}(\mathbf{R}^n)$, since then $\varphi u = \psi_k u \in H^{t+k}(\mathbf{R}^n) = H^{s+2}(\mathbf{R}^n)$ will complete the proof.

We only need to prove that if $\varphi, \psi \in \mathcal{D}(\Omega)$ are such that

$$\operatorname{supp} \varphi \prec \psi \text{ and } \psi u \in H^t(\mathbf{R}^n),$$

then $\varphi u \in H^{t+1}(\mathbf{R}^n)$.

From the definition of L and from the condition $\operatorname{supp} \varphi \prec \psi$,

$$[L, \varphi]u = L(\varphi u) - \varphi Lu = \sum_{j=1}^{n} \left((\partial_j^2 \varphi)u + 2(\partial_j \varphi)\partial_j u\right)$$

is a differential operator of order 1 with smooth coefficients and satisfies $[L, \varphi]u = [L, \varphi](\psi u)$. Hence $L(\varphi u) = [L, \varphi](\psi u) + \varphi Lu$ with $[L, \varphi](\psi u) \in H^{t+1}(\mathbf{R}^n)$ and $\varphi Lu \in \mathcal{D}(\mathbf{R}^n)$, and we conclude that

$$L(\varphi u) \in H^{t+1}(\mathbf{R}^n).$$

Therefore also

$$(\Delta - 1)(\varphi u) = L(\varphi u) - (\lambda - 1)\varphi u \in H^{t+1}(\mathbf{R}^n)$$

and, since $\Delta - 1$ is a bijective operator from $H^{t-1}(\mathbf{R}^n)$ to $H^{t+1}(\mathbf{R}^n)$, we conclude that $\varphi u \in H^{t-1}(\mathbf{R}^n)$. $\square$

For a more complete analysis concerning the Dirichlet problem we refer the reader to Brezis, "Analyse fonctionelle" [**5**], and Folland, "Introduction to Partial Differential Equations" [**14**].

7.8. Exercises

Exercise 7.1. Calculate the Fourier integral of the following functions on $\mathbf{R}$:

(a) $f_1(t) = te^{-t^2}$.

(b) $f_2(t) = \chi_{(a,b)}$.

(c) $f_3(t) = e^{-|t|}$.

(d) $f_4(t) = (1+t^2)^{-1}$.

Exercise 7.2. Assuming that $1 \le p \le \infty$ and $f \in L^p(\mathbf{R}^n)$, check that the Gauss-Weierstrass kernel

$$W_t(x) := \frac{1}{(4\pi t)^{n/2}} e^{-|x|^2/4t} \qquad (h > 0)$$

is a summability kernel in $\mathcal{S}(\mathbf{R}^n)$, and prove that

$$u(t,x) := \frac{1}{(4\pi t)^{n/2}} \int_{\mathbf{R}^n} e^{-|y|^2/4t} f(x-y)\, dy \qquad (t > 0)$$

defines a solution of the heat equation

$$\partial_t u - \Delta u = 0$$

on $(0,\infty) \times \mathbf{R}^n$.

If $p < \infty$, show that $\lim_{t\downarrow 0} u(t,\cdot) = f$ in $L^p(\mathbf{R}^n)$. If f is bounded and continuous, prove that u has an extension to a continuous function on $[0,\infty) \times \mathbf{R}^n$ such that $u(0,x) = f(x)$ for all $x \in \mathbf{R}^n$.

Exercise 7.3. The heat flow in an infinitely long road, given an initial temperature f, is described as the solution of the problem

$$\partial_t u(x,t) = \partial_x^2 u(x,t), \quad u(x,0) = f(x).$$

Prove that if $f \in \mathcal{C}_0(\mathbf{R})$ is integrable, then the unique bounded classical solution is

$$u(x,t) = \int_{\mathbf{R}} \widehat{f}(\xi) e^{-4\pi^2\xi^2 t} e^{2\pi i \xi x}\, d\xi = (f * K_t)(x)$$

where

$$K_t(x) = \frac{1}{(4\pi t)^{1/2}} e^{-\frac{x^2}{4t}}.$$

Exercise 7.4. Find the norm of the Fourier transform $\mathcal{F} : L^1(\mathbf{R}^n) \to L^\infty(\mathbf{R}^n)$.

Exercise 7.5. Is it true that $f, g \in L^1(\mathbf{R})$ and $f * g = 0$ imply either $f = 0$ or $g = 0$?

Exercise 7.6. Let $1 \leq p \leq \infty$. Prove that $\varphi \in \mathcal{E}(\mathbf{R}^n)$ is in $\mathcal{S}(\mathbf{R}^n)$ if and only if the functions $x^\beta D^\alpha \varphi(x)$ are all in $L^p(\mathbf{R}^n)$. Show that the inclusion $\mathcal{S}(\mathbf{R}^n) \hookrightarrow L^p(\mathbf{R}^n)$ is continuous.

Exercise 7.7. Show that if a rational function f belongs to $\mathcal{S}(\mathbf{R}^n)$, then $f = 0$.

Exercise 7.8. If f is a function on $\mathbf{R}^n$ such that $f\varphi \in \mathcal{S}(\mathbf{R}^n)$ for all $\varphi \in \mathcal{S}(\mathbf{R}^n)$, prove that the pointwise multiplication $f\cdot$ is a continuous linear operator on $\mathcal{S}(\mathbf{R}^n)$.

Exercise 7.9. Suppose $f \in \mathcal{S}(\mathbf{R})$ and $\widehat{f} \in \mathcal{D}_{[-R,R]}$, and let $1 \leq p < \infty$. Prove that there is a sequence of constants $C_n \geq 0$ $(n = 0, 1, 2, \ldots)$, which depend only on p, such that

$$\|f^{(n)}\|_p \leq C_n R^n \|f\|_p \qquad (n \in \mathbf{N}).$$

Exercise 7.10 (Hausdorff-Young). Show that, if $1 \leq p \leq 2$ and $f \in L^p(\mathbf{R}^n)$, then $\widehat{f} \in L^\infty(\mathbf{R}^n) + L^2(\mathbf{R}^n)$ and $\mathcal{F} : L^p(\mathbf{R}^n) \to L^{p'}(\mathbf{R}^n)$.

Exercise 7.11. Show that $\operatorname{sinc} \in L^2(\mathbf{R}) \backslash L^1(\mathbf{R})$, and find $\|\operatorname{sinc}\|_2$, $\mathcal{F}(\operatorname{sinc})$, and $\widetilde{\mathcal{F}}(\operatorname{sinc})$.

Exercise 7.12. If u is a solution of the Dirichlet problem (7.11) on a half-plane, find another solution by adding to u an appropriate harmonic function.

Exercise 7.13. Show that $\ell^p \subset \ell$ $(1 \leq p \leq \infty)$ and that the injective linear map $\ell^p \hookrightarrow \mathcal{S}'$ such that $x[k] \mapsto \sum_{k=-\infty}^{\infty} x[k]\delta_k(t)$ is continuous.

Exercise 7.14 (Hausdorff-Young). Show that $(\sum_{k=-\infty}^{\infty} |c_k(f)|^{p'})^{1/p'} \leq \|f\|_p$, with the usual change if $p' = \infty$, if $1 \leq p \leq 2$ and $f \in L^p(\mathbf{T})$.

Exercise 7.15 (Poisson summation formula). Prove the following facts:

(a) If $\varphi \in \mathcal{S}(\mathbf{R})$, $\varphi_1(t) = \sum_{k=-\infty}^{+\infty} \varphi(t-k)$ is uniformly convergent.

(b) The Fourier series of φ_1 is also uniformly convergent.

(c) $\sum_{k=-\infty}^{+\infty} \varphi(k) = \sum_{k=-\infty}^{+\infty} \widehat{\varphi}(k)$, with absolute convergence.

(d) For the Dirac comb, $\widehat{\text{Ш}} = \text{Ш}$.

Exercise 7.16. If $1 \leq q \leq \infty$, prove that

$$\|f\| := \|\{\|D^\alpha f\|_p\}_{|\alpha| \leq m}\|_q$$

defines on $W^{p,m}(\Omega)$ a norm which is equivalent to $\|\cdot\|_{(m,p)}$.

Exercise 7.17. Every $u \in W^{1,p}(a, \infty)$ is uniformly continuous.

Exercise 7.18. For a half-line (a, ∞) prove a similar result to Theorem 7.25, now about the continuity of $W^{1,p}(a,\infty) \hookrightarrow \mathcal{C}[a,\infty) \cap L^\infty(a,\infty)$.

Exercise 7.19. Every $u \in W^{1,p}(0,\infty)$ can be extended to $Ru \in W^{1,p}(\mathbf{R})$ so that $Ru(t) = u(-t)$ if $t < 0$ and $Ru(x) - u(0) = \int_0^x v(t)\,dt$ for all $x \in \mathbf{R}$. Moreover, $R : W^{1,p}(0,\infty) \to W^{1,p}(\mathbf{R})$ is linear and continuous.

Exercise 7.20. The extension by zero, $P : H_0^1(\Omega) \to H^1(\mathbf{R}^n)$, is a continuous operator.

Exercise 7.21. If $\partial\Omega$ has zero measure, the extension by zero operator, P, satisfies $\partial_j Pu = P\partial_j u$ for every $u \in H_0^1(\Omega)$.

Exercise 7.22. If $u \in H^1(-1,1)$, its extension by zero, u^o, is not always in $H^1(\mathbf{R})$.

Exercise 7.23. If $s - k > n/2$ ($s \in \mathbf{R}$) and $m - k > n/2$ ($m \in \mathbf{N}$), prove that the inclusions $H^s(\mathbf{R}^n) \hookrightarrow \mathcal{E}^k(\mathbf{R}^n)$ and $H^m(\Omega) \hookrightarrow \mathcal{E}^k(\Omega)$ of the Sobolev Theorem 7.28 are continuous.

Exercise 7.24. Let $u(x) = e^{-2\pi|x|}$ as in Example 7.12. Prove the following facts:

(a) $u, u' \in L^2(\mathbf{R})$ (distributional derivative), and $u \in H^s(\mathbf{R})$ if $s < 3/2$.

(b) $u \notin H^{3/2}(\mathbf{R})$.

Exercise 7.25. Prove that the Dirichlet problem

$$-\Delta u + u = f, \quad u = 0 \text{ on } \partial\Omega \quad (f \in L^2(\Omega))$$

has a unique weak solution by applying the Lax-Milgram theorem to the sesquilinear form

$$B(u,v) := \int_\Omega \big(\nabla u(x) \cdot \nabla \bar{v}(x) + u(x)\bar{v}(x)\big)\,dx$$

on $H_0^1(\Omega) \times H_0^1(\Omega)$.

References for further reading:

R. A. Adams, *Sobolev Spaces.*

H. Brezis, *Analyse fonctionelle: Théorie et applications.*

G. B. Folland, *Introduction to Partial Differential Equations.*

I. M. Gelfand and G. E. Chilov, *Generalized Functions.*

D. Gilbarg and N. S. Trudinger, *Elliptic Partial Differential Equations of Second Order.*

L. Hörmander, *Linear Partial Differential Operators.*

E. H. Lieb and M. Loss, *Analysis.*

V. Mazya, *Sobolev Spaces.*

W. Rudin, *Functional Analysis.*

E. M. Stein and G. Weiss, *Introduction to Fourier Analysis on Euclidean Spaces.*

Chapter 8

Banach algebras

Some important Banach spaces are equipped in a natural way with a continuous product that determines a Banach algebra structure.[1] Two basic examples are $\mathcal{C}(K)$ with the pointwise multiplication and $\mathcal{L}(E)$ with the product of operators if E is a Banach space. It can be useful for the reader to retain $\mathcal{C}(K)$ as a simple reference model.

The first work devoted to concrete Banach algebras is contained in some papers by J. von Neumann and beginning in 1930. The advantage of considering algebras of operators was clear in his contributions, but it was the abstract setting of Banach algebras which proved to be convenient and which allowed the application of similar ideas in many directions.

The main operator on these algebras is the Gelfand transform[2] $\mathcal{G} : a \mapsto \widehat{a}$, which maps a unitary commutative Banach algebra A on $\mathbf{C}$ to the space $\mathcal{C}(\Delta)$ of all complex continuous functions on the spectrum Δ of A, which is the set of all nonzero elements $\chi \in A'$ that are multiplicative. Here Δ is endowed with the restriction of the w^*-topology and it is compact. As seen in Example 8.14, Δ is the set of all the evaluations δ_t $(t \in K)$ if $A = \mathcal{C}(K)$, and $\widehat{f}(\delta_t) = \delta_t(f) = f(t)$, so that in this case one can consider $\widehat{f} = f$.

But we will be concerned with the spectral theory of operators in a complex Hilbert space H. If T is a bounded normal operator in H, so that

---

[1] Banach algebras were first introduced in 1936 with the name of "linear metric rings" by the Japanese mathematician Mitio Nagumo. He extended Cauchy's function theory to the functions with values in such an algebra to study the resolvent of an operator. They were renamed "Banach algebras" by Charles E. Rickart in 1946.

[2] Named after the Ukrainian mathematician Israel Moiseevich Gelfand, who is considered the creator, in 1941, of the theory of commutative Banach algebras. Gelfand and his colleagues created this theory which included the spectral theory of operators and proved to be an appropriate setting for harmonic analysis.

T and the adjoint T^* commute, then the closed Banach subalgebra $A = \langle T \rangle$ of $\mathcal{L}(H)$ generated by I, T, and T^* is commutative.

It turns out that the Gelfand theory of commutative Banach algebras is especially well suited in this setting. Through the change of variables $z = \widehat{T}(\chi)$ one can consider $\sigma(T) \equiv \Delta$, and the Gelfand transform is a bijective mapping that allows us to define a functional calculus $g(T)$ by $\widehat{g(T)} = g(\widehat{T})$ if g is a continuous function on the spectrum of T.

For this continuous functional calculus there is a unique operator-valued measure E on $\sigma(T)$ such that

$$g(T) = \int_{\sigma(T)} g(\lambda)\, dE(\lambda),$$

and the functional calculus is extended by

$$f(T) = \int_{\sigma(T)} f(\lambda)\, dE(\lambda)$$

to bounded measurable functions f.

The Gelfand transform, as a kind of abstract Fourier operator, is also a useful tool in harmonic analysis and in function theory. The proof of Wiener's 1932 lemma contained in Exercise 8.15 is a nice unexpected application discovered by Gelfand in 1941, and generalizations of many theorems of Tauberian type and applications to the theory of locally compact groups have also been obtained with Gelfand's methods. We refer the reader to the book by I. M. Gelfand, D. A. Raikov and G. E. Chilov [**16**] for more information.

8.1. Definition and examples

We say that A is a complex Banach algebra or, simply, a **Banach algebra** if it is a complex Banach space with a bilinear multiplication and the norm satisfies

$$\|xy\| \leq \|x\|\|y\|,$$

so that the multiplication is continuous since, if $(x_n, y_n) \to (x, y)$, then

$$\|xy - x_n y_n\| \leq \|x\|\|y - y_n\| + \|x - x_n\|\|y_n\| \to 0.$$

Real Banach algebras are defined similarly.

The Banach algebra A is said to be **unitary** if it has a unit, which is an element e such that $xe = ex = x$ for all $x \in A$ and $\|e\| = 1$. This unit is unique since, if also $e'x = xe' = x$, then $e = ee' = e'$.

We will only consider unitary Banach algebras. As a matter of fact, every Banach algebra can be embedded in a unitary Banach algebra, as shown in Exercise 8.1.

Example 8.1. (a) If X is a nonempty set, $\mathbf{B}(X)$ will denote the unitary Banach algebra of all complex bounded functions on X, with the pointwise multiplication and the uniform norm $\|f\|_X := \sup_{x\in X} |f(x)|$. The unit is the constant function 1.

(b) If K is a compact topological space, then $\mathcal{C}(K)$ is the closed subalgebra of $\mathbf{B}(K)$ that contains all the continuous complex functions on K. It is a unitary Banach subalgebra of $\mathbf{B}(K)$, since $1 \in \mathcal{C}(K)$.

(c) The **disc algebra** is the unitary Banach subalgebra $A(D)$ of $\mathcal{C}(\bar{D})$. Since the uniform limits of analytic functions are also analytic, $A(D)$ is closed in $\mathcal{C}(\bar{D})$.

Example 8.2. If Ω is a σ-finite measure space, $L^\infty(\Omega)$ denotes the unitary Banach algebra of all measurable complex functions on Ω with the usual norm $\|\cdot\|_\infty$ of the essential supremum. As usual, two functions are considered equivalent when they are equal a.e.

Example 8.3. Let E be any nonzero complex Banach space. The Banach space $\mathcal{L}(E) = \mathcal{L}(E; E)$ of all bounded linear operators on E, endowed with the usual product of operators, is a unitary Banach algebra. The unit is the identity map I.

8.2. Spectrum

Throughout this section, A denotes a unitary Banach algebra, possibly not commutative. An example is $\mathcal{L}(E)$, if E is a complex Banach space.

A **homomorphism** between A and a second unitary Banach algebra B is a homomorphism of algebras $\Psi : A \to B$ such that $\Psi(e) = e$ if e denotes the unit both in A and in B.

The notion of the spectrum of an operator is extended to any element of A:

The **spectrum** of $a \in A$ is the subset of $\mathbf{C}$

$$\sigma_A(a) = \sigma(a) := \{\lambda \in \mathbf{C};\ \lambda e - a \notin G(A)\},$$

where $G(A)$ denotes the multiplicative group of all invertible elements of A.

Note that, if B is a unitary Banach subalgebra of A and $b \in B$, an inverse of $\lambda e - b$ in B is also an inverse in A, so that $\sigma_A(b) \subset \sigma_B(b)$.

Example 8.4. If E is a complex Banach space and $T \in \mathcal{L}(E)$, we denote $\sigma(T) = \sigma_{\mathcal{L}(E)}(T)$. Thus, $\lambda \in \sigma(T)$ if and only if $T - \lambda I$ is not bijective, by the Banach-Schauder theorem. Recall that the eigenvalues of T, and also the approximate eigenvalues, are in $\sigma(T)$. Cf. Subsection 4.4.2.

Example 8.5. If E is an infinite-dimensional Banach space and $T \in \mathcal{L}(E)$ is compact, the Riesz-Fredholm theory shows that $\sigma(T) \setminus \{0\}$ can be arranged in a sequence of nonzero eigenvalues (possibly finite), all of them with finite multiplicity, and $0 \in \sigma(T)$, by the Banach-Schauder theorem.

Example 8.6. The spectrum of an element f of the Banach algebra $\mathcal{C}(K)$ is its image $f(K)$.

Indeed, the continuous function $f - \lambda$ has an inverse if it has no zeros, that is, if $f(t) \neq \lambda$ for all $t \in K$. Hence, $\lambda \in \sigma(f)$ if and only if $\lambda \in f(K)$.

Let us consider again a general unitary Banach algebra A.

Theorem 8.7. *If $p(\lambda) = \sum_{n=0}^{N} c_n \lambda^n$ is a polynomial and $a \in A$, then $\sigma(p(a)) = p(\sigma(a))$.*

Proof. We assume that $p(a) = c_0 e + c_1 a + \cdots + c_N a^N$, and we exclude the trivial case of a constant polynomial $p(\lambda) \equiv c_0$.

For a given $\mu \in \mathbf{C}$, by division we obtain $p(\mu) - p(\lambda) = (\mu - \lambda) q(\lambda)$ and $p(\mu)e - p(a) = (\mu e - a) q(a)$. If $\mu e - a \notin G(A)$, then also $p(\mu)e - p(a) \notin G(A)$. Hence, $p(\sigma(a)) \subset \sigma(p(a))$.

Conversely, if $\mu \in \sigma(p(a))$, by factorization we can write

$$\mu - p(\lambda) = \alpha(\lambda_1 - \lambda) \cdot \ldots \cdot (\lambda_N - \lambda)$$

with $\alpha \neq 0$. Then $\mu e - p(a) = \alpha(\lambda_1 e - a) \cdot \ldots \cdot (\lambda_N e - a)$, where $\mu e - p(a) \notin G(A)$, so that $\lambda_i e - a \notin G(A)$ for some $1 \leq i \leq N$. Thus, $\lambda_i \in \sigma(a)$ and we have $p(\lambda_i) = \mu$, which means that $\mu \in p(\sigma(a))$. □

The **resolvent** of an element $a \in A$ is the function $R_a : \sigma(a)^c \to A$ such that $R_a(\lambda) = (\lambda e - a)^{-1}$. It plays an important role in spectral theory.

Note that, if $\lambda \neq 0$,

$$R_a(\lambda) = -(a - \lambda e)^{-1} = \lambda^{-1}(e - \lambda^{-1} a)^{-1}.$$

To study the basic properties of R_a, we will use some facts from function theory.

As in the numerical case and with the same proofs, a vector-valued function $F : \Omega \to A$ on an open subset Ω of $\mathbf{C}$ is said to be analytic or holomorphic if every point $z_0 \in \Omega$ has a neighborhood where F is the sum of a convergent power series:

$$F(z) = \sum_{n=0}^{\infty} (z - z_0)^n a_n \qquad (a_n \in A).$$

The series is absolutely convergent at every point of the convergence disc, which is the open disc in $\mathbf{C}$ with center z_0 and radius

$$R = \frac{1}{\limsup_n \|a_n\|^{1/n}} > 0.$$

The Cauchy theory remains true without any change in this setting, and F is analytic if and only if, for every $z \in \Omega$, the complex derivative

$$F'(z) = \lim_{h\to 0} \frac{F(z+h) - F(z)}{h}$$

exists.

We will show that $\sigma(a)$ is closed and bounded and, to prove that R_a is analytic on $\sigma(a)^c$, we will see that $\mathbf{R}'_a(\lambda)$ exists whenever $\lambda \notin \sigma(a)$.

Let us first show that R_a is analytic on $|\lambda| > \|a\|$.

Theorem 8.8. (a) *If* $\|a\| < 1$, $e - a \in G(A)$ *and*

$$(e-a)^{-1} = \sum_{n=0}^{\infty} a^n \qquad (a^0 := e).$$

(b) *If* $|\lambda| > \|a\|$, *then* $\lambda \notin \sigma(a)$ *and*

$$R_a(\lambda) = \sum_{n=0}^{\infty} \lambda^{-n-1} a^n.$$

(c) *Moreover,*

$$\|R_a(\lambda)\| \leq \frac{1}{|\lambda| - \|a\|},$$

and $\lim_{|\lambda|\to\infty} R_a(\lambda) = 0$.

Proof. (a) As in (2.7), the Neumann series $\sum_{n=0}^{\infty} a^n$ is absolutely convergent ($\|a^m\| \leq \|a\|^m$ and $\|a\| < 1$), so that $z = \sum_{n=0}^{\infty} a^n \in A$ exists, and it is easy to check that z is the right and left inverse of $e - a$. For instance,

$$\lim_{N\to\infty} (e-a) \sum_{n=0}^{N} a^n = (e-a)z,$$

since the multiplication by $e - a$ is linear and continuous, so that

$$(e-a)\sum_{n=0}^{N} a^n = \sum_{n=0}^{N} a^n - \sum_{n=1}^{N+1} a^n = e - a^{N+1} \to e \quad \text{if } N \to \infty.$$

(b) Note that

$$R_a(\lambda) = \lambda^{-1}(e - \lambda^{-1}a)^{-1}$$

and, if $\|\lambda^{-1}a\| < 1$, we obtain the announced expansion from (a).

(c) Finally,

$$\|R_a(\lambda)\| = |\lambda|^{-1}\|\sum_{n=0}^{\infty} \lambda^{-n}a^n\| \leq \frac{1}{|\lambda| - \|a\|}.$$

□

The **spectral radius** of $a \in A$ is the number

$$r(a) := \sup\{|\lambda|;\ \lambda \in \sigma(a)\}.$$

From Theorem 8.8 we have that $r(a) \leq \|a\|$, an inequality that can be strict.

The following estimates are useful.

Lemma 8.9. (a) *If* $\|a\| < 1$,

$$\|(e-a)^{-1} - e - a\| \leq \frac{\|a\|^2}{1 - \|a\|}.$$

(b) *If* $x \in G(A)$ *and* $\|h\| < 1/(2\|x^{-1}\|)$, *then* $x + h \in G(A)$ *and*

$$\|(x+h)^{-1} - x^{-1} + x^{-1}hx^{-1}\| \leq 2\|x^{-1}\|^3\|h\|^2.$$

Proof. To check (a), we only need to sum the right-hand side series in

$$\|(e-a)^{-1} - e - a\| = \|\sum_{n=2}^{\infty} a^n\| \leq \sum_{n=2}^{\infty} \|a\|^n.$$

To prove (b) note that $x + h = x(e + x^{-1}h)$, and we have

$$\|x^{-1}h\| \leq \|x^{-1}\|\|h\| < 1/2.$$

If we apply (a) to $a = -x^{-1}h$, since $\|a\| < 1/2$, we obtain that $x+h \in G(A)$, and

$$\|(x+h)^{-1} - x^{-1} + x^{-1}hx^{-1}\| \leq \|(e-a)^{-1} - e - a\|\|x^{-1}\|$$

with $\|(e-a)^{-1} - e - a\| \leq \|x^{-1}h\|^2/(1 - \|a\|) \leq 2\|x^{-1}h\|^2$. □

Theorem 8.10. (a) $G(A)$ *is an open subset of* A *and* $x \in G(A) \mapsto x^{-1} \in G(A)$ *is continuous.*

(b) R_a *is analytic on* $\sigma(a)^c$ *and zero at infinity.*

(c) $\sigma(a)$ *is a nonempty subset of* **C** *and*[3]

$$r(a) = \lim_{n\to\infty} \|a^n\|^{1/n} = \inf_n \|a^n\|^{1/n}.$$

[3]This spectral radius formula and the analysis of the resolvent have a precedent in the study by Angus E. Taylor (1938) of operators which depend analytically on a parameter. This formula was included in the 1941 paper by I. Gelfand on general Banach algebras.

Proof. (a) According to Lemma 8.9(b), for every $x \in G(A)$,

$$B\big(x, \frac{1}{2\|x^{-1}\|}\big) \subset G(A)$$

and $G(A)$ is an open subset of A.

Moreover

$$\|(x+h)^{-1} - x^{-1}\| \le \|(x+h)^{-1} - x^{-1} + x^{-1}hx^{-1}\| + \|x^{-1}hx^{-1}\| \to 0$$

if $\|h\| \to 0$, and $x \in G(A) \mapsto x^{-1} \in G(A)$ is continuous.

(b) On $\sigma(a)^c$,

$$R_a'(\lambda) = \lim_{\mu \to 0} \mu^{-1}[((\lambda + \mu)e - a)^{-1} - (\lambda e - a)^{-1}] = -R_a(\lambda)^2$$

follows from an application of Lemma 8.9(b) to $x = \lambda e - a$ and $h = \mu e$. In this case $x^{-1}hx^{-1} = \mu x^{-1}x^{-1}$ and, writing $x^{-2} = x^{-1}x^{-1}$, we obtain

$$\mu^{-1}[(x+\mu e)^{-1} - x^{-1}] = \mu^{-1}[(x+\mu e)^{-1} - x^{-1} + x^{-1}hx^{-1}] - x^{-2} \to x^{-2}$$

as $\mu \to 0$, since

$$\|\mu^{-1}[(x+\mu e)^{-1} - x^{-1} + x^{-1}hx^{-1}]\| \le |\mu|^{-1} 2\|x^{-1}\|^3 |\mu|^2 \to 0.$$

By Theorem 8.8(b), $\|R_a(\lambda)\| \le 1/(|\lambda| - \|a\|) \to 0$ if $|\lambda| \to \infty$.

(c) Recall that $\sigma(a) \subset \{\lambda;\, |\lambda| \le r(a)\}$ and $r(a) \le \|a\|$. This set is closed in $\mathbf{C}$, since $\sigma(a)^c = F^{-1}(G(A))$ with $F(\lambda) := \lambda e - x$, which is a continuous function from $\mathbf{C}$ to A, and $G(A)$ is an open subset of A. Hence $\sigma(a)$ is a compact subset of $\mathbf{C}$.

If we suppose that $\sigma(a) = \emptyset$, we will arrive at a contradiction. The function R_a would be entire and bounded, with $\lim_{|\lambda| \to \infty} R_a(\lambda) = 0$, and the Liouville theorem is also true in the vector-valued case: for every $u \in A'$, $u \circ R_a$ would be an entire complex function and $\lim_{|\lambda| \to \infty} u(R_a(\lambda)) = 0$, so that $u(R_a(\lambda)) = 0$ and by the Hahn-Banach theorem $R_a(\lambda) = 0$, a contradiction to $R_a(\lambda) \in G(A)$.

Let us calculate the spectral radius. Since

$$R_a(\lambda) = \lambda^{-1} \sum_{n=0}^{\infty} \lambda^{-n} a^n$$

if $|\lambda| > r(a)$, the power series $\sum_{n=0}^{\infty} z^n a^n$ is absolutely convergent when $|z| = |\lambda|^{-1} < 1/r(a)$, and the convergence radius of $\sum_{n=0}^{\infty} \|a^n\| |z|^n$ is

$$R = (\limsup_{n \to \infty} \|a^n\|^{1/n})^{-1} \ge 1/r(a).$$

Then, $r(a) \ge \limsup_{n \to \infty} \|a^n\|^{1/n}$.

Conversely, if $\lambda \in \sigma(a)$, then $\lambda^n \in \sigma(a^n)$ by Theorem 8.7, so that $|\lambda^n| \leq \|a^n\|$ and

$$|\lambda| \leq \inf_n \|a^n\|^{1/n} \leq \liminf \|a^n\|^{1/n}.$$

Then it follows that $r(a) = \lim_{n\to\infty} \|a^n\|^{1/n} = \inf_n \|a^n\|^{1/n}$. □

As an important application of these results, let us show that $\mathbf{C}$ is the unique Banach algebra which is a field, in the sense that if A is a field, then $\lambda \mapsto \lambda e$ is an isometric isomorphism from $\mathbf{C}$ onto A. The inverse isometry is the **canonical isomorphism**:

Theorem 8.11 (Gelfand-Mazur[4]). *If every nonzero element of the unitary Banach algebra A is invertible (i.e., $G(A) = A \setminus \{0\}$), then $A = \mathbf{C}e$, and $\lambda \mapsto \lambda e$ is the unique homomorphism of unitary algebras between $\mathbf{C}$ and A.*

Proof. Let $a \in A$ and $\lambda \in \sigma(a)$ ($\sigma(a) \neq \emptyset$). Then $\lambda e - a \notin G(A)$ and it follows from the hypothesis that $a = \lambda e$. A homomorphism $\mathbf{C} \to A = \mathbf{C}e$ maps $1 \to e$ and necessarily $\lambda \to \lambda e$. □

8.3. Commutative Banach algebras

In this section A **represents a commutative unitary Banach algebra**. Some examples are $\mathbf{C}$, $\mathbf{B}(X)$, $\mathcal{C}(K)$, and $L^\infty(\Omega)$. Recall that $\mathcal{L}(E)$ (if $\dim E > 1$) is not commutative, and the convolution algebra $L^1(\mathbf{R})$ is not unitary.

8.3.1. Maximal ideals, characters, and the Gelfand transform.

A **character** of A is a homomorphism $\chi : A \to \mathbf{C}$ of unitary Banach algebras (hence $\chi(e) = 1$). We use $\Delta(A)$, or simply Δ, to denote the set of all characters of A. It is called **the spectrum of** A.

An **ideal**, J, of A is a linear subspace such that $AJ \subset J$ and $J \neq A$. It cannot contain invertible elements, since $x \in J$ invertible would imply $e = xx^{-1} \in J$ and then $A = Ae \subset J$, a contradiction to $J \neq A$.

Note that, if J is an ideal, then $\bar{J}$ is also an ideal, since it follows from $J \cap G(A) = \emptyset$ that $e \notin \bar{J}$ and $\bar{J} \neq A$. The continuity of the operations implies that $\bar{J} + \bar{J} \subset \bar{J}$ and $A\bar{J} \subset \bar{J}$.

This shows that every maximal ideal is closed.

[4]According to a result announced in 1938 by Stanislaw Mazur, a close collaborator of Banach who made important contributions to geometrical methods in linear and nonlinear functional analysis, and proved by Gelfand in 1941.

Theorem 8.12. (a) *The kernel of every character is a maximal ideal and the map $\chi \mapsto \operatorname{Ker}\chi$ between characters and maximal ideals of A is bijective.*

(b) *Every character $\chi \in \Delta(A)$ is continuous and*

$$\|\chi\| = \sup_{\|a\|_A \leq 1} |\chi(a)| = 1.$$

(c) *An element $a \in A$ is invertible if and only if $\chi(a) \neq 0$ for every $\chi \in \Delta$.*

(d) $\sigma(a) = \{\chi(a);\ \chi \in \Delta(A)\}$, *and* $r(a) = \sup_{\chi\in\Delta} |\chi(a)|$.

Proof. (a) The kernel M of any $\chi \in \Delta(A)$ is an ideal and, as the kernel of a nonzero linear functional, it is a hyperplane; that is, the complementary subspaces of M in A are one-dimensional, since χ is bijective on them, and M is maximal.

If M is a maximal ideal, the quotient space A/M has a natural structure of unitary Banach algebra, and it is a field. Indeed, if $\pi : A \to A/M$ is the canonical mapping and $\pi(x) = \widetilde{x}$ is not invertible in A/M, then $J = \pi(xA) \neq A/M$ is an ideal of A/M, and $\pi^{-1}(J) \neq A$ is an ideal of A which is contained in a maximal ideal that contains M. Thus, $\pi^{-1}(J) = M$, so that $\pi(xA) \subset \pi(M) = \{0\}$ and $\widetilde{x} = 0$.

Let $\widetilde{\chi} : A/M = \mathbf{C}\tilde{e} \to \mathbf{C}$ be the canonical isometry, so that M is the kernel of the character $\chi_M := \widetilde{\chi} \circ \pi_M$. Any other character χ_1 with the same kernel M factorizes as a product of π_M with a bijective homomorphism between A/M and $\mathbf{C}$ which has to be the canonical mapping $\mathbf{C}\tilde{e} \to \mathbf{C}$, and then $\chi_1 = \chi_M$.

(b) If $\chi = \chi_M \in \Delta(A)$, then $\|\chi\| \leq \|\pi_M\|\|\widetilde{\chi}\| = \|\pi_M\| \leq 1$ and $\|\chi\| \geq \chi(e) = 1$.

(c) If $x \in G(A)$, we have seen that it does not belong to any ideal. If $x \notin G(A)$, then xA does not contain e and is an ideal, and by Zorn's lemma every ideal is contained in a maximal ideal. So $x \in G(A)^c$ if and only if x belongs to a maximal ideal or, equivalently, $\chi(x) \neq 0$ for every character χ.

(d) Finally, $\lambda e - a \notin G(A)$ if and only if $\chi(\lambda e - a) = 0$, that is , $\lambda = \chi(a)$ for some $\chi \in \Delta(A)$. □

We associate to every element a of the unitary commutative algebra A the function $\widehat{a}$ which is the restriction of $\langle a, \cdot\rangle$ to the characters,[5] so that

$$\widehat{a} : \Delta(A) \to \mathbf{C}$$

is such that $\widehat{a}(\chi) = \chi(a)$. On $\Delta(A) \subset \bar{B}_{A'}$ we consider the **Gelfand topology**, which is the restriction of the weak-star topology $w^* = \sigma(A', A)$ of A'.

[5]Recall that $\langle a, u\rangle = u(a)$ was defined for every u in the dual A' of A as a Banach space.

In this way, $\widehat{a} \in \mathcal{C}(\Delta(A))$, and

$$\mathcal{G} : a \in A \mapsto \widehat{a} \in \mathcal{C}(\Delta(A))$$

is called the **Gelfand transform**.

Theorem 8.13. *Endowed with the Gelfand topology, $\Delta(A)$ is compact and the Gelfand transform $\mathcal{G} : A \mapsto \mathcal{C}(\Delta(A))$ is a continuous homomorphism of commutative unitary Banach algebras.*

Moreover $\|\widehat{a}\| = r(a) \leq \|a\|$ and $\mathcal{G}e = 1$, so that $\|\mathcal{G}\| = 1$.

Proof. For the first part we only need to show that $\Delta \subset \bar{B}_{A'}$ is weakly closed, since $\bar{B}_{A'}$ is weakly compact, by the Alaoglu theorem. But

$$\Delta = \Big\{\xi \in \bar{B}_{A'};\, \xi(e) = 1,\, \xi(xy) = \xi(x)\xi(y)\,\forall x, y \in A\Big\}$$

is the intersection of the weakly closed sets of $\bar{B}_{A'}$ defined by the conditions $\langle xy, \cdot\rangle - \langle x, \cdot\rangle\langle y, \cdot\rangle = 0$ $(x, y \in A)$ and $\langle e, \cdot\rangle = 1$.

It is clear that it is a homomorphism of commutative unitary Banach algebras. For instance, $\widehat{e}(\chi) = \chi(e) = 1$ and $\widehat{xy}(\chi) = \chi(x)\chi(y) = \widehat{x}(\chi)\widehat{y}(\chi)$.

Also, $\|\widehat{a}\| = \sup_{\chi\in\Delta} |\chi(a)| \leq \|a\|$, according to Theorem 8.12(d). □

Example 8.14. If K is a compact topological space, then $\mathcal{C}(K)$ is a unitary commutative Banach algebra whose characters are the evaluation maps δ_t at the different points $t \in K$, and $t \in K \mapsto \delta_t \in \Delta$ is a homeomorphism.

Obviously $\delta_t \in \Delta$. Conversely, if $\chi = \chi_M \in \Delta$, we will show that there is a common zero for all $f \in M$. If not, for every $t \in K$ there would exist some $f_t \in M$ such that $f_t(t) \neq 0$, and $|f_t| \geq \varepsilon_t > 0$ on a neighborhood $U(t)$ of this point t. Then, $K = U(t_1) \cup \cdots \cup U(t_N)$, and the function

$$f = |f_{t_1}|^2 + \cdots + |f_{t_N}|^2 = f_{t_1}\overline{f_{t_1}} + \cdots + f_{t_N}\overline{f_{t_N}},$$

which belongs to M, would be invertible, since it has no zeros.

Hence, there exists some $t \in K$ such that $f(t) = 0$ for every $f \in M$. But M is maximal and contains all the functions $f \in \mathcal{C}(K)$ such that $f(t) = 0$.

Both K and Δ are compact spaces and $t \in K \mapsto \delta_t \in \Delta$, being continuous, is a homeomorphism.

8.3.2. Algebras of bounded analytic functions. Suppose that Ω is a bounded domain of $\mathbf{C}$ and denote by $H^\infty(\Omega)$ the algebra of bounded analytic functions in Ω, which is a commutative Banach algebra under the uniform norm

$$\|f\|_\infty = \sup_{z\in\Omega} |f(z)|.$$

It is a unitary Banach subalgebra of $\mathbf{B}(\Omega)$.

The Gelfand transform $\mathcal{G} : H^\infty(\Omega) \to \mathcal{C}(\Delta)$ is an isometric isomorphism, since $\|f^2\| = \|f\|^2$ and $r(f) = \|f\|$ for every $f \in A$, and we can see $H^\infty(\Omega)$ is a unitary Banach subalgebra of $\mathcal{C}(\Delta)$ (cf. Exercise 8.18).

For every $\zeta \in \Omega$, the evaluation map δ_ζ is the character of $H^\infty(\Omega)$ uniquely determined by $\delta_\zeta(z) = \zeta$, where z denotes the coordinate function.

Indeed, if $\chi \in \Delta$ satisfies the condition $\chi(z) = \zeta$ and if $f \in H^\infty(\Omega)$, then

$$f(z) = f(\zeta) + \frac{f(z) - f(\zeta)}{z - \zeta}(z - \zeta)$$

and

$$\chi(f) = f(\zeta) + \chi\Big(\frac{f - f(\zeta)}{z - \zeta}\Big)\chi(z - \zeta) = f(\zeta).$$

It can be shown that the embedding $\Omega \hookrightarrow \Delta$ such that $\zeta \mapsto \delta_\zeta$ is a homeomorphism from Ω onto an open subset of Δ (see Exercise 8.14 where we consider the case $\Omega = U$, the unit disc) and, for every $f \in H^\infty(\Omega)$, it is convenient to write $\widehat{f}(\delta_\zeta) = f(\zeta)$ if $\zeta \in \Omega$.

Suppose now that $\xi \in \partial\Omega$, a boundary point. Note that $z - \xi$ is not invertible in $H^\infty(\Omega)$, so that

$$\Delta_\xi := \Big\{\chi \in \Delta;\ \chi(z - \xi) = 0\Big\} = \Big\{\chi \in \Delta;\ \chi(z) = \xi\Big\} = (\widehat{z})^{-1}(\xi)$$

is not empty.

For every $\chi \in \Delta$, $\chi(z - \chi(z)) = 0$ and $z - \chi(z)$ is not invertible, so that $\chi(z) \in \bar{\Omega}$ and $\chi \in \Omega$ or $\chi \in \Omega_\xi$ for some $\xi \in \partial\Omega$. That is,

$$\Delta = \Omega \cup \Big(\bigcup_{\xi \in \partial\Omega} \Delta_\xi\Big) \tag{8.1}$$

and we can imagine Δ as the domain Ω with a compact fiber $\Delta_\xi = (\widehat{z})^{-1}(\xi)$ lying above every $\xi \in \partial\Omega$.

The corona problem asks whether Ω is dense in Δ for the Gelfand topology, and it admits a more elementary equivalent formulation in terms of function theory:

Theorem 8.15. *For the Banach algebra $H^\infty(\Omega)$, the domain Ω is dense in Δ if and only if the following condition holds:*

If $f_1, \ldots, f_n \in H^\infty(\Omega)$ and if

$$|f_1(\zeta)| + \cdots + |f_n(\zeta)| \geq \delta > 0 \tag{8.2}$$

for every $\zeta \in \Omega$, then there exist $g_1, \ldots, g_n \in H^\infty(\Omega)$ such that

$$f_1 g_1 + \cdots + f_n g_n = 1. \tag{8.3}$$

Proof. Suppose that Ω is dense in Δ. By continuity, if $|f_1| + \cdots + |f_n| \geq \delta$ on Ω, then also $|\widehat{f_1}| + \cdots + |\widehat{f_n}| \geq \delta$ on Δ, so that $\{f_1, \dots, f_n\}$ is contained in no maximal ideal and

$$1 \in H^\infty(\Omega) = f_1 H^\infty(\Omega) + \cdots + f_n H^\infty(\Omega).$$

Conversely, suppose Ω is not dense in Δ and choose $\chi_0 \in \Delta$ with a neighborhood V disjoint from Ω. The Gelfand topology is the w^*-topology and this neighborhood has the form

$$V = \Big\{\chi;\ \max_{j=1,\dots,n} |\chi(h_j) - \chi_0(h_j)| < \delta,\ h_1, \dots, h_n \in H^\infty(\Omega)\Big\}.$$

The functions $f_j = h_j - \chi_0(h_j)$ are in V and they satisfy (8.2) because $\delta_\zeta \notin V$ and then $|f_j(\zeta)| \geq \delta$. But (8.3) is not possible because $f_1, \dots, f_n \in \operatorname{Ker} \chi_0$ and $\chi_0(1) = 1$. □

Starting from the above equivalence, in 1962 Carleson[6] solved the corona problem for the unit disc, that is, D is dense in $\Delta(H^\infty(D))$.

The version of the corona theorem for the disc algebra is much easier. See Exercise 8.3.

8.4. C^*-algebras

We are going to consider a class of algebras whose Gelfand transform is a bijective and isometric isomorphism. Gelfand introduced his theory to study these algebras.

8.4.1. Involutions.

A **C^*-algebra** is a unitary Banach algebra with an **involution**, which is a mapping $x \in A \mapsto x^* \in A$ that satisfies the following properties:

(a) $(x + y)^* = x^* + y^*$,

(b) $(\lambda x)^* = \bar{\lambda} x^*$,

(c) $(xy)^* = y^* x^*$,

(d) $x^{**} = x$, and

(e) $e^* = e$

for any $x, y \in A$ and $\lambda \in \mathbf{C}$, and such that $\|x^* x\| = \|x\|^2$ for every $x \in A$.

---

[6] The Swedish mathematician Lennart Carleson, awarded the Abel Prize in 2006, has solved some outstanding problems such as the corona problem (1962) and the almost everywhere convergence of Fourier series of any function in $L^2(\mathbf{T})$ (1966) and in complex dynamics. To quote Carleson, "The corona construction is widely regarded as one of the most difficult arguments in modern function theory. Those who take the time to learn it are rewarded with one of the most malleable tools available. Many of the deepest arguments concerning hyperbolic manifolds are easily accessible to those who understand well the corona construction."

An involution is always bijective and it is its own inverse. It is isometric, since $\|x\|^2 = \|x^*x\| \le \|x^*\|\|x\|$, so that $\|x\| \le \|x^*\|$ and $\|x^*\| \le \|x^{**}\| = \|x\|$.

Throughout this section, A **will be a** C^***-algebra**.

If H is a complex Hilbert space, $\mathcal{L}(H)$ is a C^*-algebra with the involution $T \mapsto T^*$, where T^* denotes the adjoint of T. It has been proved in Theorem 4.4 that $\|T^*T\| = \|TT^*\| = \|T\|^2$.

Let A and B be two C^*-algebras. A **homomorphism of** C^***-algebras** is a homomorphism $\Psi : A \to B$ of unitary Banach algebras such that $\Psi(x^*) = \Psi(x)^*$ (and, of course, $\Psi(e) = e$).

We say that $a \in A$ is **hermitian** or **self-adjoint** if $a = a^*$. The orthogonal projections of H are hermitian elements of $\mathcal{L}(H)$. We say that $a \in A$ is **normal** if $aa^* = a^*a$.

Example 8.16. If $a \in A$ is normal and $\langle a \rangle$ denotes the closed subalgebra of A generated by a, a^*, and e, then $\langle a \rangle$ contains all elements of A that can be obtained as the limits of sequences of polynomials in a, a^* and e. With the restriction of the involution of A, $\langle a \rangle$ is a commutative C^*-algebra.

Lemma 8.17. *Assume that A is commutative.*

(a) *If $a = a^* \in A$, then $\sigma_A(a) \subset \mathbf{R}$.*

(b) *For every $a \in A$ and $\chi \in \Delta(A)$, $\chi(a^*) = \overline{\chi(a)}$.*

Proof. If $t \in \mathbf{R}$, since $\|\chi\| = 1$,

$$\begin{aligned} |\chi(a + ite)|^2 &\le \|a + ite\|^2 = \|(a + ite)^*(a + ite)\| \\ &= \|(a - ite)(a + ite)\| = \|a^2 + t^2 e\| \le \|a\|^2 + t^2. \end{aligned}$$

Let $\chi(a) = \alpha + i\beta$ ($\alpha, \beta \in \mathbf{R}$). Then

$$\|a\|^2 + t^2 \ge |\alpha + i\beta + it|^2 = \alpha^2 + \beta^2 + 2\beta t + t^2,$$

i.e., $\|a\|^2 \ge \alpha^2 + \beta^2 + 2\beta t$, and it follows that $\beta = 0$ and $\chi_A(a) = \alpha \in \mathbf{R}$.

For any $a \in A$, if $x = (a + a^*)/2$ and $y = (a - a^*)/2i$, we obtain $a = x + iy$ with x, y hermitian, $\chi(x), \chi(y) \in \mathbf{R}$, and $a^* = x - iy$. Hence, $\chi(a) = \chi(x) + i\chi(y)$ and $\chi(a^*) = \chi(x) - i\chi(y) = \overline{\chi(a)}$. □

Theorem 8.18. *If B is a closed unitary subalgebra of A such that $b^* \in B$ for every $b \in B$, then $\sigma_B(b) = \sigma_A(b)$ for every $b \in B$.*

Proof. First let $b^* = b$. From Lemma 8.17 we know that $\sigma_{\langle b \rangle} \subset \mathbf{R}$ and, obviously,

$$\sigma_A(b) \subset \sigma_B(b) \subset \sigma_{\langle b \rangle}(b) = \partial\sigma_{\langle b \rangle}(b).$$

To prove the inverse inclusions, it is sufficient to show that $\partial\sigma_{\langle b \rangle}(b) \subset \sigma_A(b)$. Let $\lambda \in \partial\sigma_{\langle b \rangle}(b)$ and suppose that $\lambda \notin \sigma_A(b)$. There exists $x \in A$ so that $x(b-\lambda e) = (b-\lambda e)x = e$ and the existence of $\lambda_n \notin \sigma_{\langle b\rangle}(b)$ such that $\lambda_n \to \lambda$ follows from $\lambda \in \partial\sigma_{\langle b\rangle}(b)$. Thus we have

$$(b-\lambda_n e)^{-1} \in \langle b\rangle \subset A, \quad b-\lambda_n e \to b-\lambda e, \text{ and } (b-\lambda_n e)^{-1} \to (b-\lambda e)^{-1} = x.$$

Hence $x \in \langle b\rangle$, in contradiction to $\lambda \in \sigma_{\langle b\rangle}(b)$.

In the general case we only need to prove that if $x \in B$ has an inverse y in A, then $y \in B$ also. But it follows from $xy = e = yx$ that $(x^*x)(yy^*) = e = (yy^*)(x^*x)$, and x^*x is hermitian. In this case we have seen above that x^*x has its unique inverse in B, so that $yy^* = (x^*x)^{-1} \in B$ and $y = y(y^*x^*) = (yy^*)x^* \in B$. □

8.4.2. The Gelfand-Naimark theorem and functional calculus. We have proved in Theorem 8.13 that the Gelfand transform satisfies $\|\widehat{a}\|_\Delta = r(a) \leq \|a\|$, but in the general case it may not be injective. This is not the case for C^*-algebras.

Theorem 8.19 (Gelfand-Naimark). *If A is a commutative C^*-algebra, then the Gelfand transform $\mathcal{G} : A \to \mathcal{C}(\Delta(A))$ is a bijective isometric isomorphism of C^*-algebras.*

Proof. We have $\widehat{a^*}(\chi) = \overline{\chi(a)} = \overline{\widehat{a}(\chi)}$ and $\mathcal{G}(a^*) = \overline{\mathcal{G}(a)}$.

If $x^* = x$, then $r(x) = \lim_n \|x^{2^n}\|^{1/2^n} = \|x\|$, since $\|x^2\| = \|xx^*\| = \|x\|^2$ and, by induction, $\|x^{2^{(n+1)}}\| = \|(x^{2^n})^2\| = (\|x\|^{2^n})^2 = \|x\|^{2^{(n+1)}}$.

If we take $x = a^*a$, then $\|\widehat{a^*a}\|_\Delta = \|a^*a\|$, so

$$\|a\|^2 = \|a^*a\| = \|\widehat{a^*a}\|_\Delta = \|\overline{\widehat{a}}\widehat{a}\|_\Delta = \|\widehat{a}\|_\Delta^2$$

and $\|a\| = \|\widehat{a}\|_\Delta$.

Since $\mathcal{G}$ is an isometric isomorphism, $\mathcal{G}(A)$ is a closed subalgebra of $\mathcal{C}(\Delta(A))$. This subalgebra contains the constant functions ($\widehat{e} = 1$) and it is self-conjugate and separates points (if $\chi_1 \neq \chi_2$, there exists $a \in A$ such that $\chi_1(a) \neq \chi_2(a)$, i.e., $\widehat{a}(\chi_1) \neq \widehat{a}(\chi_2)$). By the complex form of the Stone-Weierstrass theorem, the image is also dense, so $\mathcal{G}(A) = \mathcal{C}(\Delta(A))$ and $\mathcal{G}$ is bijective. □

Theorem 8.20. *Let a be a normal element of the C^*-algebra A, let $\Delta = \Delta\langle a\rangle$ be the spectrum of the subalgebra $\langle a\rangle$, and let $\mathcal{G} : \langle a\rangle \to \mathcal{C}(\Delta)$ be the Gelfand transform. The function $\widehat{a} : \Delta \to \sigma_A(a) = \sigma_{\langle a\rangle}(a)$ is a homeomorphism.*

Proof. We know $\sigma(a) = \widehat{a}(\Delta)$. If $\chi_1, \chi_2 \in \Delta$, from $\widehat{a}(\chi_1) = \widehat{a}(\chi_2)$ we obtain $\chi_1(a) = \chi_2(a)$, $\chi_1(a^*) = \overline{\chi_1(a)} = \overline{\chi_2(a)} = \chi_2(a^*)$, and $\chi_1(e) = 1 = \chi_2(e)$, so that $\chi_1(x) = \chi_2(x)$ for all $x \in \langle a\rangle$; hence, $\chi_1 = \chi_2$ and $\widehat{a} : \Delta \to \sigma(a)$ is bijective and continuous between two compact spaces, and then the inverse is also continuous. □

The homeomorphism $\widehat{a} : \Delta \to \sigma(a)$ $(\lambda = \widehat{a}(\chi))$ allows us to define the isometric isomorphism of C^*-algebras $\tau = \circ\widehat{a} : \mathcal{C}(\sigma(a)) \to \mathcal{C}(\Delta)$ such that $[g(\lambda)] \mapsto [G(\chi)] = [g(\widehat{a}(\chi))]$.

By Theorem 8.19, the composition

$$\Phi_a = \mathcal{G}^{-1} \circ \tau : \mathcal{C}(\sigma(a)) \to \mathcal{C}(\Delta) \to \langle a \rangle \subset A,$$

such that $g \in \mathcal{C}(\sigma(a)) \mapsto \mathcal{G}^{-1}(g(\widehat{a})) \in \langle a \rangle$, is also an isometric isomorphism of C^*-algebras. If $g \in \mathcal{C}(\sigma(a))$, then the identity $\widehat{\Phi_a(g)} = g \circ \widehat{a} = g(\widehat{a})$ suggests that we may write $g(a) := \Phi_a(g)$.

So, we have the isometric isomorphism of C^*-algebras

$$\Phi_a : g \in \mathcal{C}(\sigma(a)) \mapsto g(a) \in \langle a \rangle \subset A$$

such that, if $g_0(\lambda) = \lambda$ is the identity on $\sigma(a)$, then $\widehat{\Phi_a(g_0)} = \widehat{a}$ and $g_0(a) = a$, since $\tau(g_0) = \widehat{a} = \mathcal{G}(a)$. Also $\bar{g}_0(a) = a^*$ and

$$p(a) = \sum_{0 \le j,k \le N} c_{j,k} a^j (a^*)^k \text{ if } p(z) = \sum_{0 \le j,k \le N} c_{j,k} z^j \bar{z}^k. \tag{8.4}$$

We call Φ_a the **functional calculus with continuous functions**. It is the unique homomorphism $\Phi : \mathcal{C}(\sigma(a)) \to A$ of C^*-algebras such that

$$\Phi(p) = \sum_{0 \le j,k \le N} c_{j,k} a^j (a^*)^k (a)$$

if $p(z) = \sum_{0 \le j,k \le N} c_{j,k} z^j \bar{z}^k$.

Indeed, it follows from the Stone-Weierstrass theorem that the subalgebra $\mathcal{P}$ of all polynomials $p(z)$ considered in (8.4) is dense in $\mathcal{C}(\sigma(a))$ and, if $g = \lim_n p_n$ in $\mathcal{C}(\sigma(a))$ with $p_n \in \mathcal{P}$, then

$$\Phi(g) = \lim_n p_n(a) = \Phi_a(g).$$

These facts are easily checked and justify the notation $g(a)$ for $\Phi_a(g)$.

8.5. Spectral theory of bounded normal operators

In this section we are going to consider normal operators $T \in \mathcal{L}(H)$. By Theorem 8.18,

$$\sigma(T) = \sigma_{\mathcal{L}(H)}(T) = \sigma_{\langle T \rangle}(T)$$

and it is a nonempty compact subset of $\mathbf{C}$.

From now on, by $\mathbf{B}(\sigma(T))$ we will denote the C^*-algebra of all bounded Borel measurable functions $f : \sigma(T) \to \mathbf{C}$, endowed with the involution $f \mapsto \bar{f}$ and with the uniform norm. Obviously $\mathcal{C}(\sigma(T))$ is a closed unitary subalgebra of $\mathbf{B}(\sigma(T))$.

An application of the Gelfand-Naimark theorem to the commutative C^*-algebra $\langle T \rangle$ gives an isometric homomorphism from $\langle T \rangle$ onto $\mathcal{C}(\Delta\langle T \rangle)$.

The composition of this homomorphism with the change of variables $\lambda = \widehat{T}(\chi)$ ($\lambda \in \sigma(T)$ and $\chi \in \Delta\langle T\rangle$) defines the functional calculus with continuous functions on $\sigma(T)$, $g \in \mathcal{C}(\sigma(T)) \mapsto g(T) \in \langle T\rangle \subset \mathcal{L}(H)$, which is an isometric homomorphism of C^*-algebras.

If $x, y \in H$ are given, then

$$u_{x,y}(g) := (g(T)x, y)_H$$

defines a continuous linear form on $\mathcal{C}(\sigma(T))$ and, by the Riesz-Markov representation theorem,

$$(g(T)x, y)_H = u_{x,y}(g) = \int_{\sigma(T)} g \, d\mu_{x,y}$$

for a unique complex Borel measure $\mu_{x,y}$ on $\sigma(T)$.

We will say that $\{\mu_{x,y}\}$ is **the family of complex spectral measures of** T. For any bounded Borel measurable function f on $\sigma(T)$, we can define

$$u_{x,y}(f) := \int_{\sigma(T)} f \, d\mu_{x,y}$$

and in this way we extend $u_{x,y}$ to a linear form on these functions. Note that $|u_{x,y}(g)| = |(g(T)x, y)_H| \le \|x\|_H \|y\|_H \|g\|_{\sigma(T)}$.

8.5.1. Functional calculus of normal operators. Now our goal is to show that it is possible to define $f(T) \in \mathcal{L}(H)$ for every f in the C^*-algebra $\mathbf{B}(\sigma(T))$ of all bounded Borel measurable functions on $\sigma(T) \subset \mathbf{C}$, equipped with the uniform norm and with the involution $f \mapsto \bar{f}$, so that

$$(f(T)x, y)_H = u_{x,y}(f) = \int_{\sigma(T)} f \, d\mu_{x,y}$$

in the hope of obtaining a functional calculus $f \mapsto f(T)$ for bounded but not necessarily continuous functions.

Theorem 8.21. *Let $T \in \mathcal{L}(H)$ be a normal operator ($TT^* = T^*T$) and let $\{\mu_{x,y}\}$ be its family of complex spectral measures. Then there exists a unique homomorphism of C^*-algebras*

$$\Phi_T : \mathbf{B}(\sigma(T)) \to \mathcal{L}(H)$$

such that

$$(\Phi_T(f)x, y)_H = \int_{\sigma(T)} f \, d\mu_{x,y} \qquad (x, y \in H).$$

It is an extension of the continuous functional calculus $g \mapsto g(T)$, and $\|\Phi_T(f)\| \le \|f\|_{\sigma(T)}$.

Proof. Note that, if μ_1 and μ_2 are two complex Borel measures on $\sigma(T)$ and if $\int g\,d\mu_1 = \int g\,d\mu_2$ for all real $g \in \mathcal{C}(\sigma(T))$, then $\mu_1 = \mu_2$, by the uniqueness in the Riesz-Markov representation theorem.

If $g \in \mathcal{C}(\sigma(T))$ is a real function, then $g(T)$ is self-adjoint, since $g(T)^* = \bar{g}(T)$. Hence, $(g(T)x, y)_H = \overline{(g(T)y, x)_H}$ and then

$$\int_{\sigma(T)} g\,d\mu_{x,y} = \overline{\int_{\sigma(T)} g\,d\mu_{y,x}} = \int_{\sigma(T)} g\,d\bar{\mu}_{y,x},$$

so that

$$\mu_{x,y} = \bar{\mu}_{y,x}.$$

Obviously, $(x, y) \mapsto \int_{\sigma(T)} g\,d\mu_{x,y} = (g(T)x, y)_H$ is a continuous sesquilinear form and, from the uniqueness in the Riesz-Markov representation theorem, the map $(x, y) \mapsto \mu_{x,y}(B)$ is also sesquilinear, for any Borel set $B \subset \sigma(T)$. For instance, $\mu_{x,\lambda y} = \bar{\lambda}\mu_{x,y}$, since for continuous functions we have

$$\int_{\sigma(T)} g\,d\mu_{x,\lambda y} = \bar{\lambda}(g(T)x, y)_H = \int_{\sigma(T)} g\,d\bar{\lambda}\mu_{x,y}.$$

With the extension $u_{x,y}(f) := \int_{\sigma(T)} f\,d\mu_{x,y}$ of $u_{x,y}$ to functions f in $\mathbf{B}(\sigma(T))$, it is still true that

$$|u_{x,y}(f)| \le \|x\|_H \|y\|_H \|f\|_{\sigma(T)}.$$

For every $f \in \mathbf{B}(\sigma(T))$,

$$(x, y) \mapsto B_f(x, y) := \int_{\sigma(T)} f\,d\mu_{x,y}$$

is a continuous sesquilinear form on $H \times H$ and $B_f(y, x) = \overline{B_f(x, y)}$, since

$$\overline{\int_{\sigma(T)} f\,d\mu_{x,y}} = \int_{\sigma(T)} f\,d\mu_{y,x}$$

($\overline{\mu_{x,y}(B)} = \mu_{y,x}(B)$ extends to simple functions). Let us check that an application of the Riesz representation theorem produces a unique operator $\Phi_T(f) \in \mathcal{L}(H)$ such that $B_f(x, y) = (\Phi_T(f)x, y)_H$.

Note that $B_f(\cdot, x) \in H'$ and there is a unique $\Phi_T(f)x \in H$ so that $B_f(y, x) = (y, \Phi_T(f)x)_H$ for all $y \in H$. Then

$$(\Phi_T(f)x, y)_H = \overline{B_f(y, x)} = B_f(x, y) = \int_{\sigma(T)} f\,d\mu_{x,y} \qquad (x, y \in H).$$

It is clear that $B_f(x, y)$ is linear in f and that we have defined a bounded linear mapping $\Phi_T : \mathbf{B}(\sigma(T)) \to \mathcal{L}(H)$ such that

$$|(\Phi_T(f)x, y)_H| \le \|f\|_{\sigma(T)} \|x\|_H \|y\|_H$$

and $\|\Phi_T(f)\| \leq \|f\|_{\sigma(T)}$. Moreover, with this definition, Φ_T extends the functional calculus with continuous functions, $g \mapsto g(T)$.

To prove that Φ_T is a continuous homomorphism of C^*-algebras, all that remains is to check its behavior with the involution and with the product.

If f is real, then from $\mu_{x,y} = \bar{\mu}_{y,x}$ we obtain $(\Phi_T(f)x, y)_H = \overline{(\Phi_T(f)y, x)}_H$ and $\Phi_T(f)^* = \Phi_T(f)$. In the case of a complex function, f, $\Phi_T(f)^* = \Phi_T(\bar{f})$ follows by linearity.

Finally, to prove that $\Phi_T(f_1 f_2) = \Phi_T(f_1)\Phi_T(f)(f_2)$, we note that, on continuous functions,

$$\int_{\sigma(T)} hg\, d\mu_{x,y} = (h(T)g(T)x, y)_H = \int_{\sigma(T)} h\, d\mu_{g(T)x,y}$$

and $g\, d\mu_{x,y} = d\mu_{g(T)x,y}$ $(x, y \in H)$. Hence, also

$$\int_{\sigma(T)} f_1 g\, d\mu_{x,y} = \int_{\sigma(T)} f_1\, d\mu_{g(T)x,y}$$

if f_1 is bounded, and then

$$\begin{aligned}\int_{\sigma(T)} f_1 g\, d\mu_{x,y} &= (\Phi_T(f_1)g(T)x, y)_H = (g(T)x, \Phi_T(f_1)^* y)_H \\ &= \int_{\sigma(T)} g\, d\mu_{x,\Phi(f_1)^* y}.\end{aligned}$$

Again $f_1 d\mu_{x,y} = d\mu_{x,\Phi_T(f_1)^* y}$, and also $\int_{\sigma(T)} f_1 f_2\, d\mu_{x,y} = \int_{\sigma(T)} f_2\, d\mu_{x,f_1(T)^* y}$ if f_1 and f_2 are bounded. Thus,

$$\begin{aligned}(\Phi_T(f_1 f_2)x, y)_H &= \int_{\sigma(T)} f_1 f_2\, d\mu_{x,y} \\ &= \int_{\sigma(T)} f_2\, d\mu_{x,\Phi_T(f_1)^* y} = (\Phi_T(f_1)\Phi_T(f_2)x, y)_H\end{aligned}$$

and $\Phi_T(f)$ is multiplicative. □

As in the case of the functional calculus for continuous functions, if $f \in \mathbf{B}(\sigma(T))$, we will denote the operator $\Phi_T(f)$ by $f(T)$; that is,

$$(f(T)x, y)_H = \int_{\sigma(T)} f\, d\mu_{x,y} \qquad (x, y \in H).$$

8.5.2. Spectral measures. For a given Hilbert space, H, a **spectral measure**, or a **resolution of the identity**, on a locally compact subset K of $\mathbf{C}$ (or of $\mathbf{R}^n$), is an operator-valued mapping defined on the Borel σ-algebra $\mathcal{B}_K$ of K,

$$E : \mathcal{B}_K \longrightarrow \mathcal{L}(H),$$

that satisfies the following conditions:

(1) Each $E(B)$ is an orthogonal projection.

(2) $E(\emptyset) = 0$ and $E(K) = I$, the identity operator.

(3) If $B_n \in \mathcal{B}_K$ $(n \in \mathbf{N})$ are disjoint, then

$$E(\biguplus_{n=1}^{\infty} B_n)x = \sum_{n=1}^{\infty} E(B_n)x$$

for every $x \in H$, and it is said that

$$E(\biguplus_{n=1}^{\infty} B_n) = \sum_{n=1}^{\infty} E(B_n)$$

for the strong convergence, or that E is strongly σ-additive.

Note that E also has the following properties:

(4) If $B_1 \cap B_2 = \emptyset$, then $E(B_1)E(B_2) = 0$ (orthogonality).

(5) $E(B_1 \cap B_2) = E(B_1)E(B_2) = E(B_2)E(B_1)$ (multiplicativity).

(6) If $B_1 \subset B_2$, then $\operatorname{Im} E(B_1) \subset \operatorname{Im} E(B_2)$ (usually represented by $E(B_1) \leq E(B_2)$).

(7) If $B_n \uparrow B$ or $B_n \downarrow B$, then $\lim_n E(B_n)x = E(B)x$ for every $x \in H$ (it is said that $E(B_n) \to E(B)$ strongly).

Indeed, to prove (4), if $y = E(B_2)x$, the equality

$$(E(B_1) + E(B_2))^2 = E(B_1 \uplus B_2)^2 = E(B_1) + E(B_2)$$

and the condition $B_1 \cap B_2 = \emptyset$ yield

$$E(B_1)E(B_2)x + E(B_2)E(B_1)x = 0,$$

that is, $E(B_1)y + y = 0$ and, applying $E(B_1)$ to both sides, $E(B_1)y = 0$.

Now (5) follows from multiplying the equations

$$E(B_1) = E(B_1 \cap B_2) + E(B_1 \setminus B_1 \cap B_2), \quad E(B_2) = E(B_1 \cap B_2) + E(B_2 \setminus B_1 \cap B_2)$$

and taking into account (4).

If $B_n \uparrow B$, then $\lim_n E(B_n)x = E(B)x$ follows from (3), since

$$B = B_1 \uplus (B_2 \setminus B_1) \uplus (B_3 \setminus B_2) \uplus \cdots.$$

The decreasing case $B_n \downarrow B$ reduces to $K \setminus B_n \uparrow K \setminus B$.

It is also worth noticing that the spectral measure E generates the family of complex measures $E_{x,y}$ $(x, y \in H)$ defined as

$$E_{x,y}(B) := (E(B)x, y)_H.$$

If $x \in H$, then $E_x(B) := E(B)x$ defines a vector measure $E_x : \mathcal{B}_K \to H$, i.e., $E_x(\emptyset) = 0$ and $E_x(\biguplus_{n=1}^{\infty} B_n) = \sum_{n=1}^{\infty} E_x(B_n)$ in H.

Note that for every $x \in H$, $E_{x,x}$ is a (positive) measure such that

$$E_{x,x}(B) = (E(B)x, y)_H = \|E(B)x\|_H^2, \quad E_{x,x}(K) = \|x\|_H^2,$$

a probability measure if $\|x\|_H = 1$, and that operations with the complex measures $E_{x,y}$, by polarization, reduce to operations with positive measures:

$$E_{x,y}(B) = \frac{1}{4}\Big(E_{x+y,x+y}(B) - E_{x-y,x-y}(B) + iE_{x+iy,x+iy}(B) - iE_{x-iy,x-iy}(B)\Big).$$

The notions E**-almost everywhere** (E-a.e.) and E**-essential supremum** have the usual meaning. In particular, if f is a real measurable function,

$$E\text{-}\sup f = \inf\{M \in \mathbf{R};\ f \leq M\ E\text{-a.e.}\}.$$

Note that $E(B) = 0$ if and only if $E_{x,x}(B) = 0$ for every $x \in H$. Thus, if $B_1 \subset B_2$ and $E(B_2) = 0$, then $E(B_1) = 0$, and the class of E-null sets is closed under countable unions.

The **support of a spectral measure** E on K is defined as the least closed set $\operatorname{supp} E$ such that $E(K \setminus \operatorname{supp} E) = 0$. The support consists precisely of those points in K for which every neighborhood has nonzero E-measure and $E(B) = E(B \cap \operatorname{supp} E)$ for every Borel set $B \subset K$.

The existence of the support is proved by considering the union V of the open sets V_α of K such that $E(V_\alpha) = 0$. Since there is a sequence V_n of open sets in K such that $V_\alpha = \bigcup_{\{n;\, V_n \subset V_\alpha\}} V_n$, then also $V = \bigcup_{\{n;\, V_n \subset V\}} V_n$ and $E(V) = 0$. Then $\operatorname{supp} E = K \setminus V$.

We write

$$R = \int_K f\, dE$$

to mean that

$$(Rx, y)_H = \int_K f\, dE_{x,y} \qquad (x, y \in H).$$

It is natural to ask whether the family $\{\mu_{x,y}\}$ of complex measures associated to a normal operator T is generated by a single spectral measure E associated to T. The next theorem shows that the answer is affirmative, allowing us to rewrite the functional calculus of Theorem 8.21 as $f(T) = \int_{\sigma(T)} f\, dE$.

Theorem 8.22 (Spectral resolution[7]). *If $T \in \mathcal{L}(H)$ is a normal operator, then there exists a unique spectral measure $E : \mathcal{B}_{\sigma(T)} \to \mathcal{L}(H)$ which satisfies*

$$T = \int_{\sigma(T)} \lambda \, dE(\lambda). \tag{8.5}$$

Furthermore,

$$f(T) = \int_{\sigma(T)} f(\lambda) \, dE(\lambda) \qquad (f \in \mathbf{B}(\sigma(T))), \tag{8.6}$$

and

$$E(B) = \chi_B(T) \qquad (B \in \mathcal{B}_{\sigma(T)}). \tag{8.7}$$

Proof. If $\Phi_T : \mathbf{B}(\sigma(T)) \to \mathcal{L}(H)$ is the homomorphism that defines the functional calculus, then to obtain (8.6) we must define E by condition (8.7),

$$E(B) := \Phi_T(\chi_B) \qquad (B \in \mathcal{B}_{\sigma(T)}),$$

and then check that E is a spectral measure with the convenient properties.

Obviously, $E(B) = E(B)^2$ and $E(B)^* = E(B)$ (χ_B is real), so that $E(B)$ is an orthogonal projection (Theorem 3.13).

Moreover, it follows from the properties of the functional calculus for continuous functions that $E(\sigma(T)) = \Phi(1) = 1(T) = I$ and $E(\emptyset) = \Phi(0) = 0$ and, since Φ is linear, E is additive. Also, from

$$(E(B)x, y)_H = (\Phi(\chi_B)x, y)_H = \mu_{x,y}(B),$$

we obtain that

$$\int_{\sigma(T)} g \, dE = g(T), \quad \int_{\sigma(T)} f \, dE = \Phi(f) \qquad \Big(g \in \mathcal{C}(\sigma(T)),\ f \in \mathbf{B}(\sigma(T))\Big).$$

Finally, E is strongly σ-additive since, if B_n ($n \in \mathbf{N}$) are disjoint Borel sets, $E(B_n)E(B_m) = 0$ if $n \neq m$, so that the images of the projections $E(B_n)$ are mutually orthogonal (if $y = E(B_m)x$, we have $y \in \operatorname{Ker} E(B_n)$ and $y \in E(B_n)(H)^\perp$) and then, for every $x \in H$, $\sum_n E(B_n)x$ is convergent to some $Px \in H$ since

$$\sum_n \|E(B_n)x\|_H^2 \leq \|x\|_H^2,$$

this being true for partial sums, $\|E(\biguplus_{n=1}^N B_n)x\|_H^2 \leq \|x\|_H^2$.

---

[7] In their work on integral equations, D. Hilbert for a self-adjoint operator on ℓ^2 and F. Riesz on L^2 used the Stieltjes integral

$$\int_{\sigma(T)} t \, dE(t) = \lim \sum_k t_k (E(t_k) - E(t_{k-1})) \qquad (\text{here } E(t_k) - E(t_{k-1}) = E(t_{k-1}, t_k])$$

to obtain this spectral theorem.

But then

$$(Px, y)_H = \sum_{n=1}^{\infty} (E(B_n)x, y)_H = \mu_{x,y}\Big(\biguplus_{n=1}^{\infty} B_n\Big) = \Big(E\Big(\biguplus_{n=1}^{\infty} B_n\Big)x, y\Big)_H$$

and $\sum_n E(B_n)x = Px = E\big(\biguplus_n B_n\big)x$.[8]

The uniqueness of E follows from the uniqueness for the functional calculus for continuous functions Φ_T and from the uniqueness of the measures $E_{x,y}$ in the Riesz-Markov representation theorem. □

Remark 8.23. A more general spectral theorem due to John von Neumann in 1930 can also be obtained from the Gelfand-Naimark theorem: any commutative family of normal operators admits a single spectral measure which simultaneously represents all operators of the family as integrals $\int_K g\, dE$ for various functions g.

8.5.3. Applications. There are two special instances of normal operators that we are interested in: self-adjoint operators and unitary operators.

Recall that an operator $U \in \mathcal{L}(H)$ is said to be **unitary** if it is a bijective isometry of H. This means that

$$UU^* = U^*U = I$$

since $U^*U = I$ if and only if $(Ux, Uy)_H = (x, y)$ and U is an isometry. If it is bijective, then $(U^{-1}x, U^{-1}y)_H = (x, y)_H$ and $((U^{-1})^*U^{-1}x, y)_H = (x, y)_H$, where $(U^{-1})^*U^{-1}x = (U^*)^{-1}U^{-1}x = (UU^*)^{-1}x$ and then $((UU^*)^{-1}x, y)_H = (x, y)_H$, so that $UU^* = I$. Conversely, if $UU^* = I$, then U is exhaustive.

The Fourier transform is an important example of a unitary operator of $L^2(\mathbf{R}^n)$.

Knowing the spectrum allows us to determine when a normal operator is self-adjoint or unitary:

Theorem 8.24. *Let $T \in \mathcal{L}(H)$ be a normal operator.*

(a) *T is self-adjoint if and only if $\sigma(T) \subset \mathbf{R}$.*

(b) *T is unitary if and only if $\sigma(T) \subset \mathbf{S} = \{\lambda;\ |\lambda| = 1\}$.*

Proof. We will apply the continuous functional calculus Φ_T for T to the identity function $g(\lambda) = \lambda$ on $\sigma(T)$, so that $g(T) = T$ and $\bar{g}(T) = T^*$.

From the injectivity of Φ_T, $T = T^*$ if and only if $g = \bar{g}$, meaning that $\lambda = \bar{\lambda} \in \mathbf{R}$ for every $\lambda \in \sigma(T)$.

Similarly, T is unitary if and only if $TT^* = T^*T = I$, i.e., when $g\bar{g} = 1$, which means that $|\lambda| = 1$ for all $\lambda \in \sigma(T)$. □

[8] See also Exercise 8.21.

Positivity can also be described through the spectrum:

Theorem 8.25. *Suppose $T \in \mathcal{L}(H)$. Then*

$$(Tx, x)_H \geq 0 \qquad (x \in H) \tag{8.8}$$

if and only if

$$T = T^* \text{ and } \sigma(T) \subset [0, \infty). \tag{8.9}$$

Such an operator is said to be positive.

Proof. It follows from (8.8) that $(Tx, x)_H \in \mathbf{R}$ and then

$$(Tx, x)_H = (x, Tx)_H = (T^*x, x)_H.$$

Let us show that then $S := T - T^* = 0$.

Indeed, $(Sx, y)_H + (Sy, x) = 0$ and, replacing y by iy, $-i(Sx, y)_H + i(Sy, x) = 0$. Now we multiply by i and add to obtain $(Sx, y)_H = 0$ for all $x, y \in H$, so that $S = 0$.

Thus $\sigma(T) \subset \mathbf{R}$. To prove that $\lambda < 0$ cannot belong to $\sigma(T)$, we note that the condition (8.8) allows us to set

$$\|(T - \lambda I)x\|_H^2 = \|Tx\|_H^2 - 2\lambda(Tx, x)_H + \lambda^2\|x\|_H^2.$$

This shows that $T_\lambda := T - \lambda I : H \to F = \Im(T - \lambda I)$ has a continuous inverse with domain F, which is closed. This operator is easily extended to a left inverse R of T_λ by defining $R = 0$ on $F^\perp$. But T_λ is self-adjoint and $RT_\lambda = I$ also gives $T_\lambda R^* = I$, T_λ is also right invertible, and $\lambda \notin \sigma(T)$.

Suppose now that $T = T^*$ and $\sigma(T) \subset [0, \infty)$. In the spectral resolution

$$(Tx, x)_H = \int_{\sigma(T)} \lambda \, dE_{x,x}(\lambda) \geq 0,$$

since $E_{x,x}$ is a positive measure and $\lambda \geq 0$ on $\sigma(T) \subset [0, \infty)$. □

Let us now give an application of the functional calculus with bounded functions:

Theorem 8.26. *If $T = \int_{\sigma(T)} \lambda \, dE(\lambda)$ is the spectral resolution of a normal operator $T \in \mathcal{L}(H)$ and if $\lambda_0 \in \sigma(T)$, then*

$$\operatorname{Ker}(T - \lambda_0 I) = \operatorname{Im} E\{\lambda_0\},$$

so that λ_0 is an eigenvalue of T if and only if $E(\{\lambda_0\}) \neq 0$.

Proof. The functions $g(\lambda) = \lambda - \lambda_0$ and $f = \chi_{\{\lambda_0\}}$ satisfy $fg = 0$ and $g(T)f(T) = 0$. Since $f(T) = E(\{\lambda_0\})$,

$$\operatorname{Im} E(\{\lambda_0\}) \subset \operatorname{Ker}(g(T)) = \operatorname{Ker}(T - \lambda_0 I).$$

Conversely, let us take

$$G = \sigma(T) \setminus \{\lambda_0\} = \biguplus_n B_n$$

with $d(\lambda_0, B_n) > 0$ and define the bounded functions

$$f_n(\lambda) = \frac{\chi_{B_n}(\lambda)}{\lambda - \lambda_0}.$$

Then $f_n(T)(T - \lambda_0 I) = E(B_n)$, and $(T - \lambda_0 I)x = 0$ implies $E(B_n)x = 0$ and $E(G)x = \sum_n E(B_n)x = 0$. Hence, $x = E(G)x + E(\{\lambda_0\})x = E(\{\lambda_0\})x$, i.e., $x \in \operatorname{Im} E(\{\lambda_0\})$. □

As shown in Section 4.4, if T is compact, then every nonzero eigenvalue has finite multiplicity and $\sigma(T) \setminus \{0\}$ is a finite or countable set of eigenvalues with finite multiplicity with 0 as the only possible accumulation point. If T is normal, the converse is also true:

Theorem 8.27. *If $T \in \mathcal{L}(T)$ is a normal operator such that $\sigma(T)$ has no accumulation point except possibly* 0 *and* $\dim \operatorname{Ker}(T - \lambda I) < \infty$ *for every $\lambda \neq 0$, then T is compact.*

Proof. Let $\sigma(T) \setminus \{0\} = \{\lambda_1, \lambda_2, \ldots\}$ and $|\lambda_1| \geq |\lambda_2| \geq \cdots$. We apply the functional calculus to the functions g_n defined as

$$g_n(\lambda) = \lambda \text{ if } \lambda = \lambda_k \text{ and } k \leq n$$

and $g_n(\lambda) = 0$ at the other points of $\sigma(T)$ to obtain the compact operator with finite-dimensional range

$$g_n(T) = \sum_{k=1}^{n} \lambda_k E(\{\lambda_k\}).$$

Then

$$\|T - g_n(T)\| \leq \sup_{\lambda \in \sigma(T)} |\lambda - g_n(\lambda)| \leq |\lambda_n|$$

and $|\lambda_n| \to 0$ as $n \to \infty$ if $\sigma(T) \setminus \{0\}$ is an infinite set. This shows that T is compact as a limit of compact operators. □

8.6. Exercises

Exercise 8.1. Show that every Banach algebra A without a unit element can be considered as a Banach subalgebra of a unitary Banach algebra A_1 constructed in the following fashion. On $A_1 = A \times \mathbf{C}$, which is a vector space, define the multiplication $(a, \lambda)(b, \mu) := (ab + \lambda b + \mu a, \lambda\mu)$ and the norm $\|(a, \lambda)\| := \|a\| + |\lambda|$. The unit is $\delta = (0, 1)$.

The map $a \mapsto (a, 0)$ is an isometric homomorphism (i.e., linear and multiplicative), so that we can consider A as a closed subalgebra of A_1. By denoting $a = (a, 0)$ if $a \in A$, we can write $(a, \lambda) = a + \lambda\delta$ and the projection $\chi_0(a + \lambda\delta) := \lambda$ is a character $\chi_0 \in \Delta(A_1)$.

If A is unitary, then the unit e cannot be the unit δ of A_1.

Exercise 8.2. Suppose that $A = \mathbf{C}_1$ is the commutative unitary Banach algebra obtained by adjoining a unit to $\mathbf{C}$ as in Exercise 8.1. Describe $\Delta(A)$ and the corresponding Gelfand transform.

Exercise 8.3. (a) Prove that the polynomials $P(z) = \sum_{n=0}^{N} c_n z^n$ are dense in the disc algebra $A(D)$ by showing that, if $f \in A(D)$ and

$$f_n(z) := f\Big(\frac{nz}{n+1}\Big),$$

then $f_n \to f$ uniformly on $\bar{D}$ and, if $\|f - f_n\| \leq \varepsilon/2$, there is a Taylor polynomial P of f_n such that $\|f_n - P\| \leq \varepsilon/2$.

Hence, polynomials P are not dense in $\mathcal{C}(\bar{D})$. Why is this not in contradiction to the Stone-Weierstrass theorem?

(b) Prove that the characters of $A(D)$ are the evaluations δ_z $(|z| \leq 1)$ and that $z \in \bar{D} \mapsto \delta_z \in \Delta(A(D))$ is a homeomorphism.

(c) If $f_1, \ldots, f_n \in A(D)$ have no common zeros, prove that there exist $g_1, \ldots, g_n \in A(D)$ such that $f_1 g_1 + \cdots + f_n g_n = 1$.

Exercise 8.4. Show that, with the convolution product,

$$f * g(x) := \int_{\mathbf{R}} f(x - y) g(y)\, dy,$$

the Banach space $L^1(\mathbf{R})$ becomes a nonunitary Banach algebra.

Exercise 8.5. Show also that $L^1(\mathbf{T})$, the Banach space of all complex 1-periodic functions that are integrable on $(0, 1)$, with the convolution product

$$f * g(x) := \int_0^1 f(x - y) g(y)\, dy,$$

and the usual L^1 norm, is a nonunitary Banach algebra.

Exercise 8.6. Show that $\ell^1(\mathbf{Z})$, with the discrete convolution,

$$(u * v)[k] := \sum_{m=-\infty}^{+\infty} u[k - m] v[m],$$

is a unitary Banach algebra.

Exercise 8.7. Every unitary Banach algebra, A, can be considered a closed subalgebra of $\mathcal{L}(A)$ by means of the isometric homomorphism $a \mapsto L_a$, where $L_a(x) := ax$.

Exercise 8.8. In this exercise we want to present the Fourier transform on $L^1(\mathbf{R})$ as a special case of the Gelfand transform. To this end, consider the unitary commutative Banach algebra $L^1(\mathbf{R})_1$ obtained as in Exercise 8.1 by adjoining the unit to $L^1(\mathbf{R})$, which is a nonunitary convolution Banach algebra $L^1(\mathbf{R})$ (see Exercise 8.4).

(a) Prove that, if $\chi \in \Delta(L^1(\mathbf{R})_1) \setminus \{\chi_0\}$ and $\chi(u) = 1$ with $u \in L^1(\mathbf{R})$, then $\gamma_\chi(\alpha) := \chi(\tau_{-\alpha}u) = \chi([u(t+\alpha)])$ defines a function $\gamma_\chi : \mathbf{R} \to \mathbf{T} \subset \mathbf{C}$ which is continuous and such that $\gamma_\chi(\alpha+\beta) = \gamma_\chi(\alpha)\gamma_\chi(\beta)$.

(b) Prove that there exists a uniquely determined number $\xi_\chi \in \mathbf{R}$ such that $\gamma_\chi(\alpha) = e^{2\pi i \xi_\chi \alpha}$.

(c) Check that, if $\mathcal{G}f$ denotes the Gelfand transform of $f \in L^1(\mathbf{R})$, then $\mathcal{G}f(\chi) = \int_{\mathbf{R}} f(\alpha) e^{-2\pi i \alpha \xi_\chi}\, d\alpha = \mathcal{F}f(\xi_\chi)$.

Exercise 8.9. Let us consider the unitary Banach algebra $L^\infty(\Omega)$ of Example 8.2. The **essential range**, $f[\Omega]$, of $f \in L^\infty(\Omega)$ is the complement of the open set $\bigcup\{G;\ G \text{ open},\ \mu(f^{-1}(G)) = 0\}$. Show that $f[\Omega]$ is the smallest closed subset F of $\mathbf{C}$ such that $\mu(f^{-1}(F^c)) = 0$, $\|f\|_\infty = \max\{|\lambda|;\ \lambda \in f[\Omega]\}$, and $f[\Omega] = \sigma(f)$.

Exercise 8.10. The algebra of **quaternions**, $\mathbf{H}$, is the real Banach space $\mathbf{R}^4$ endowed with the distributive product such that

$$1x = x,\ ij = -ji = k,\ jk = -kj = i,\ ki = -ik = j,\ i^2 = j^2 = k^2 = -1$$

if $x \in \mathbf{H}$, $1 = (1,0,0,0)$, $i = (0,1,0,0)$, $j = (0,0,1,0)$, and $k = (0,0,0,1)$, so that one can write $(a,b,c,d) = a + bi + cj + dk$.

Show that $\mathbf{H}$ is an algebra such that $\|xy\| = \|x\|\|y\|$ and that every nonzero element of A has an inverse.

Remark. It can be shown that every real Banach algebra which is a field is isomorphic to the reals, the complex numbers or the quaternions (cf. Rickart, *General Theory of Banach Algebras*, [**35**, 1.7]). Hence, $\mathbf{C}$ is the only (complex) Banach algebra which is a field and $\mathbf{H}$ is the only real Banach algebra which is a noncommutative field.

Exercise 8.11. If $\chi : A \to \mathbf{C}$ is linear such that $\chi(ab) = \chi(a)\chi(b)$ and $\chi \neq 0$, then prove that $\chi(e) = 1$, so that χ is a character.

Exercise 8.12. Prove that, if $\mathcal{T}$ is a compact topology on $\Delta(A)$ and every function $\widehat{a}$ $(a \in A)$ is $\mathcal{T}$-continuous, then $\mathcal{T}$ is the Gelfand topology.

Exercise 8.13. Prove that the Gelfand transform is an isometric isomorphism from $\mathcal{C}(K)$ onto $\mathcal{C}(\Delta)$.

Exercise 8.14. Let U be the open unit disc of $\mathbf{C}$ and suppose Δ is the spectrum of $H^\infty(U)$. Prove that, through the embedding $U \hookrightarrow \Delta$, U is an open subset of Ω. Write

$$\Delta = D \cup \Big(\bigcup_{\xi \in \partial D} \Delta_\xi \Big),$$

as in (8.1), and prove that the fibers Δ_ξ ($|\xi| = 1$) are homeomorphic to one another.

Exercise 8.15 (Wiener algebra)**.** Show that the set of all 2π-periodic complex functions on $\mathbf{R}$

$$f(t) = \sum_{k=-\infty}^{+\infty} c_k e^{ikt} \qquad \Big(\sum_{k=-\infty}^{+\infty} |c_k| < \infty\Big),$$

with the usual operations and the norm $\|f\|_W := \sum_{k=-\infty}^{+\infty} |c_k|$, is a commutative unitary Banach algebra, W. Moreover prove that the characters of W are the evaluations δ_t on the different points $t \in \mathbf{R}$ and that, if $f \in W$ has no zeros, then $1/f \in W$.

Exercise 8.16. Every $f \in W$ is 2π-periodic and it can be identified as the function F on $\mathbf{T}$ such that $f(t) = F(e^{it})$. If $f(t) = \sum_k c_k e^{ikt}$, $F(z) = \sum_k c_k z^k$. In Exercice 8.15 we have seen that the δ_t ($t \in \mathbf{R}$) are the characters of W, but show that $\mathcal{G} : W \to \mathcal{C}(\mathbf{T})$, one-to-one and with $\|\widehat{f}\| \le \|f\|_W$, is not an isometry and it is not exhaustive.

Exercise 8.17. Suppose A is a unitary Banach algebra and $a \in A$, and denote $M(U) = \sup_{\lambda \in U^c} \|R_a(\lambda)\|$. Prove that, if $U \subset \mathbf{C}$ is an open set and $\sigma_A(a) \subset U$, then $\sigma_A(b) \subset U$ whenever $\|b - a\| < \delta$ if $\delta \le 1/M(U)$ (upper semi-continuity of σ_A).

Exercise 8.18. Let A be a commutative unitary Banach algebra. Prove that the Gelfand transform $\mathcal{G} : A \to \mathcal{C}(\Delta)$ is an isometry if and only if $\|a^2\| = \|a\|^2$ for every $a \in A$.

Show that in order for $\|a\|$ to coincide with the spectral radius $r(a)$, the condition $\|a^2\| = \|a\|^2$ is necessary and sufficient.

Remark. This condition characterizes when a Banach algebra A is a **uniform algebra**, meaning that A is a closed unitary subalgebra of $\mathcal{C}(K)$ for some compact topological space K.

Exercise 8.19. In the definition of an involution, show that property (e), $e^* = e$, is a consequence of (a)–(d). If $x \in A$ is invertible, prove that $(x^*)^{-1} = (x^{-1})^*$.

Exercise 8.20. With the involution $f \mapsto \bar{f}$, where $\bar{f}$ is the complex conjugate of f, show that $\mathcal{C}(K)$ is a commutative C^*-algebra. Similarly, show that $L^\infty(\Omega)$, with the involution $f \mapsto \bar{f}$, is also a commutative C^*-algebra.

Exercise 8.21. If $\{P_n\}_{n=1}^{\infty}$ is a sequence of orthogonal projections and their images are mutually orthogonal, then the series $\sum_{n=1}^{\infty} P_n$ is strongly convergent to the orthogonal projection on the closed linear hull $\bigoplus P_n(H)$ of the images of the projections P_n.

Exercise 8.22. Let $Ax = \sum_{k=1}^{\infty} \lambda_k (x, e_k)_H e_k$ be the spectral representation of a self-adjoint compact operator of H, and let $\{P_n\}_{n=1}^{\infty}$ be the sequence of the orthogonal projections on the different eigensubspaces

$$H_1 = [e_1, \dots, e_{k(1)}], \ \dots, \ H_n = [e_{k(n-1)+1}, \dots, e_{k(n)}], \ \dots$$

for the eigenvalues $\alpha_n = \lambda_{k(n-1)+1} = \dots = \lambda_{k(n)}$ of A.

Show that we can write

$$A = \sum_{n=1}^{\infty} \alpha_n P_n$$

and prove that

$$E(B) = \sum_{\alpha_n \in B} P_n$$

is the resolution of the identity of the spectral resolution of A.

Exercise 8.23. Let μ be a Borel measure on a compact set $K \subset \mathbf{C}$ and let $H = L^2(\mu)$. Show that multiplication by characteristic functions of Borel sets in K, $E(B) := \chi_B \cdot$, is a spectral measure $E : \mathcal{B}_K \to L^2(\mu)$.

Exercise 8.24. If $E : \mathcal{B}_K \to \mathcal{L}(H)$ is a spectral measure, show that the null sets for the spectral measure have the following desirable properties:

(a) If $E(B_n) = 0$ $(n \in \mathbf{N})$, then $E(\bigcup_{n=1}^{\infty} B_n) = 0$.

(b) If $E(B_1) = 0$ and $B_2 \subset B_1$, then $E(B_2) = 0$.

Exercise 8.25. Show that the equivalences of Theorem 8.25 are untrue on the real Hilbert space $\mathbf{R}^2$.

Exercise 8.26. Show that every positive $T \in \mathcal{L}(H)$ in the sense of Theorem 8.25 has a unique positive square root.

Exercise 8.27. With the functional calculus, prove also that, if $T \in \mathcal{L}(H)$ is normal, then it can be written as

$$T = UP$$

with U unitary and P positive. This is the **polar decomposition** of a bounded normal operator in a complex Hilbert space.

References for further reading:

I. M. Gelfand, D. A. Raikov and G. E. Chilov, *Commutative Normed Rings.*

E. Hille and R. S. Phillips, *Functional Analysis and Semigroups.*

T. Kato, *Perturbation Theory for Linear Operators.*

P. D. Lax, *Functional Analysis.*

M. A. Naimark, *Normed Rings.*

M. Reed and B. Simon, *Methods of Modern Mathematical Physics.*

C. E. Rickart, *General Theory of Banach Algebras.*

F. Riesz and B. Sz. Nagy, *Leçons d'analyse fonctionelle.*

W. Rudin, *Functional Analysis.*

A. E. Taylor and D. C. Lay, *Introduction to Functional Analysis.*

K. Yosida, *Functional Analysis.*

Chapter 9

Unbounded operators in a Hilbert space

Up to this moment all of our linear operators have been bounded, but densely defined unbounded operators also occur naturally in connection with the foundations of quantum mechanics.

When in 1927 J. von Neumann[1] introduced axiomatically Hilbert spaces, he recognized the need to extend the spectral theory of self-adjoint operators from the bounded to the unbounded case and immediately started to obtain this extension, which was necessary for his presentation of the transformation theory of quantum mechanics created in 1925–1926 by Heisenberg and Schrödinger.[2]

The definition of unbounded self-adjoint operators on a Hilbert space requires a precise selection of the domain, the symmetry condition $(x, Ax)_H = (Ax, x)_H$ for a densely defined operator not being sufficient for A to be self-adjoint, since its spectrum has to be a subset of $\mathbf{R}$. The creators of quantum

[1] The Hungarian mathematician János (John) von Neumann is considered one of the foremost mathematicians of the 20th century: he was a pioneer of the application of operator theory to quantum mechanics, a member of the Manhattan Project, and a key figure in the development of game theory and of the concepts of cellular automata. Between 1926 and 1930 he taught in the University of Berlin. In 1930 he emigrated to the USA where he was invited to Princeton University and was one of the first four people selected for the faculty of the Institute for Advanced Study (1933–1957).

[2] The German physicist Werner Karl Heisenberg, in Göttingen, was one of the founders of quantum mechanics and the head of the German nuclear energy project; with Max Born and Pascual Jordan, Heisenberg formalized quantum mechanics in 1925 using matrix transformations. The Austrian physicist Erwin Rudolf Josef Alexander Schrödinger, while in Zurich, in 1926 derived what is now known as the Schrödinger wave equation, which is the basis of his development of quantum mechanics. Based on the Born statistical interpretation of quantum theory, P. Dirac and Jordan unified "matrix mechanics" and "wave mechanics" with their "transformation theory".

mechanics did not care about this and it was von Neumann himself who clarified the difference between a self-adjoint operator and a symmetric one.

In this chapter, with the Laplacian as a reference example, we include the Rellich theorem, showing that certain perturbations of self-adjoint operators are still self-adjoint, and the Friedrichs method of constructing a self-adjoint extension of many symmetric operators.

Then the spectral theory of bounded self-adjoint operators on a Hilbert space is extended to the unbounded case by means of the Cayley transform, which changes a self-adjoint operator T into a unitary operator U. The functional calculus of this operator allows us to define the spectral resolution of T.

We include a very short introduction on the principles of quantum mechanics, where an observable, such as position, momentum, and energy, is an unbounded self-adjoint operator, their eigenvalues are the observable values, and the spectral representing measure allows us to evaluate the observable in a given state in terms of the probability of belonging to a given set.[3]

Von Neumann's text "Mathematical Foundations of Quantum Mechanics" [**43**] is strongly recommended here for further reading: special attention is placed on motivation, detailed calculations and examples are given, and the thought processes of a great mathematician appear in a very transparent manner. More modern texts are available, but von Neumann's presentation contains in a lucid and very readable way the germ of his ideas on the subject.

In that book, for the first time most of the modern theory of Hilbert spaces is defined and elaborated, as well as "quantum mechanics in a unified representation which ... is mathematically correct". The author explains that, just as Newton mechanics was associated with infinitesimal calculus, quantum mechanics relies on the Hilbert theory of operators.

With von Neumann's work, quantum mechanics is Hilbert space analysis and, conversely, much of Hilbert space analysis is quantum mechanics.

9.1. Definitions and basic properties

Let H denote a complex linear space. We say that T is an operator on H if it is a linear mapping $T : \mathcal{D}(T) \to H$, defined on a linear subspace $\mathcal{D}(T)$ of H, which is called the **domain** of the operator.

[3]Surprisingly, in this way the atomic spectrum appears as Hilbert's spectrum of an operator. Hilbert himself was extremely surprised to learn that his spectrum could be interpreted as an atomic spectrum in quantum mechanics.

Example 9.1. The derivative operator $D : f \mapsto f'$ (distributional derivative) on $L^2(\mathbf{R})$ has

$$\mathcal{D}(D) = \left\{ f \in L^2(\mathbf{R});\ f' \in L^2(\mathbf{R}) \right\}$$

as its domain, which is the Sobolev space $H^1(\mathbf{R})$. This domain is dense in $L^2(\mathbf{R})$, since it contains $\mathcal{D}(\mathbf{R})$.

Example 9.2. As an operator on $L^2(\mathbf{R})$, the domain of **the position operator**, $Q : f(x) \mapsto xf(x)$, is

$$\mathcal{D}(Q) = \left\{ f \in L^2(\mathbf{R});\ [xf(x)] \in L^2(\mathbf{R}) \right\}.$$

It is unbounded, since $\|\chi_{(n,n+1)}\|_2 = 1$ and $\|Q\chi_{(n,n+1)}\|_2 \geq n$.

Recall that $f \in L^2(\mathbf{R})$ if and only if $\widehat{f} \in L^2(\mathbf{R})$ and both f and $x\widehat{f}(x)$ are in $L^2(\mathbf{R})$ if and only if $f, f' \in L^2(\mathbf{R})$. Thus, the Fourier transform is a unitary operator which maps $\mathcal{D}(Q)$ onto $H^1(\mathbf{R}) = \mathcal{D}(D)$ and changes $2\pi iQ$ into D. Conversely, $2\pi iQ = \mathcal{F}^{-1}D\mathcal{F}$ on $\mathcal{D}(Q)$.

Under these conditions it is said that $2\pi iD$ and Q are **unitarily equivalent**. Unitarily equivalent operators have the same spectral properties.

Of course, it follows that D is also unbounded (see Exercise 9.3).

We are interested in the **spectrum** of T. If for a complex number λ the operator $T - \lambda I : \mathcal{D}(T) \to H$ is bijective and $(T - \lambda I)^{-1} : H \to \mathcal{D}(T) \subset H$ is continuous, then we say that λ is a **regular point** for T.

The spectrum $\sigma(T)$ is the subset of $\mathbf{C}$ which consists of all nonregular points, that is, all complex numbers λ for which $T - \lambda I : \mathcal{D}(T) \to H$ does not have a continuous inverse. Thus $\lambda \in \sigma(T)$ when it is in one of the following disjoint sets:

(a) The **point spectrum** $\sigma_p(T)$, which is the set of the eigenvalues of T. That is, $\lambda \in \sigma_p(T)$ when $T - \lambda I : \mathcal{D}(T) \to H$ is not injective. In this case $(T - \lambda I)^{-1}$ does not exist.

(b) The **continuous spectrum** $\sigma_c(T)$, the set of all $\lambda \in \mathbf{C} \setminus \sigma_p(T)$ such that $T - \lambda I : \mathcal{D}(T) \to H$ is not exhaustive but $\overline{\operatorname{Im}(T - \lambda I)} = H$ and $(T - \lambda I)^{-1}$ is unbounded.

(c) The **residual spectrum** $\sigma_r(T)$, which consists of all $\lambda \in \mathbf{C} \setminus \sigma_p(T)$ such that $\overline{\operatorname{Im}(T - \lambda I)} \neq H$. Then $(T - \lambda I)^{-1}$ exists but is not densely defined.

The set $\sigma(T)^c$ of all regular points is called the **resolvent set**. Thus, $\lambda \in \sigma(T)^c$ when we have $(T - \lambda I)^{-1} \in \mathcal{L}(H)$.

The **resolvent** of T is again the function

$$R_T : \sigma(T)^c \to \mathcal{L}(H), \quad R_T(\lambda) := (T - \lambda I)^{-1}.$$

The spectrum of T is not necessarily a bounded subset of $\mathbf{C}$, but it is still closed and the resolvent function is analytic:

Theorem 9.3. *The set $\sigma(T)^c$ is an open subset of $\mathbf{C}$, and every point $\lambda_0 \in \sigma(T)^c$ has a neighborhood where*

$$R_T(\lambda) = -\sum_{k=0}^{\infty} (\lambda - \lambda_0)^k R_T(\lambda_0)^{k+1},$$

the sum of a convergent Neumann series.

Proof. Let us consider $\lambda = \lambda_0 + \mu$ such that $|\mu| < \|R_T(\lambda_0)\|$. The sum of the Neumann series

$$S(\mu) := \sum_{k=0}^{\infty} \mu^k R_T(\lambda_0)^{k+1} \qquad (|\mu| < 1/\|R_T(\lambda_0)\|)$$

will be the bounded inverse of $T - \lambda I$.

The condition $\|\mu R_T(\lambda_0)\| < 1$ ensures that the series is convergent, and it is easily checked that $(T - \lambda I) S(\mu) = I$:

$$(T - \lambda_0 I - \mu I) \sum_{k=0}^{N} \mu^k ((T - \lambda_0 I)^{-1})^{k+1} = I - (\mu R_T(\lambda_0))^{N+1} \to I,$$

and $S(\mu)$ commutes with T. □

A **graph** is a linear subspace $F \subset H \times H$ such that, for every $x \in H$, the section $F_x := \{y;\ (x, y) \in F\}$ has at most one point, y, so that the first projection $\pi_1(x, y) = x$ is one-to-one on F. This means that $x \mapsto y$ ($y \in F_x$) is an operator T_F on H with $\mathcal{D}(T_F) = \{x \in H;\ F_x \neq \emptyset\}$ and $\mathcal{G}(T_F) = F$.

We write $S \subset T$ if the operator T is an extension of another operator S, that is, if $\mathcal{D}(S) \subset \mathcal{D}(T)$ and $T_{|\mathcal{D}(S)} = S$ or, equivalently, if $\mathcal{G}(S) \subset \mathcal{G}(T)$.

If $\mathcal{G}(T)$ is closed in $H \times H$, then we say that T is a **closed operator**. Also, T is said to be **closable** if it has a closed extension $\tilde{T}$. This means that $\overline{\mathcal{G}(T)}$ is a graph, since, if $\tilde{T}$ is a closed extension of T, $\overline{\mathcal{G}(T)} \subset \mathcal{G}(\tilde{T})$ and ψ_1 is one-to-one on $\mathcal{G}(\bar{T})$, so that it is also one-to-one on $\overline{\mathcal{G}(T)}$. Conversely, if $\overline{\mathcal{G}(T)}$ is a graph, it is the graph of a closed extension of T, since $\mathcal{G}(T) \subset \overline{\mathcal{G}(T)}$.

If T is closable, then $\bar{T}$ will denote the closure of T; that is, $\bar{T} = T_{\overline{\mathcal{G}(T)}}$.

When defining operations with unbounded operators, the domains of the new operators are the intersections of the domains of the terms. Hence

$$\mathcal{D}(S \pm T) = \mathcal{D}(S) \cap \mathcal{D}(T) \ \text{ and } \ \mathcal{D}(ST) = \Big\{x \in \mathcal{D}(T);\ Tx \in \mathcal{D}(S)\Big\}.$$

Example 9.4. The domain of the commutator $[D, Q] = DQ - QD$ of the derivation operator with the position operator on $L^2(\mathbf{R})$ is $\mathcal{D}(DQ) \cap \mathcal{D}(QD)$, which contains $\mathcal{D}(\mathbf{R})$, a dense subspace of $L^2(\mathbf{R})$.

Since $D(xf(x)) - xDf(x) = f(x)$, the commutator $[D, Q]$ coincides with the identity operator on its domain, so that we simply write $[D, Q] = I$ and consider it as an operator on $L^2(\mathbf{R})$.

9.1.1. The adjoint.

We will only be interested in **densely defined** operators, which are the operators T such that $\overline{\mathcal{D}(T)} = H$.

If T is densely defined, then every bounded linear form on $\mathcal{D}(T)$ has a unique extension to H, and from the Riesz representation Theorem 4.1 we know that it is of the type $(\cdot, z)_H$. This fact allows us to define the **adjoint** T^* of T. Its domain is defined as

$$\mathcal{D}(T^*) = \{y \in H;\ x \mapsto (Tx, y)_H \text{ is bounded on } \mathcal{D}(T)\}$$

and, if $y \in \mathcal{D}(T^*)$, $T^*y \in H$ is the unique element such that

$$(Tx, y)_H = (x, T^*y)_H \qquad (x \in \mathcal{D}(T)).$$

Hence, $y \in \mathcal{D}(T^*)$ if and only if $(Tx, y)_H = (x, y^*)_H$ for some $y^* \in H$, for all $x \in \mathcal{D}(T)$, and then $y^* = T^*y$.

Theorem 9.5. *Let T be densely defined. Then the following properties hold:*

(a) $(\lambda T)^* = \bar{\lambda} T^*$.

(b) $(I + T)^* = I + T^*$.

(c) *T^* is closed.*

(d) *If $T : \mathcal{D}(T) \to H$ is one-to-one with dense image, then T^* is also one-to-one and densely defined, and $(T^{-1})^* = (T^*)^{-1}$.*

Proof. Both (a) and (b) are easy exercises.

To show that the graph of T^* is closed, suppose that $(y_n, T^*y_n) \to (y, z)$ ($y_n \in \mathcal{D}(T^*)$). Then $(x, T^*y_n)_H \to (x, z)_H$ and $(Tx, y_n)_H \to (Tx, y)_H$ for every $x \in \mathcal{D}(T)$, with $(x, T^*y_n)_H = (Tx, y_n)_H$. Hence $(x, z)_H = (Tx, y)_H$ and $z = T^*y$, so that $(y, z) \in \mathcal{G}(T^*)$.

In (d) the inverse $T^{-1} : \operatorname{Im} T \to \mathcal{D}(T)$ is a well-defined operator with dense domain and image. We need to prove that $(T^*)^{-1}$ exists and coincides with $(T^{-1})^*$.

First note that $T^*y \in \mathcal{D}((T^{-1})^*)$ for every $y \in \mathcal{D}(T^*)$, since the linear form $x \mapsto (T^{-1}x, T^*y)_H = (x, y)_H$ on $\mathcal{D}(T^{-1})$ is bounded and T^*y is well-defined. Moreover $(T^{-1})^*T^*y = y$, so that $(T^{-1})^*T^* = I$ on $\mathcal{D}(T^*)$, $(T^*)^{-1} : \operatorname{Im} T^* \to \mathcal{D}(T^*)$, and

$$(T^*)^{-1} \subset (T^{-1})^*$$

since, for $y = (T^*)^{-1}z$ in $(T^{-1})^*T^*y = y$, we have $(T^{-1})^*z = (T^*)^{-1}z$.

To also prove that $(T^{-1})^* \subset (T^*)^{-1}$, let $x \in \mathcal{D}(T)$ and $y \in \mathcal{D}((T^*)^{-1})$. Then $Tx \in \operatorname{Im}(T) = \mathcal{D}(T^{-1})$ and

$$(Tx, (T^{-1})^* y)_H = (x, y)_H, \quad (Tx, (T^{-1})^* y)_H = (x, T^*(T^{-1})^* y)_H.$$

Thus, $(T^{-1})^* y \in \mathcal{D}(T^*)$ and $T^*(T^{-1})^* y = y$, so that $T^*(T^{-1})^* = I$ on $\mathcal{D}((T^*)^{-1}) = \operatorname{Im} T^*$, and $(T^*)^{-1} : \operatorname{Im}(T^*) \to \mathcal{D}(T^*)$ is bijective. □

It is useful to consider the "rotation operator" $G : H \times H \to H \times H$, such that $G(x, y) = (-y, x)$. It is an isometric isomorphism with respect to the norm $\|(x, y)\| := (\|x\|_H^2 + \|y\|_H^2)^{1/2}$ associated to the scalar product

$$((x, y), (x', y'))_{H \times H} := (x, x')_H + (y, y')_H,$$

which makes $H \times H$ a Hilbert space. Observe that $G^2 = -I$.

Theorem 9.6. *If T is closed and densely defined, then*

$$H \times H = G(\mathcal{G}(T)) \oplus \mathcal{G}(T^*) = \mathcal{G}(T) \oplus G(\mathcal{G}(T^*)),$$

orthogonal direct sums, T^* is also closed and densely defined, and $T^{**} = T$.

Proof. Let us first prove that $\mathcal{G}(T^*) = G(\mathcal{G}(T))^\perp$, showing the first equality, and that T^* is closed. Since $(y, z) \in \mathcal{G}(T^*)$ if and only if $(Tx, y)_H = (x, z)_H$ for every $x \in \mathcal{D}(T)$, we have

$$(G(x, Tx), (y, z))_{H \times H} = ((-Tx, x), (y, z))_{H \times H} = 0,$$

and this means that $(y, z) \in G(\mathcal{G}(T))^\perp$, so that $\mathcal{G}(T^*) = G(\mathcal{G}(T))^\perp$.

Also, since $G^2 = -I$,

$$H \times H = G(G(\mathcal{G}(T)) \oplus \mathcal{G}(T^*)) = \mathcal{G}(T) \oplus G(\mathcal{G}(T^*)).$$

If $(z, y)_H = 0$ for all $y \in \mathcal{D}(T^*)$, then $((0, z), (-T^* y, y))_{H \times H} = 0$. Hence, $(0, z) \in G(\mathcal{G}(T^*))^\perp = \mathcal{G}(T)$ and it follows that $z = T0 = 0$. Thus, $\mathcal{D}(T^*)$ is dense in H.

Finally, since also $H \times H = G(\mathcal{G}(T^*)) \oplus \mathcal{G}(T^{**})$ and $\mathcal{G}(T)$ is the orthogonal complement of $G(\mathcal{G}(T^*))$, we obtain the identity $T = T^{**}$. □

9.2. Unbounded self-adjoint operators

$T : \mathcal{D}(T) \subset H \to H$ is still a possibly unbounded linear operator on the complex Hilbert space H.

9.2.1. Self-adjoint operators. We say that the operator T is **symmetric** if it is densely defined and

$$(Tx, y)_H = (x, Ty)_H \qquad (x, y \in \mathcal{D}(T)).$$

Note that this condition means that $T \subset T^*$.

Theorem 9.7. *Every symmetric operator T is closable and its closure is T^{**}.*

Proof. Since T is symmetric, $T \subset T^*$ and, $\mathcal{G}(T^*)$ being closed,

$$\mathcal{G}(T) \subset \overline{\mathcal{G}(T)} \subset \mathcal{G}(T^*).$$

Hence $\overline{\mathcal{G}(T)}$ is a graph, T is closable, and $\overline{\mathcal{G}(T)}$ is the graph of $\bar{T}$. As a consequence, let us show that the domain of T^* is dense.

According to Theorem 9.6, $(x, y) \in \mathcal{G}(T^*)$ if and only if $(-y, x) \in \mathcal{G}(T)^\perp$, in $H \times H$. Hence,

$$\overline{\mathcal{G}(T)} = \mathcal{G}(T)^{\perp\perp} = \{(T^*x, -x);\, x \in \mathcal{D}(T^*)\}^\perp.$$

This subspace is not a graph if and only if $(y, z_1), (y, z_2) \in \overline{\mathcal{G}(T)}$ for two different points $z_1, z_2 \in H$; that is, $(0, z) \in \{(T^*x, -x);\, x \in \mathcal{D}(T^*)\}^\perp$ for some $z \neq 0$. Then $(z, x)_H = 0$ for all $x \in \mathcal{D}(T^*)$ which means that $0 \neq z \in \mathcal{D}(T^*)^\perp$, and it follows that $\overline{\mathcal{D}(T^*)} \neq H$.

Since $\overline{\mathcal{D}(T^*)} = H$, T^{**} is well-defined. We need to prove that

$$\overline{\mathcal{G}(T)} = \mathcal{G}(\bar{T}) = \{(T^*x, -x);\, x \in \mathcal{D}(T^*)\}^\perp$$

is $\mathcal{G}(T^{**})$. But $(v, u) \in \overline{\mathcal{G}(T)}$ if and only if $(T^*x, v)_H - (x, u)_H = 0$ for all $x \in \mathcal{D}(T^*)$; that is, $v \in \mathcal{D}(T^{**})$ and $u = T^{**}v$, which means that $(v, u) \in \mathcal{G}(T^{**})$. □

The operator T is called **self-adjoint** if it is densely defined and $T = T^*$, i.e., if it is symmetric and

$$\mathcal{D}(T^*) \subset \mathcal{D}(T),$$

this inclusion meaning that the existence of $y^* \in H$ such that $(Tx, y)_H = (x, y^*)_H$ for all $x \in \mathcal{D}(T)$ implies $y^* = Tx$.

Theorem 9.8. *If T is self-adjoint and S is a symmetric extension of T, then $S = T$. Hence T does not have any strict symmetric extension; it is "maximally symmetric".*

Proof. It is clear that $T = T^* \subset S$ and $S \subset S^*$, since S is symmetric. It follows from the definition of a self-adjoint operator that $T \subset S$ implies $S^* \subset T^*$. From $S \subset S^* \subset T \subset S$ we obtain the identity $S = T$. □

We are going to show that, in the unbounded case, the spectrum of a self-adjoint operator is also real. This property characterizes the closed symmetric operators that are self-adjoint.

First note that, if $T = T^*$, the point spectrum is real, since if $Tx = \lambda x$ and $0 \neq x \in \mathcal{D}(T)$, then

$$\bar{\lambda}(x,x)_H = (x,Tx)_H = (Tx,x)_H = \lambda(x,x)_H$$

and $\bar{\lambda} = \lambda$.

Theorem 9.9. *Suppose that T is self-adjoint. The following properties hold:*

(a) $\lambda \in \sigma(T)^c$ *if and only if* $\|Tx - \lambda x\|_H \geq c\|x\|_H$ *for all* $x \in \mathcal{D}(T)$, *for some constant* $c > 0$.

(b) *The spectrum* $\sigma(T)$ *is real and closed.*

(c) $\lambda \in \sigma(T)$ *if and only if* $Tx_n - \lambda x_n \to 0$ *for some sequence* $\{x_n\}$ *in* $\mathcal{D}(T)$ *such that* $\|x_n\|_H = 1$ (λ *is an* **approximate eigenvalue**).

(d) *The inequality* $\|R_T(\lambda)\| \leq 1/|\Im\lambda|$ *holds.*

Proof. (a) If $\lambda \in \sigma(T)^c$, then $R_T(\lambda) \in \mathcal{L}(H)$ and

$$\|x\|_H \leq \|R_T(\lambda)\|\|(T-\lambda I)x\|_H = c^{-1}\|(T-\lambda I)x\|_H.$$

Suppose now that $\|Tx - \lambda x\|_H \geq c\|x\|_H$ and let $M = \operatorname{Im}(T-\lambda I)$, so that we have $T - \lambda I : \mathcal{D}(T) \to M$ with continuous inverse. To prove that $M = H$, let us first show that M is dense in H.

If $z \in M^\perp$, then for every $Tx - \lambda x \in M$ we have

$$0 = (Tx - \lambda x, z)_H = (Tx,z)_H - \lambda(x,z)_H.$$

Hence $(Tx,z)_H = (x,\bar{\lambda}z)_H$ if $x \in \mathcal{D}(T)$, and then $z \in \mathcal{D}(T^*) = \mathcal{D}(T)$ and $Tz = \bar{\lambda}z$. Suppose $z \neq 0$, so that $\bar{\lambda} = \lambda$ and we arrive at $Tz - \lambda z = 0$ and $0 \neq z \in M$, a contradiction. Thus, $M^\perp = 0$ and M is dense.

To prove that M is closed in H, let $M \ni y_n = Tx_n - \lambda x_n \to y$. Then $\|x_p - x_q\| \leq c^{-1}\|y_p - y_q\|_H$, and there exist $x = \lim x_n \in H$ and $\lim_n Tx_n = y + \lambda x$. But T is closed, so that $Tx = y + \lambda x$ and $y \in M$.

(b) To show that every $\lambda = \alpha + i\beta \in \sigma(T)$ is real, observe that, if $x \in \mathcal{D}(T)$,

$$(Tx-\lambda x, x)_H = (Tx,x)_H - \lambda(x,x)_H, \quad \overline{(Tx-\lambda x, x)}_H = (Tx,x)_H - \bar{\lambda}(x,x)_H,$$

since $(Tx,x)_H \in \mathbf{R}$. Subtracting,

$$\overline{(Tx-\lambda x, x)}_H - (Tx-\lambda x,x)_H = 2i\beta\|x\|_H^2,$$

where $\overline{(Tx-\lambda x, x)}_H - (Tx-\lambda x,x)_H = -2i\operatorname{Im}(Tx-\lambda x,x)_H$. Hence,

$$|\beta|\|x\|_H^2 = |\operatorname{Im}(Tx-\lambda x)_H| \leq |(Tx-\lambda x,x)_H| \leq \|Tx-\lambda x\|_H\|x\|_H$$

and then $|\beta|\|x\|_H \le \|Tx - \lambda x\|_H$ if $x \in \mathcal{D}(T)$. As seen in the proof of (a), the assumption $\beta \neq 0$ would imply $\lambda \in \sigma(T)^c$.

(c) If $\lambda \in \sigma(T)$, the estimate in (a) does not hold and then, for every $c = 1/n$, we can choose $x_n \in \mathcal{D}(T)$ with norm one such that $\|Tx_n - \lambda x_n\|_H \le 1/n$ and λ is an approximate eigenvalue. Every approximate eigenvalue λ is in $\sigma(T)$, since, if $(T-\lambda I)^{-1}$ were bounded on H, then it would follow from $Tx_n - \lambda x_n \to 0$ that $x_n = (T-\lambda I)^{-1}(Tx_n - \lambda x_n) \to 0$, a contradiction to $\|x_n\|_H = 1$.

(d) If $y \in \mathcal{D}(T)$ and $\lambda = \Re\lambda + i\Im\lambda \notin \mathbf{R}$, then it follows that

$$\|(T-\lambda I)y\|_H^2 = (Ty - \lambda y, Ty - \lambda y)_H \ge ((\Im\lambda)y, (\Im\lambda)y)_H = |\Im\lambda|^2\|y\|_H^2.$$

If $x = (T-\lambda I)y \in H$, then $y = R_T(\lambda)x$ and $\|x\|_H^2 \ge |\Im\lambda|^2\|R_T(\lambda)x\|_H^2$; thus $|\Im\lambda|\|R_T(\lambda)\| \le 1$. □

The condition $\sigma(T) \subset \mathbf{R}$ is sufficient for a symmetric operator to be self-adjoint. In fact we have more:

Theorem 9.10. *Suppose that T is symmetric. If there exists $z \in \mathbf{C}\setminus\mathbf{R}$ such that $z, \bar{z} \in \sigma(T)^c$, then T is self-adjoint.*

Proof. Let us first show that $((T-zI)^{-1})^* = (T-\bar{z}I)^{-1}$, that is,

$$((T-zI)^{-1}x_1, x_2)_H = (x_1, (T-\bar{z}I)^{-1}x_2)_H.$$

We denote $(T-zI)^{-1}x_1 = y_1$ and $(T-\bar{z}I)^{-1}x_2 = y_2$. The desired identity means that $(y_1, (T-\bar{z}I)y_2)_H = ((T-zI)y_1, y_2)_H$ and it is true if $y_1, y_2 \in \mathcal{D}(T)$, since T is symmetric. But the images of $T - zI$ and $T - \bar{z}I$ are both the whole space H, so that $((T-zI)^{-1}x_1, x_2)_H = (x_1, (T-\bar{z}I)^{-1}x_2)_H$ holds for any $x_1, x_2 \in H$.

Now we can prove that $\mathcal{D}(T^*) \subset \mathcal{D}(T)$. Let $v \in \mathcal{D}(T^*)$ and $w = T^*v$, i.e.,

$$(Ty_1, v)_H = (y_1, w)_H \qquad (\forall y_1 \in \mathcal{D}(T)).$$

We subtract $z(y_1, v)_H$ to obtain

$$((T-zI)y_1, v)_H = (y_1, w - \bar{z}v)_H.$$

Still with the notation $(T-zI)^{-1}x_1 = y_1$ and $(T-\bar{z}I)^{-1}x_2 = y_2$, but now with $x_2 = w - \bar{z}v$, since $(x_1, v)_H = ((T-zI)y_1, v)_H = (y_1, w-\bar{z}v)_H$,

$$(x_1, v)_H = \big((T-zI)^{-1}x_1, w-\bar{z}v\big)_H = \big(x_1, (T-\bar{z}I)^{-1}(w-\bar{z}v)\big)_H \quad (\forall x_1 \in H).$$

Thus, $v = (T-\bar{z}I)^{-1}(w-\bar{z}v)$ and $v \in \operatorname{Im}(T-\bar{z})^{-1} = \mathcal{D}(A)$. □

In the preceding proof, we have only needed the existence of $z \notin \mathbf{R}$ such that $(T-zI)^{-1}$ and $(T-\bar{z}I)^{-1}$ are defined on H, but not their continuity.

Example 9.11. The position operator Q of Example 9.2 is self-adjoint and $\sigma(Q) = \mathbf{R}$, but it does not have an eigenvalue.

Obviously Q is symmetric, since, if $f, g, [xf(x)], [xg(x)] \in L^2(\mathbf{R})$, then

$$(Qf, g)_2 = \int_{\mathbf{R}} xf(x)\overline{g(x)}\,dx = \int_{\mathbf{R}} f(x)\overline{xg(x)}\,dx = (f, Qg)_2.$$

Let us show that $\mathcal{D}(Q^*) \subset \mathcal{D}(Q)$. If $g \in \mathcal{D}(Q^*)$, then there exists $g^* \in L^2(\mathbf{R})$ such that $(Qf, g)_2 = (f, g^*)_2$ for all $f \in \mathcal{D}(Q)$. So $\int_{\mathbf{R}} \varphi(x)\overline{xg(x)}\,dx = \int_{\mathbf{R}} \varphi(x)\overline{g^*(x)}\,dx$ if $\varphi \in \mathcal{D}(\mathbf{R})$, and then $g^*(x) = xg(x)$ a.e. on $\mathbf{R}$. I.e., $xg(x)$ is in $L^2(\mathbf{R})$ and $g \in \mathcal{D}(Q)$. Hence Q is self-adjoint, since it is symmetric and $\mathcal{D}(Q^*) \subset \mathcal{D}(Q)$.

If $\lambda \notin \sigma(Q)$, then $T = (Q - \lambda I)^{-1} \in \mathcal{L}(H)$ and, for every $g \in L^2(\mathbf{R})$, the equality $(Q - \lambda I)Tg = g$ implies that $(Tg)(x) = g(x)/(x - \lambda) \in L^2(\mathbf{R})$ and $\lambda \notin \mathbf{R}$, since, when $\lambda \in \mathbf{R}$ and $g := \chi_{(\lambda,b)}$, $g(x)/(x - \lambda) \notin L^2(\mathbf{R})$.

Example 9.12. The adjoint of the derivative operator D of Example 9.1 is $-D$, and iD is self-adjoint. The spectrum of iD is also $\mathbf{R}$ and it has no eigenvalues.

By means of the Fourier transform we can transfer the properties of Q. If $f, g \in H^1(\mathbf{R})$, then $(Df, g)_2 = (\widehat{Df}, \widehat{g})_2$. From $(Q\widehat{f}, \widehat{g})_2 = (\widehat{f}, Q\widehat{g})_2$ and $\widehat{Df}(x) = 2\pi i t\widehat{f}(x) = 2\pi i(Q\widehat{f})(x)$ we obtain

$$(Df, g)_2 = (2\pi i Q\widehat{f}, \widehat{g})_2 = (\widehat{f}, -2\pi i Q\widehat{g})_2 = (f, -Dg)_2$$

and $-D \subset D^*$. As in the case of Q, also $\mathcal{D}(D^*) \subset \mathcal{D}(D)$.

Furthermore, $(iD)^* = -iD^* = iD$ and $\sigma(iD) \subset \mathbf{R}$. If $T = (iD - \lambda I)^{-1}$ and $g \in L^2$, then an application of the Fourier transform to $(iD - \lambda I)Tg = g$ gives $-2\pi x\widehat{Tg}(x) - \lambda\widehat{Tg}(x) = \widehat{g}(x)$, and

$$\widehat{Tg}(x) = -\frac{\widehat{g}(x)}{2\pi x + \lambda}$$

has to lie in $L^2(\mathbf{R})$. So we arrive to $\lambda \notin \mathbf{R}$ by taking convenient functions $\widehat{g} = \chi_{(a,b)}$.

Example 9.13. The Laplace operator Δ of $L^2(\mathbf{R}^n)$ with domain $H^2(\mathbf{R}^n)$ is self-adjoint. Its spectrum is $\sigma(\Delta) = [0, \infty)$.

Recall that

$$\begin{aligned} H^2(\mathbf{R}^n) &= \Big\{u \in L^2(\mathbf{R}^n);\ D^\alpha u \in L^2(\mathbf{R}^n),\ |\alpha| \le 2\Big\} \\ &= \Big\{u \in L^2(\mathbf{R}^n);\ \int_{\mathbf{R}^n} |(1 + |\xi|^2)\widehat{u}(\xi)|^2\,d\xi < \infty\Big\} \end{aligned}$$

and Δ, with this domain, is symmetric: $(\Delta u, v)_2 = (u, \Delta v)_2$ follows from the Fourier transforms, since

$$\int_{\mathbf{R}^n} \widehat{u}(\xi)|\xi|^2\overline{\widehat{v}(\xi)}\, d\xi = \int_{\mathbf{R}^n} \widehat{u}(\xi)\overline{|\xi|^2\widehat{v}(\xi)}\, d\xi.$$

To prove that it is self-adjoint, let $u \in \mathcal{D}(\Delta^*) \subset L^2(\mathbf{R}^n)$. If $w \in L^2(\mathbf{R}^n)$ is such that

$$(\Delta v, u)_2 = (v, w)_2 \qquad (v \in H^2(\mathbf{R}^n)),$$

then, up to a nonzero multiplicative constant,

$$\int_{\mathbf{R}^n} \widehat{v}(\xi)|\xi|^2\overline{\widehat{u}(\xi)}\, d\xi = \int_{\mathbf{R}^n} \widehat{v}(\xi)\overline{\widehat{w}(\xi)}\, d\xi$$

for every $\widehat{v} \in H^2(\mathbf{R}^n)$, a dense subspace of $L^2(\mathbf{R}^n)$, and $|\xi|^2\widehat{u}(\xi) = c\widehat{w}(\xi)$, in $L^2(\mathbf{R}^n)$. Hence, $\int_{\mathbf{R}^n} |(1+|\xi|^2)\widehat{u}(\xi)|^2\, d\xi < \infty$ and $u \in H^1(\mathbf{R}^n) = \mathcal{D}(\Delta)$. Thus, $\mathcal{D}(\Delta^*) \subset \mathcal{D}(\Delta)$.

The Fourier transform, $\mathcal{F}$, is a unitary operator of $L^2(\mathbf{R}^n)$, so that the spectrum of Δ is the same as the spectrum of the multiplication operator $\mathcal{F}\Delta\mathcal{F}^{-1} = 4\pi^2|\xi|^2\cdot$, which is self-adjoint with domain

$$\Big\{f \in L^2(\mathbf{R}^n);\ \int_{\mathbf{R}^n} |(1+|\xi|^2)f(\xi)|^2\, d\xi < \infty\Big\},$$

and $\lambda \in \sigma(\mathcal{F}\Delta\mathcal{F}^{-1})^c$ if and only if the multiplication by $4\pi^2|\xi|^2 - \lambda$ has a continuous inverse on $L^2(\mathbf{R}^n)$, the multiplication by $1/(4\pi^2|\xi|^2 - \lambda)$. This means that $\lambda \neq 4\pi^2|\xi|^2$ for every $\xi \in \mathbf{R}^n$, i.e., $\lambda \notin [0, \infty)$.

An application of Theorem 9.10 shows that a perturbation of a self-adjoint operator with a "small" symmetric operator is still self-adjoint. For a more precise statement of this fact, let us say that an operator S is **relatively bounded**, with constant α, with respect to another operator A if $\mathcal{D}(A) \subset \mathcal{D}(S)$ and there are two constants $\alpha, c \geq 0$ such that

$$\|Sx\|_H^2 \leq \alpha^2\|Ax\|_H^2 + c^2\|x\|_H^2 \qquad (x \in \mathcal{D}). \tag{9.1}$$

Let us check that this kind of estimate is equivalent to

$$\|Sx\|_H \leq \alpha'\|Ax\|_H + c'\|x\|_H \qquad (x \in \mathcal{D}) \tag{9.2}$$

and that we can take $\alpha' < 1$ if $\alpha < 1$ and $\alpha < 1$ if $\alpha' < 1$.

By completing the square, it is clear that (9.2) follows from (9.1) with $\alpha = \alpha'$ and $c = c'$. Also, from (9.2) we obtain (9.1) with $\alpha^2 = (1+\varepsilon^{-1})\alpha'^2$ and $c^2 = (1+\varepsilon)c'^2$, for any $\varepsilon > 0$, since $2\alpha'\|Ax\|_H c'\|x\|_H \leq \varepsilon^{-1}\alpha'^2\|Ax\|_H^2 + \varepsilon c'^2\|x\|_H^2$ and an easy substitution shows that

$$(\alpha'\|Ax\|_H + c'\|x\|_H)^2 \leq \alpha^2\|Ax\|_H^2 + c^2\|x\|_H^2.$$

Theorem 9.14 (Rellich[4]). *Let A be a self-adjoint operator and let S be symmetric, with the same domain $D \subset H$. If S is relatively bounded with constant α with respect to A, then $T = A+S$ is also self-adjoint with domain D.*

Proof. Let us first check that the symmetric operator T is closed. If $(x, y) \in \overline{\mathcal{G}(T)}$, then we choose $x_n \in D$ so that $x_n \to x$ and $Tx_n \to y$. From the hypothesis we have

$$\|Ax_n - Ax_m\|_H \leq \|Tx_n - Tx_m\|_H + \alpha\|Ax_n - Ax_m\|_H + c\|x_n - x_m\|_H,$$

which implies

$$\|Ax_n - Ax_m\|_H \leq \frac{1}{1-\alpha}\|Tx_n - Tx_m\|_H + \frac{c}{1-\alpha}\|x_n - x_m\|_H,$$

and there exists $z = \lim_n Ax_n$. But A is closed and $z = Ax$ with $x \in D$.

Moreover $\|Sx_n - Sx\|_H \leq \alpha\|A(x_n - x)\|_H + c\|x_n - x\|_H$ and $Sx_n \to Sx$. Hence $y = \lim_n Tx_n = Tx$ and $(x, y) \in \mathcal{G}(T)$.

The operators $T - zI$ ($z \in \mathbf{C}$), with domain D, are also closed. With Theorem 9.10 in hand, we only need to check that $\pm\lambda i \in \sigma(T)^c$ when $\lambda \in \mathbf{R}$ is large enough ($|\lambda| > c$).

To show that $T - \lambda iI$ is one-to-one if $\lambda \neq 0$, let

$$(T - \lambda iI)x = y \qquad (x \in D)$$

and note that the absolute values of the imaginary parts of both sides of

$$(Tx, x)_H - \lambda i(x, x)_H = (y, x)_H$$

are equal, so that $|\lambda|\|x\|_H^2 = |\Im(y, x)_H| \leq \|y\|_H\|x\|_H$ and

$$\|x\|_H \leq |\lambda|^{-1}\|y\|_H \qquad (x \in D).$$

Thus, $y = 0$ implies $x = 0$.

Let us prove now that $T - \lambda iI$ has a closed image. Let $y_n \to y$ with $y_n = (T - \lambda iI)x_n$. Then $\|x_n - x_m\|_H \leq |\lambda|^{-1}\|y_n - y_m\|_H$ and the limit $x = \lim x_n \in H$ exists. Since $(T - \lambda iI)x_n \to y$ and the graph of $T - \lambda iI$ is closed, $x \in D$ and $y = (T - \lambda iI)x \in \operatorname{Im}(T - \lambda iI)$.

Let us also show that $\operatorname{Im}(T - \lambda iI) = H$ by proving that the orthogonal is zero. Let $v \in H$ be such that

$$(Ax + Sx - \lambda ix, v)_H = 0 \qquad (x \in D).$$

[4]F. Rellich worked on the foundations of quantum mechanics and on partial differential equations, and his most important contributions, around 1940, refer to the perturbation of the spectrum of self-adjoint operators $A(\epsilon)$ which depend on a parameter ϵ. See also footnote 15 in Chapter 7.

Then $(A - \lambda i)(D) = H$, since $\lambda i \in \sigma(A)^c$. If $(A - \lambda i I)u = v$, let $x = u$, and then we obtain that

$$((A - \lambda i)u, (A - \lambda i)u)_H + (Su, (A - \lambda i)u)_H = 0.$$

From the Cauchy-Schwarz inequality, $\|Au - \lambda iu\|_H^2 \leq \|Su\|_H \|Au - \lambda iu\|_H$ and

$$\|Au - \lambda iu\|_H \leq \|Su\|_H.$$

Since A is symmetric, $(Ay - \lambda iy, Ay - \lambda iy)_H = \|Ay\|_H^2 + \lambda^2 \|y\|_H^2$ and

$$\|Au\|^2 + \lambda^2 \|u\|_H^2 = \|Au - \lambda iu\|_H^2 \leq \|Su\|_H^2 \leq \alpha^2 \|Au\|_H^2 + c^2 \|u\|_H^2.$$

But, if $|\lambda| > c$, the condition $\alpha^2 < 1$ implies $u = 0$, and then $v = 0$.

We have proved that $(T - \lambda i I)^{-1} : H \to H$ is well-defined and closed, i.e., it is bounded. Hence, $\pm i\lambda \in \sigma(T)^c$ and the symmetric operator T is self-adjoint. □

Example 9.15. The operator $H = -\Delta - |x|^{-1}$ on $L^2(\mathbf{R}^3)$, with domain $H^2(\mathbf{R}^3)$, is self-adjoint.

Let $-|x|^{-1} = V_0(x) + V_1(x)$ with $V_0(x) := \chi_B(x)V(x)$ ($B = \{|x| \leq 1\}$). Multiplication by the real function $|x|^{-1}$ is a symmetric operator whose domain contains $H^2(\mathbf{R}^3)$, the domain of $-\Delta$, since $V_0 u \in L^2(\mathbf{R}^3)$ if $u \in H^2(\mathbf{R}^3)$, with

$$\|V_0 u\|_2 \leq \|V_0\|_\infty \|u\|_2 = \|u\|_2$$

and $\|V_1 u\|_2 \leq \|V_1\|_2 \|u\|_\infty$, where $\|u\|_\infty \leq \|\widehat{u}\|_1$. To apply Theorem 9.14, we will show that multiplication by $-|x|^{-1}$ is relatively bounded with respect to $-\Delta$.

From the Cauchy-Schwarz inequality and from the relationship between the Fourier transform and the derivatives, we obtain

$$\left(\int_{\mathbf{R}^3} |\widehat{u}(\xi)|\, d\xi \right)^2 \leq \int_{\mathbf{R}^3} \frac{d\xi}{(|\frac{2\pi}{\xi}|^2 + \beta^2)^2} \|(-\Delta + \beta^2 I)u\|_2^2 = \frac{2\pi^3}{\xi} \|(-\Delta + \beta^2 I)u\|_2^2.$$

From the inversion theorem we obtain that u is bounded and continuous, since it is the Fourier co-transform of the integrable function $\widehat{u}$. Then

$$\|V_1 u\|_2 \leq c(\beta^{-1/2}\| - \Delta u\|_2 + \beta^{3/2}\|u\|_2) \qquad (u \in H^2(\mathbf{R}^3))$$

so that

$$\|Vu\|_2 \leq c\beta^{-1/2}\| - \Delta u\|_2 + (c\beta^{3/2} + 1)\|u\|_2$$

and $c\beta^{-1/2} < 1$ if β is large.

It follows from the Rellich theorem that H is self-adjoint with domain $H^2(\mathbf{R}^3)$.

9.2.2. Essentially self-adjoint operators. Very often, operators appear to be symmetric but they are not self-adjoint, and in order to apply the spectral theory, it will be useful to know whether they have a self-adjoint extension. Recall that a symmetric operator is closable and that a self-adjoint operator is always closed and maximally symmetic.

A symmetric operator is said to be **essentially self-adjoint** if its closure is self-adjoint. In this case, the closure is the unique self-adjoint extension of the operator.

Example 9.16. It follows from Example 9.13 that the Laplacian Δ, as an operator on $L^2(\mathbf{R}^n)$ with domain $\mathcal{S}(\mathbf{R}^n)$, is essentially self-adjoint. Its closure is again Δ, but with domain $H^2(\mathbf{R}^n)$.

Theorem 9.17. *If T is symmetric and a sequence $\{u_n\}_{n\in\mathbf{N}} \subset \mathcal{D}(T)$ is an orthonormal basis of H such that $Tu_n = \lambda_n u_n$ $(n \in \mathbf{N})$, then T is essentially self-adjoint and the spectrum of its self-adjoint extension $\bar{T}$ is $\sigma(\bar{T}) = \overline{\{\lambda_n;\, n \in \mathbf{N}\}}$.*

Proof. The eigenvalues λ_n are all real. Define

$$\mathcal{D}(\bar{T}) := \Big\{ \sum_{n=1}^{\infty} \alpha_n u_n;\, \sum_{n=1}^{\infty} |\alpha_n|^2 + \sum_{n=1}^{\infty} \lambda_n^2 |\alpha_n|^2 < \infty \Big\},$$

a linear subspace of H that contains $\mathcal{D}(T)$, since, if $x = \sum_{n=1}^{\infty} \alpha_n u_n \in \mathcal{D}(T)$ and $Tx = \sum_{n=1}^{\infty} \beta_n u_n \in H$, the Fourier coefficients α_n and β_n satisfy

$$\beta_n = (Tx, u_n)_H = (x, Tu_n)_H = \lambda_n (x, u_n)_H = \lambda_n \alpha_n$$

and $\{\alpha_n\}, \{\lambda_n \alpha_n\} \in \ell^2$.

We can define the operator $\bar{T}$ on $\mathcal{D}(\bar{T})$ by

$$\bar{T}\Big(\sum_{n=1}^{\infty} \alpha_n u_n\Big) := \sum_{n=1}^{\infty} \lambda_n \alpha_n u_n.$$

Let us show that $\bar{T}$ is a self-adjoint extension of T.

It is clear that $\bar{T}$ is symmetric, $T \subset \bar{T}$, and every λ_n is an eigenvalue of $\bar{T}$, so that $\overline{\{\lambda_n;\, n \in \mathbf{N}\}} \subset \sigma(\bar{T})$.

If $\lambda \notin \overline{\{\lambda_n;\, n \in \mathbf{N}\}}$, so that $|\lambda - \lambda_n| \geq \delta > 0$, then it follows that $\lambda \notin \sigma(\bar{T})$ since we can construct the inverse of

$$(T - \lambda I)\Big(\sum_{n=1}^{\infty} \alpha_n u_n\Big) = \sum_{n=1}^{\infty} \alpha_n (\lambda_n - \lambda) u_n$$

by defining

$$R\Big(\sum_{n=1}^{\infty} \alpha_n u_n\Big) := \sum_{n=1}^{\infty} \frac{\alpha_n}{\lambda_n - \lambda} u_n.$$

Indeed, we obtain an operator $R \in \mathcal{L}(H)$ ($\|R\| \leq 1/\delta$) which obviously is one-to-one and its image is $\mathcal{D}(\bar{T})$, since

$$\sum_{n=1}^{\infty} \left| \frac{\alpha_n}{\lambda - \lambda_n} \right|^2 \lambda_n^2 \leq |\alpha_n|^2 \Big(1 + \frac{\lambda}{\delta}\Big)^2 < \infty$$

and, moreover, we can associate to every $x = \sum_{n=1}^{\infty} \alpha_n u_n \in \mathcal{D}(\bar{T})$ the element

$$y = \sum_{n=1}^{\infty} \beta_n u_n = \sum_{n=1}^{\infty} \alpha_n (\lambda_n - \lambda) u_n \in H$$

such that $Ry = x$.

Also,

$$(\bar{T} - \lambda I) R \Big(\sum_{n=1}^{\infty} \alpha_n u_n \Big) = \sum_{n=1}^{\infty} - \frac{\alpha_n}{\lambda - \lambda_n} (\bar{T} - \lambda I) u_n = \sum_{n=1}^{\infty} \alpha_n u_n$$

and $R = (\bar{T} - \lambda I)^{-1}$.

To prove that $\bar{T}$ is self-adjoint, we need to see that $\mathcal{D}((\bar{T})^*) \subset \mathcal{D}(\bar{T})$. If $x = \sum_{n=1}^{\infty} \alpha_n u_n \in \mathcal{D}((\bar{T})^*)$ and $y = (\bar{T})^* x$, then, for every $n \in \mathbf{N}$,

$$(y, u_n)_H = (x, \bar{T} u_n)_H = \lambda_n (x, u_n)_H = \lambda_n \alpha_n$$

and $\sum_{n=1}^{\infty} |\lambda_n \alpha_n|^2 < \infty$, i.e., $x \in \mathcal{D}(T)$.

Finally, to prove that $\bar{T}$ is the closure of T, consider

$$(x, \bar{T} y) = \Big(\sum_{n=1}^{\infty} \alpha_n u_n, \sum_{n=1}^{\infty} \lambda_n \alpha_n u_n\Big) \in \mathcal{G}(\bar{T}).$$

Then $x_N := \sum_{n=1}^{N} \alpha_n u_n \in \mathcal{D}(T)$ and

$$(x_N, T x_N) = \Big(\sum_{n=1}^{N} \alpha_n u_n, \sum_{n=1}^{N} \lambda_n \alpha_n u_n\Big) \to (x, \bar{T} x)$$

in $H \times H$, since $\{\alpha_n\}, \{\lambda_n \alpha_n\} \in \ell^2$. □

Remark 9.18. A symmetric operator T may have no self-adjoint extensions at all, or many self-adjoint extensions. According to Theorems 9.7 and 9.8, if T is essentially self-adjoint, T^{**} is the unique self-adjoint extension of T.

9.2.3. The Friedrichs extensions. A sufficient condition for a symmetric operator T to have self-adjoint extensions, known as the **Friedrichs extensions**, concerns the existence of a lower bound for the quadratic form $(Tx, x)_H$.

We say that T, symmetric, is **semi-bounded**[5] with constant c, if

$$c := \inf_{x\in\mathcal{D}(T),\,\|x\|_H=1} (Tx,x)_H > -\infty,$$

so that $(Tx,x)_H \geq c\|x\|_H^2$ for all $x \in \mathcal{D}(T)$.

In this case, for any $c' \in \mathbf{R}$, $T-c'I$ is also symmetric on the same domain and semi-bounded, with constant $c+c'$. If $\widetilde{T}$ is a self-adjoint extension of T, then $\widetilde{T}-c'I$ is a self-adjoint extension of $T-c'I$, and we will choose a convenient constant in our proofs. Let us denote

$$(x,y)_T := (Tx,y)_H,$$

a sesquilinear form on $\mathcal{D}(T)$ such that $(y,x)_T = \overline{(x,y)_T}$. If $c>0$, then we have an inner product.

Theorem 9.19 (Friedrichs-Stone[6]). *If T is a semi-bounded symmetric operator, with constant c, then it has a self-adjoint extension $\widetilde{T}$ such that $(\widetilde{T}x,x)_H \geq c\|x\|_H^2$, if $x\in\mathcal{D}(\widetilde{T})$.*

Proof. We can suppose that $c=1$, and then $(x,y)_T$ is a scalar product on $D=\mathcal{D}(T)$ which defines a norm $\|x\|_T = (x,x)_T^{1/2} \geq \|x\|_H$.

Let D_T be the $\|\cdot\|_T$-completion of D. Since $\|x\|_H \leq \|x\|_T$, every $\|\cdot\|_T$-Cauchy sequence $\{x_n\}\subset D$, which represents a point $\widetilde{x}\in D_T$, has a limit x in H, and we have a natural mapping $J: D_T \to H$, such that $J\widetilde{x} = x$.

This mapping J is one-to-one, since, if $Jy=0$ and $x_n \to y$ in D_T, $\{x_n\}\subset D$ is also a Cauchy sequence in H and there exists $x=\lim x_n$ in H. Then $x=Jy=0$ and, from the definition of $(y,x)_T$ and by the continuity of the scalar product, it follows that, for every $v\in D$,

$$(v,y)_T = \lim_n (v,x_n)_T = \lim_n (Tv,x_n)_H = (Tv,x)_H = 0.$$

But D is dense in D_T and $y=0$.

We have $D=\mathcal{D}(T)\subset D_T \hookrightarrow H$ and, to define the Friedrichs extension $\widetilde{T}$ of T, we observe that, for every $u=(\cdot,y)_H \in H'$,

$$|u(x)| \leq \|x\|_H\|y\|_H \leq \|x\|_T\|y\|_H \qquad (x\in D_T)$$

and there exists a unique element $w\in D_T$ such that $u=(\cdot,w)_T$ on D_T. We define $\mathcal{D}(\widetilde{T})$ as the set of all these elements,

$$\mathcal{D}(\widetilde{T}) = \Big\{w\in D_T;\ (\cdot,w)_T = (\cdot,y)_H \text{ on } D_T \text{ for some } y\in H\Big\},$$

[5]In 1929 J. von Neumann and also A. Wintner identified this class of operators that admit self-adjoint extensions.

[6]Kurt Otto Friedrichs (1901–1982) made contributions to the theory of partial differential equations, operators in Hilbert space, perturbation theory, and bifurcation theory. He published his extension theorem in Göttingen in 1934, and M. Stone did the same in New York in 1932.

i.e., $\mathcal{D}(\widetilde{T}) = \mathcal{D}(T^*) \cap D_T$ and $T^*w = y$ for a unique $w \in \mathcal{D}(\widetilde{T})$, for every $y \in H$. Next we define

$$\widetilde{T}w = y \text{ if } (\cdot, y)_H = (\cdot, w)_T \text{ over } D_T \qquad (w \in \mathcal{D}(\widetilde{T})),$$

so that $\widetilde{T}$ is the restriction of T^* to $\mathcal{D}(\widetilde{T}) = \mathcal{D}(T^*) \cap D_T$.

This new operator is a linear extension of T, since, for all $v \in D_T$,

$$(9.3) \qquad (v, w)_T = (v, \widetilde{T}w)_H \qquad (w \in \mathcal{D}(\widetilde{T}))$$

and, if $y = Tx \in H$ with $x \in D$,

$$(v, y)_H = (v, Tx)_H = (Tv, x)_H = (v, x)_T \qquad (v \in D).$$

Thus, $x = w$ and $\widetilde{T}w = Tx$, i.e., $D \subset \mathcal{D}(\widetilde{T})$ and $T \subset \widetilde{T}$.

To show that $\widetilde{T}$ is symmetric, apply (9.3) to $w \in \mathcal{D}(\widetilde{T}) \subset D_T$. If $v, w \in \mathcal{D}(\widetilde{T})$, then $(w, v)_T = (w, \widetilde{T}v)_H$ and the scalar product is symmetric, so that $(\widetilde{T}w, v)_H = (w, \widetilde{T}v)_H$.

Observe that $\widetilde{T} : \mathcal{D}(\widetilde{T}) \to H$ is bijective, since, in our construction, since every $y \in H$, w was the unique solution of the equation $\widetilde{T}w = y$. Moreover, the closed graph theorem shows that $A := \widetilde{T}^{-1} : H \to \mathcal{D}(\widetilde{T}) \subset H$ is a bounded operator, since $y_n \to 0$ and $\widetilde{T}^{-1}y_n \to w$ imply

$$0 = \lim_n (\widetilde{T}^{-1}x, y_n)_H = (x, \widetilde{T}^{-1}y_n)_H = (y, w)_H$$

for every $x \in H$, and then $w = 0$. This bounded operator, being the inverse of a symmetric operator, is also symmetric, i.e., it is self-adjoint. But then, every $z \in \mathbf{C} \setminus \mathbf{R}$ is in $\sigma(A)^c$.

The identity $z^{-1}I - A^{-1} = A^{-1}(A - zI)z^{-1}$ shows that $z^{-1} \in \sigma(A^{-1})^c = \sigma(\widetilde{T})^c$, if $z \notin \mathbf{R}$. By Theorem 9.10, $\widetilde{T}$ is self-adjoint. □

9.3. Spectral representation of unbounded self-adjoint operators

$T : \mathcal{D}(T) \subset H \to H$ is still a possibly unbounded linear operator.

The functional calculus for a bounded normal operator T has been based on the spectral resolution

$$T = \int_{\sigma(T)} \lambda \, dE(\lambda),$$

where E represents a spectral measure on $\sigma(T)$. If f is bounded, then this representation allows us to define

$$f(T) = \int_{\sigma(T)} f(\lambda) \, dE(\lambda).$$

This functional calculus can be extended to unbounded functions, h, and then it can be used to set a spectral theory for unbounded self-adjoint operators. The last section of this chapter is devoted to the proof of the following result:

Theorem 9.20 (Spectral theorem). *For every self-adjoint operator T on H, there exists a unique spectral measure E on $\mathbf{R}$ which satisfies*

$$T = \int_{\mathbf{R}} t\,dE(t)$$

in the sense that

$$(Tx, y)_H = \int_{-\infty}^{+\infty} t\,dE_{x,y}(t) \qquad (x \in \mathcal{D}(T),\ y \in H).$$

If f is a Borel measurable function on $\mathbf{R}$, then a densely defined operator

$$f(T) = \int_{\mathbf{R}} f(t)\,dE(t)$$

is obtained such that

$$(f(T)x, y)_H = (\Phi_E(f)x, y)_H = \int_{-\infty}^{+\infty} f(t)\,dE_{x,y}(t) \qquad (x \in \mathcal{D}(f),\ y \in H),$$

where

$$\mathcal{D}(f) = \left\{x \in H;\ \int_{-\infty}^{+\infty} |f(\lambda)|^2\,dE_{x,x} < \infty\right\}.$$

For this functional calculus,

(a) $\|f(T)x\|_H^2 = \int_{\sigma(T)} |f|^2\,dE_{x,x}$ *if* $x \in \mathcal{D}(f(T))$,

(b) $f(T)h(T) \subset (fh)(T)$, $\mathcal{D}(f(T)h(T)) = \mathcal{D}(h(T)) \cap \mathcal{D}((fh)(T))$, *and*

(c) $f(T)^* = \bar{f}(T)$ *and* $f(T)^*f(T) = |f|^2(T) = f(T)f(T)^*$.

If f is bounded, then $\mathcal{D}(f) = H$ and $f(T)$ is a bounded normal operator. If f is real, then $f(T)$ is self-adjoint.

The following example will be useful in the next section.

Example 9.21. The spectral measure of the position operator of Examples 9.2 and 9.11,

$$Q = \int_{\mathbf{R}} t\,dE(t),$$

is $E(B) = \chi_B\cdot$ and $dE_{\varphi,\psi}(t) = \varphi(t)\overline{\psi(t)}\,dt$, i.e.,

$$\int_{\mathbf{R}} t\,dE_{\varphi,\psi}(t) = (Q\varphi, \psi)_2 = \int_{\mathbf{R}} t\,\varphi(t)\overline{\psi(t)}\,dt \qquad (\varphi \in \mathcal{D}(Q)).$$

This is proved by defining $F(B)\psi := \chi_B\psi$ for every Borel set $B \subset \mathbf{R}$; that is, $F(B) = \chi_B\cdot$, a multiplication operator. It is easy to check that $F : \mathcal{B}_\mathbf{R} \to \mathcal{L}(L^2(\mathbf{R}))$ is a spectral measure, and to show that $F = E$, we only need to see that

$$\int_\mathbf{R} t\varphi(t)\overline{\psi(t)}\,dt = \int_\mathbf{R} t\,dF_{\varphi,\psi}(t),$$

where $F_{\varphi,\psi}(B) = (F(B)\varphi,\psi)_2 = \int_\mathbf{R}(F(B)\varphi)(t)\overline{\psi(t)}\,dt$.

But the integral for the complex measure $F_{\varphi,\psi}$ is a Lebesgue-Stieltjes integral with the distribution function

$$F(t) = F_{\varphi,\psi}\big((-\infty,t]\big) = (\chi_{(-\infty,t]}\varphi,\psi)_2 = \int_{-\infty}^t \varphi(s)\overline{\psi(s)}\,ds,$$

and then $dF(t) = \varphi(t)\overline{\psi(t)}\,dt$.

The spectrum $\sigma(T)$ of a self-adjoint operator can be described in terms of its spectral measure E:

Theorem 9.22. *If $T = \int_\mathbf{R} t\,dE(t)$ is the spectral representation of a self-adjoint operator T, then*

(a) $\sigma(T) = \operatorname{supp} E$,

(b) $\sigma_p(T) = \{\lambda \in \mathbf{R};\ E\{\lambda\} \neq 0\}$, *and*

(c) $\operatorname{Im} E\{\lambda\}$ *is the eigenspace of every* $\lambda \in \sigma_p(T)$.

Proof. We will use the fact that

$$\|(T-\lambda I)x\|_H^2 = \int_\mathbf{R} (t-\lambda)^2\,dE_{x,x}(t) \qquad (x \in \mathcal{D}(T),\ \lambda \in \mathbf{R}),$$

which follows from Theorem 9.20(a).

(a) If $\lambda \notin \operatorname{supp} E$, then $E_{x,x}(\lambda-\varepsilon,\lambda+\varepsilon) = 0$ for some $\varepsilon > 0$, and

$$\|(T-\lambda I)x\|_H^2 = \int_{(\lambda-\varepsilon,\lambda+\varepsilon)} (t-\lambda)^2\,dE_{x,x}(t) \geq \varepsilon^2\|x\|_H^2,$$

which means that $\lambda \notin \sigma(T)$, by Theorem 9.9.

Conversely, if $\lambda \in \operatorname{supp} E$, then $E(\lambda-1/n,\lambda+1/n) \neq 0$ for every $n > 0$ and we can choose $0 \neq x_n \in \operatorname{Im} E(\lambda-1/n,\lambda+1/n)$. Then $\operatorname{supp} E_{x_n,x_n} \subset [\lambda-1/n,\lambda+1/n]$ since it follows from $V \cap (\lambda-1/n,\lambda+1/n) = \emptyset$ that $E(V)$ and $E(\lambda-1/n,\lambda+1/n)$ are orthogonal and $E_{x,x}(V) = (E(V)x,x)_H = 0$. Thus

$$\|(T-\lambda I)x_n\|_H^2 = \int_\mathbf{R} (t-\lambda)^2\,dE_{x_n,x_n}(t) \leq \frac{1}{n^2}\|x_n\|_H^2$$

and λ is an approximate eigenvalue.

(b) $Tx = \lambda x$ for $0 \neq x \in \mathcal{D}(T)$ if and only if $\int_{\mathbf{R}}(t-\lambda)^2\,dE_{x,x}(t) = 0$, meaning that $E_{x,x}\{\lambda\} \neq 0$ and $E_{x,x}(\mathbf{R} \setminus \{\lambda\}) = 0$.

(c) The identity $E_{x,x}(\mathbf{R} \setminus \{\lambda\}) = 0$ means that $x = E\{\lambda\}(x)$ satisfies $Tx = \lambda x$.

□

Since $E(B) = E(B \cap \operatorname{supp} E)$, in the spectral representation of the self-adjoint operator, T, $\mathbf{R}$ can be changed by $\operatorname{supp} E = \sigma(T)$; that is,

$$T = \int_{\mathbf{R}} t\,dE(t) = \int_{\sigma(T)} t\,dE(t),$$

and also

$$h(T) = \int_{\mathbf{R}} h\,dE = \int_{\sigma(T)} h\,dE(t).$$

As an application, we define the square root of a positive operator:

Theorem 9.23. *A self-adjoint operator T is positive ($(Tx, x) \geq 0$ for all $x \in \mathcal{D}(T)$) if and only if $\sigma(T) \subset [0, \infty)$. In this case there exists a unique self-adjoint operator R which is also positive and satisfies $R^2 = T$, so that $R = \sqrt{T}$, the square root of T.*

Proof. If $(Tx, x)_H \geq 0$ for every $x \in \mathcal{D}(T)$ and $\lambda > 0$, we have

$$\lambda \|x\|_H^2 \leq ((T + \lambda I)x, x)_H \leq \|(T + \lambda I)x\|_H \|x\|_H,$$

so that

$$\|(T + \lambda I)x\|_H \geq \lambda \|x\|_H \qquad (x \in \mathcal{D}(T)).$$

By Theorem 9.9 there exists $(T + \lambda I)^{-1} \in \mathcal{L}(H)$ and $-\lambda \notin \sigma(T)$.

Conversely, if $\sigma(T) \subset [0, \infty)$ and $x \in \mathcal{D}(T)$, then $\int_0^\infty t\,dE_{x,x}(t) \geq 0$. Moreover

$$(Tx, y)_H = \int_0^\infty t\,dE_{x,y}(t) \qquad (x \in \mathcal{D}(T),\ y \in H).$$

Define $R = f(T)$ with $f(t) = t^{1/2}$. Then $\mathcal{D}(R) = \{x;\ \int_0^\infty t\,dE_{x,x} < \infty\}$, which contains $\mathcal{D}(T) = \{x;\ \int_0^\infty t^2\,dE_{x,x} < \infty\}$. Thus,

$$R = \sqrt{T} = \int_0^\infty t^{1/2}\,dE(t).$$

From Theorem 9.29(b), $R^2 = T$, since $\mathcal{D}(f^2) = \mathcal{D}(T) \subset \mathcal{D}(f)$.

To prove the uniqueness, suppose that we also have

$$S = \int_0^\infty t\,dF(t)$$

such that $S^2 = T$ and

$$T = \int_0^\infty t^2\,dF(t).$$

With the substitution $\lambda = t^2$ we obtain a spectral measure $E'(\lambda) = F(\lambda^{1/2})$ such that $T = \int_0^\infty \lambda \, dE'(\lambda)$. From the uniqueness of the spectral measure, $E' = E$ and then $S = R$. □

9.4. Unbounded operators in quantum mechanics

To show how unbounded self-adjoint operators are used in the fundamentals of quantum mechanics, we are going to start by studying the case of a single particle constrained to move along a line.

9.4.1. Position, momentum, and energy. In quantum mechanics, what matters about the **position** is the probability that the particle is in $[a, b] \subset \mathbf{R}$, and this probability is given by an integral

$$\int_a^b |\psi(x)|^2 \, dx.$$

The density distribution $|\psi(x)|^2$ is defined by some $\psi \in L^2(\mathbf{R})$, which is called the state function, such that $\int_{\mathbf{R}} |\psi(x)|^2 \, dx = \|\psi\|_2^2 = 1$ is the total probability. Here ψ is a complex-valued function and a complex factor α in ψ is meaningless ($|\alpha| = 1$ is needed to obtain $\|\psi\|_2 = 1$). There is a dependence on the time, t, which can be considered as a parameter.

The mean position of the particle will be

$$\mu_\psi = \int_{\mathbf{R}} x|\psi(x)|^2 \, dx = \int_{\mathbf{R}} x\psi(x)\overline{\psi(x)} \, dx = \int_{\mathbf{R}} x \, dE_{\psi,\psi}(x)$$

with $dE_{\psi,\psi} = \psi(x)\overline{\psi(x)} \, dx$.

If Q denotes the position operator, $Q\varphi(x) = x\varphi(x)$, note that $\mu_\psi = (Q\psi, \psi)_2$.

The dispersion of the position with respect to its mean value is measured by the variance,

$$\mathrm{var}_\psi = \int_{\mathbf{R}} (x-\mu_\psi)^2 |\psi(x)|^2 \, dq = \int_{\mathbf{R}} x(x-\mu_\psi)^2 \, dE_{\psi,\psi}(x) = ((Q-\mu_\psi I)^2\psi, \psi)_2.$$

Similarly, if $\int_{\mathbf{R}} |f(x)||\psi(x)|^2 \, dx < \infty$, the mathematical expectation of f is

$$\int_{\mathbf{R}} f(x)|\psi(x)|^2 \, dx = (f\psi, \psi)_2 = \int_{\mathbf{R}} f(x) \, dE_{\psi,\psi}(x). \tag{9.4}$$

The **momentum** of the particle is defined as mass × velocity:

$$p = m\dot{x}.$$

Note that, from the properties of the Fourier transform,

(9.5)
$$\int_{\mathbf{R}} \xi|\widehat{\psi}(\xi)|^2\,d\xi = \int_{\mathbf{R}} \xi\widehat{\psi}(\xi)\overline{\widehat{\psi}(\xi)}\,d\xi = \frac{1}{2\pi i}\int_{\mathbf{R}} \widehat{\psi}'(\xi)\overline{\widehat{\psi}(\xi)}\,d\xi = \frac{1}{2\pi i}(\psi',\psi)_2.$$

By assuming that the probability that $p \in [a,b]$ is given by

$$\frac{1}{h}\int_a^b \left|\widehat{\psi}\left(\frac{p}{h}\right)\right|^2 dp = \int_{a/h}^{b/h} |\widehat{\psi}(\xi)|^2\,d\xi,$$

where $h = 6.62607095(44)\cdot 10^{-34}\,J\cdot\text{seg}$ is the **Planck constant**,[7] the average value of p is

$$\frac{1}{h}\int_{\mathbf{R}} p\left|\widehat{\psi}\left(\frac{p}{h}\right)\right|^2 dp = h\int_{\mathbf{R}} \xi|\widehat{\psi}(\xi)|^2\,d\xi.$$

Here the Fourier transform can be avoided by considering the **momentum operator** P defined as

$$P = \frac{h}{2\pi i}D \qquad \left(D = \frac{d}{dx}\right),$$

since then, as noted in (9.5), this average is

$$h\int_{\mathbf{R}} \xi|\widehat{\psi}(\xi)|^2\,d\xi = (P\psi,\psi)_2. \tag{9.6}$$

If $\int_{\mathbf{R}} |f(h\xi)||\widehat{\psi}(\xi)|^2\,d\xi$, then the functional calculus gives the value

$$\mu_\psi(f) = \int_{\mathbf{R}} f(h\xi)|\widehat{\psi}(\xi)|^2\,d\xi$$

for the mathematical expectation of f, which in the case $f(p) = p^n$ is

$$\mu_\psi(p^n) = (P^n\psi,\psi)_2.$$

The **kinetic energy** is

$$T = \frac{p^2}{2m},$$

so that its mathematical expectation will be

$$\mu_\psi(T) = \frac{1}{2m}(P^2\psi,\psi)_2.$$

The potential energy is given by a real-valued function $V(x)$ and from (9.4) we obtain the value

$$\mu_\psi(V) = \int_{\mathbf{R}} V(x)|\psi(x)|^2\,dx = (V\psi,\psi)_2$$

for the mathematical expectation of V if $\int_{\mathbf{R}} |V(x)||\psi(x)|^2\,dx < \infty$.

[7]This is the value reported in October 2007 by the National Physical Laboratory for this constant, named in honor of Max Planck, considered to be the founder of quantum theory in 1901 when, in his description of the black-body radiation, he assumed that the electromagnetic energy could be emitted only in quantized form, $E = h\nu$, where ν is the frequency of the radiation.

The mathematical expectation is additive, so that the average of the total energy is

$$\mu_\psi(T+V) = \left(\frac{1}{2m}P^2\psi + V\psi, \psi\right)_2 = (H\psi, \psi)_2,$$

where

$$H = \frac{1}{2m}P^2 + V$$

is the **energy operator**, or **Hamiltonian**, of the particle.

9.4.2. States, observables, and Hamiltonian of a quantic system. As in the case of classical mechanics, the basic elements in the description of a general quantic system are those of **state** and **observable**.

Classical mechanics associates with a given system a phase space, so that for an N-particle system we have a $6N$-dimensional phase state.

Similarly, quantum mechanics associates with a given system a complex Hilbert space $\mathcal{H}$ as the **state space**, which is $L^2(\mathbf{R})$ in the case of a single particle on the line. In a quantum system the **observables** are self-adjoint operators, such as the position, momentum, and energy operators.

A quantum system, in the **Schrödinger picture**, is ruled by the following postulates:

Postulate 1: States and observables

A **state** of a physical system at time t is a line $[\psi] \subset \mathcal{H}$, which we represent by $\psi \in \mathcal{H}$ such that $\|\psi\|_{\mathcal{H}} = 1$.

A **wave function** is an $\mathcal{H}$-valued function of the time parameter $t \in \mathbf{R} \mapsto \psi(t) \in \mathcal{H}$. If $\psi(t)$ describes the state, then $c\psi(t)$, for any nonzero constant c, represents the same state.

The observable values of the system are magnitudes such as position, momentum, angular momentum, spin, charge, and energy that can be measured. They are associated to self-adjoint operators. In a quantic system, an **observable** is a time-independent[8] self-adjoint operator A on $\mathcal{H}$, which has a spectral representation

$$A = \int_{\mathbf{R}} \lambda \, dE(\lambda).$$

By the "superposition principle", all self-adjoint operators on $\mathcal{H}$ are assumed to be observable,[9] and all lines $[\psi] \subset \mathcal{H}$ are admissible states.

---

[8]In the Heisenberg picture of quantum mechanics, the observables are represented by time-dependent operator-valued functions $A(t)$ and the state ψ is time-independent.

[9]Here we are following the early assumptions of quantum mechanics, but the existence of "superselection rules" in quantum field theories indicated that this superposition principle lacks experimental support in relativistic quantum mechanics.

The elements of the spectrum, $\lambda \in \sigma(A)$, are the **observable values** of the observable A.

Postulate 2: Distribution of an observable in a given state

The values $\lambda \in \sigma(A)$ in a state ψ are observable in terms of a probability distribution P_ψ^A.

As in the case of the position operator Q for the single particle on $\mathbf{R}$, the observable $A = \int_{\mathbf{R}} \lambda \, dE(\lambda)$ on $\mathcal{H}$ is evaluated in a state ψ at a given time in terms of the probability $P_\psi^A(B)$ of belonging to a set $B \subset \mathbf{R}$ with respect to the distribution $dE_{\psi,\psi}(\lambda)$ (we are assuming that $\|\psi\|_{\mathcal{H}} = 1$), so that

$$P_\psi^A(B) = \int_B \lambda dE_{\psi,\psi}(\lambda) = (E(B)\psi, \psi)_{\mathcal{H}}$$

and the **mean value** is

$$\widehat{A}_\psi := \int_{\mathbf{R}} \lambda \, dE_{\psi,\psi}(\lambda) = (A\psi, \psi)_{\mathcal{H}}.$$

When $\psi \in \mathcal{D}(A)$, this mean value $\widehat{A}_\psi$ exists, since λ^2 is integrable with respect to the finite measure $E_{\psi,\psi}$, and also $\int_{\mathbf{R}} |\lambda| \, dE_{\psi,\psi}(\lambda) < \infty$.

In general, if f is $E_{\psi,\psi}$-integrable,

$$\widehat{f(A)}_\psi = (f(A)_\psi, \psi)_{\mathcal{H}},$$

is the **expected value** of f, the mean value with respect to $E_{\psi,\psi}$.

The **variance** of A in the state $\psi \in \mathcal{D}(A)$ is then

$$\operatorname{var}_\psi(A) = \int_{\mathbf{R}} (\lambda - \widehat{A}_\psi)^2 \, dE_{\psi,\psi}(\lambda) = ((A - \widehat{A}_\psi I)^2\psi, \psi)_{\mathcal{H}} = \|A\psi - \widehat{A}_\psi\psi\|_{\mathcal{H}}^2.$$

It is said that A **certainly takes** the value λ_0 in the state ψ if $\widehat{A}_\psi = \lambda_0$ and $\operatorname{var}_\psi(A) = 0$.

This means that ψ is an eigenvector of A with eigenvalue λ_0, since it follows from $A\psi = \lambda_0\psi$ that $\widehat{A}_\psi = (A\psi, \psi)_{\mathcal{H}} = \lambda_0$, and also

$$\operatorname{var}_\psi(A) = \|A\psi - \widehat{A}_\psi\psi\|_{\mathcal{H}}^2 = 0.$$

Conversely, $\operatorname{var}_\psi(A) = 0$ if and only if $A\psi - \widehat{A}_\psi\psi = 0$.

Postulate 3: Hamiltonians and the Schrödinger equation

There is an observable, H, the **Hamiltonian**, defining the evolution of the system

$$\psi(t) = U_t\psi_0,$$

where ψ_0 is the initial state and U_t is an operator defined as follows:

If h is the Planck constant and $g_t(\lambda) = e^{-\frac{it}{h}\lambda}$, a continuous function with its values in the unit circle, then using the functional calculus, we can define the unitary operators

$$U_t := g_t(H) \in \mathcal{L}(\mathcal{H}) \qquad (t \in \mathbf{R})$$

that satisfy the conditions

$$U_0 = I, \quad U_s U_t = U_{s+t}, \text{ and } \lim_{t \to s} \|U_t x - U_s x\|_{\mathcal{H}} = 0 \, \forall x \in \mathcal{H},$$

since $g_t \bar{g}_t = 1$, $g_0 = 1$, $g_s g_t = g_{st}$, and, if $H = \int_{\sigma(H)} \lambda \, dE(\lambda)$ is the spectral representation of H, then the continuity property

$$\|U_t \psi - U_s \psi\|_{\mathcal{H}}^2 = \int_{\sigma(H)} |e^{-\frac{it}{h}\lambda} - e^{-\frac{is}{h}\lambda}|^2 \, dE_{\psi,\psi}(\lambda) \to 0 \quad \text{as} \quad t \to s$$

follows from the dominated convergence theorem.

Such a family of operators U_t is called a **strongly continuous one-parameter group of unitary operators**, and we say that $A = -(i/h)H$ is the **infinitesimal generator**.

It can be shown (**Stone's theorem**) that the converse is also true: every strongly continuous one-parameter group of unitary operators $\{U_t\}_{t \in \mathbf{R}}$ has a self-adjoint infinitesimal generator $A = -(i/h)H$; that is, $U_t = e^{-\frac{it}{h}H}$ for some self-adjoint operator H.

It is said that

$$U_t = e^{-\frac{it}{h}H}$$

is the **time-evolution operator** of the system.

It is worth noticing that, if $\psi \in \mathcal{D}(H)$, the function $t \mapsto U_t \psi$ is differentiable and

$$\frac{d}{dt} U_t \psi = U_t A \psi = A U_t \psi$$

at every point $t \in \mathbf{R}$. Indeed,

$$\frac{1}{s}(U_s U_t \psi - U_t \psi) = U_t \frac{1}{s}(U_s \psi - \psi)$$

and

$$\lim_{s \to 0} \frac{1}{s}(U_s \psi - \psi) = A\psi = -\frac{i}{h} H\psi, \tag{9.7}$$

since

$$\left\|\frac{1}{s}(U_s \psi - \psi) + \frac{i}{h} H\psi\right\|_{\mathcal{H}}^2 = \int_{\sigma(A)} \left|\frac{e^{-is\lambda/h} - 1}{s} + \frac{i}{h}\lambda\right|^2 dE_{\psi,\psi}(\lambda) \to 0$$

as $s \to 0$, again by dominated convergence.

For a given initial state ψ_0, it is said that $\psi(t) = U_t \psi_0$ is the corresponding wave function.

If $\psi(t) \in \mathcal{D}(H)$ for every $t \in \mathbf{R}$, then the vector-valued function $t \mapsto \psi(t)$ is derivable and satisfies the **Schrödinger equation**[10]

$$ih\psi'(t) = H\psi(t),$$

since by (9.7)

$$\psi'(t) = \frac{d}{dt}(U_t\psi_0) = AU_t\psi_0 = -\frac{i}{h}H\psi(t).$$

In this way, from a given initial state, subsequent states can be calculated causally from the Schrödinger equation.[11]

9.4.3. The Heisenberg uncertainty principle and compatible observables. To illustrate the role of probabilities in the postulates, let us consider again the case of a single particle on $\mathbf{R}$. Recall that the momentum operator,

$$P\psi(q) = \frac{h}{2\pi i}\psi'(q),$$

is self-adjoint on $L^2(\mathbf{R})$ and with domain $H^1(\mathbf{R})$.

From Example 9.4 we know that the commutator of P and Q is bounded and

$$[P,Q] = PQ - QP = \frac{h}{2\pi i}I,$$

where $\mathcal{D}([P,Q]) = \mathcal{D}(PQ) \cap \mathcal{D}(QP)$, or extended to all $L^2(\mathbf{R})$ by continuity.

Lemma 9.24. *The commutator $C = [S,T] = ST - TS$ of two self-adjoint operators on $L^2(\mathbf{R})$ satisfies the estimate*

$$|\widehat{C}_\psi| \le 2\sqrt{\mathrm{var}_\psi(S)}\sqrt{\mathrm{var}_\psi(T)}$$

for every $\psi \in \mathcal{D}(C)$.

Proof. Obviously, $A = S - \widehat{S}_\psi I$ and $B = T - \widehat{T}_\psi I$ are self-adjoint (note that $\widehat{S}_\psi, \widehat{T}_\psi \in \mathbf{R}$) and $C = [A,B]$. From the definition of the expected value,

$$|\widehat{C}_\psi| \le |(B\psi, A\psi)_2| + |(A\psi, B\psi)_2| \le 2\|B\psi\|_2\|A\psi\|_2,$$

where, A being self-adjoint, $\|A\psi\|_2^2 = (A^2\psi,\psi)_2 = \mathrm{var}_\psi(S)$. Similarly, $\|B\psi\|_2^2 = \mathrm{var}_\psi(T)$. □

[10] E. Schrödinger published his equation and the spectral analysis of the hydrogen atom in a series of four papers in 1926, which where followed the same year by Max Born's interpretation of $\psi(t)$ as a probability density.

[11] We have assumed that the energy is constant and the Hamiltonian does not depend on t but, if the system interacts with another one, the Hamiltonian is an operator-valued function $H(t)$ of the time parameter. In the Schrödinger picture, all the observables except the Hamiltonian are time-invariant.

Theorem 9.25 (Uncertainty principle).

$$\sqrt{\mathrm{var}_\psi(Q)}\sqrt{\mathrm{var}_\psi(P)} \geq \frac{h}{4\pi}.$$

Proof. In the case $C = [P, Q]$, $|\widehat{C}_\psi| = |(h/2\pi i)\widehat{I}_\psi| = h/2\pi$ and we can apply Lemma 9.24. □

The standard deviations $\sqrt{\mathrm{var}_\psi(Q)}$ and $\sqrt{\mathrm{var}_\psi(P)}$ measure the uncertainties of the position and momentum, and the uncertainty principle shows that both uncertainties cannot be arbitrarily small simultaneously. Position and moment are said to be **incompatible observables**.

It is a basic principle of all quantum theories that if n observables $A_1, \ldots, A_n$ are compatible in the sense of admitting arbitrarily accurate simultaneous measurements, they must commute. However, since these operators are only densely defined, the commutators $[A_j, A_k]$ are not always densely defined. Moreover, the condition $AB = BA$ for two commuting operators is unsatisfactory; for example, taking it literally, $A0 \neq 0A$ if A is unbounded, but $A0 \subset 0A$ and $[A, 0] = 0$ on the dense domain of A.

This justifies saying that $A_j = \int_{\mathbf{R}} \lambda \, dE^j(\lambda)$ and $A_k = \int_{\mathbf{R}} \lambda \, dE^k(\lambda)$ **commute**, or that **their spectral measures commute**, if

$$[E^j(B_1), E^k(B_2)] = 0 \qquad (B_1, B_2 \in \mathcal{B}_{\mathbf{R}}). \tag{9.8}$$

If both A_j and A_k are bounded, then this requirement is equivalent to $[A_j, A_k] = 0$ (see Exercise 9.18).[12] For such commuting observables and a given (normalized) state ψ, there is a probability measure P_ψ on $\mathbf{R}^n$ so that

$$P_\psi(B_1 \times \cdots \times B_n) = (E^1(B_1) \cdots E^n(B_n)\psi, \psi)_{\mathcal{H}}$$

is the predicted probability that a measurement to determine the values $\lambda_1, \ldots, \lambda_n$ of the observables $A_1, \ldots, A_n$ will lie in $B = B_1 \times \cdots \times B_n$. See Exercise 9.19, where it is shown how a spectral measure E on $\mathbf{R}^n$ can be defined so that $dE_{\psi,\psi}$ is the distribution of this probability; with this spectral measure there is an associated functional calculus $f(A_1, \ldots, A_n)$ of n commuting observables.

9.4.4. The harmonic oscillator.

A heuristic recipe to determine a quantic system from a classical system of energy

$$T + V = \sum_{j=1}^{n} \frac{p_j^2}{2m_j} + V(q_1, \ldots, q_1)$$

[12] But E. Nelson proved in 1959 that there exist essentially self-adjoint operators A_1 and A_2 with a common and invariant domain, so that $[A_1, A_2]$ is defined on this domain and $[A_1, A_2] = 0$ but with noncommuting spectral measures.

is to make a formal substitution of the generalized coordinates q_j by the position operators Q_j (multiplication by q_j) and every p_j by the corresponding momentum P_j. Then the Hamiltonian or energy operator should be a self-adjoint extension of

$$H = \sum_{j=1}^{n} \frac{P_j^2}{2m_j} + V(Q_1, \dots, Q_1).$$

For instance, in the case of the two-body problem under Coulomb force, which derives from the potential $-e/|x|$, $n = 3$ and the energy of the system is

$$E = T + V = \frac{1}{2m}|p|^2 - \frac{e^2}{|x|}.$$

Hence, in a convenient scale, $H = -\Delta - \frac{1}{|x|}$ is the possible candidate of the Hamiltonian of the hydrogen atom. In Example 9.15 we have seen that it is a self-adjoint operator with domain $H^2(\mathbf{R}^3)$.

With the help of his friend Hermann Weyl, Schrödinger calculated the eigenvalues of this operator. The coincidence of his results with the spectral lines of the hydrogen atom was considered important evidence for the validity of Schrödinger's model for quantum mechanics.

Several problems appear with this quantization process, such as finding the self-adjoint extension of H, determining the spectrum, and describing the evolution of the system for large values of t ("scattering").

Let us consider again the simple classical one-dimensional case of a single particle with mass m, now in a Newtonian field with potential V, so that

$$-\nabla V = F = \frac{d}{dt}(m\dot{q}),$$

q denoting the position. We have the linear momentum $p = m\dot{q}$, the kinetic energy $T = (1/2)m\dot{q}^2 = p^2/2m$, and the total energy $E = T + V$.

The **classical harmonic oscillator** corresponds to the special case of the field $F(q) = -m\omega^2 q$ on a particle bound to the origin by the potential

$$V(q) = m\frac{\omega^2}{2}q^2$$

if $q \in \mathbf{R}$ is the position variable. Hence, in this case,

$$E = T + V = \frac{1}{2}m\dot{q}^2 + m\frac{\omega^2}{2}q^2 = \frac{1}{2m}p^2 + m\frac{\omega^2}{2}q^2.$$

From Newton's second law, the initial state $\dot{q}(0) = 0$ and $q(0) = a > 0$ determines the state of the system at every time,

$$q = a\cos(\omega t).$$

The state space for the **quantic harmonic oscillator** is $L^2(\mathbf{R})$, and the position $Q = q\cdot$ and the momentum P are two observables. By making the announced substitutions, we obtain as a possible Hamiltonian the operator

$$H = \frac{1}{2m}P^2 + m\frac{\omega^2}{2}Q^2.$$

On the domain $\mathcal{S}(\mathbf{R})$, which is dense in $L^2(\mathbf{R})$, it is readily checked that

$$(H\varphi, \psi)_2 = (\varphi, H\psi)_2,$$

so that H is a symmetric operator. We will prove that it is essentially self-adjoint and the Hamiltonian will be its unique self-adjoint extension $\bar{H} = H^{**}$, which is also denoted H.

In coordinates,

$$H = -\frac{h^2}{2m\cdot 4\pi^2}\frac{d^2}{dq^2} + \frac{m\omega^2}{2}q^2$$

which after the substitution $x = aq$, with $a^2 = 2\pi m\omega/h$, can be written

$$H = \frac{h\omega}{2}\Big(x^2 - \frac{d^2}{dx^2}\Big).$$

Without loss of generality, we suppose $h\omega = 1$, and it will be useful to consider the action of

$$H = \frac{1}{2}\Big(x^2 - \frac{d^2}{dx^2}\Big)$$

on

$$\mathcal{F} := \{P(x)e^{-x^2/2};\ P \text{ polynomial}\},$$

the linear subspace of $\mathcal{S}(\mathbf{R})$ that has the functions $x^n e^{-x^2/2}$ as an algebraic basis.

Since $H(x^n e^{-x^2/2}) \in \mathcal{F}$, we have $H(\mathcal{F}) \subset \mathcal{F}$. Similarly, $A(\mathcal{F}) \subset \mathcal{F}$ and $B(\mathcal{F}) \subset \mathcal{F}$ if

$$A := \frac{1}{\sqrt{2}}\Big(x + \frac{d}{dx}\Big),$$

the **annihilation operator**, and

$$B := \frac{1}{\sqrt{2}}\Big(x - \frac{d}{dx}\Big),$$

the **creation operator**.

Theorem 9.26. *The subspace $\mathcal{F}$ of $\mathcal{S}(\mathbf{R})$ is dense in $L^2(\mathbf{R})$, and the Gram-Schmidt process applied to $\{x^n e^{-x^2/2}\}_{n=0}^\infty$ generates an orthonormal basis $\{\widetilde{\psi}_n\}_{n=0}^\infty$ of $L^2(\mathbf{R})$. The functions $\widetilde{\psi}_n$ are in the domain $\mathcal{S}(\mathbf{R})$ of H and they are eigenfunctions with eigenvalues $\lambda_n = n + 1/2$. According to Theorem 9.17, the operator H is essentially self-adjoint.*

Proof. On $\mathcal{F}$, a simple computation gives

$$H = BA + \frac{1}{2}I = AB - \frac{1}{2}I;$$

hence $HB = BAB + \frac{1}{2}B$ and $BH = BAB - \frac{1}{2}B$, so that

$$[H, B] = B.$$

Then, if $H\psi = \lambda\psi$ and $B\psi \neq 0$ with $\psi \in \mathcal{F}$, it follows that $\lambda + 1$ is also an eigenvalue of H, with the eigenfunction $B\psi$, since

$$H(B\psi) = B(H\psi) + B\psi = \lambda B\psi + B\psi = (\lambda + 1)B\psi.$$

For

$$\psi_0(x) := e^{-x^2/2},$$

we have $2H\psi_0(x) = x^2e^{-x^2/2} - (e^{-x^2/2})'' = e^{-x^2/2}$, so that

$$H\psi_0 = \frac{1}{2}\psi_0$$

and ψ_0 is an eigenfunction with eigenvalue $1/2$.

We have $\sqrt{2}B\psi_0(x) = 2xe^{-x^2/2} \neq 0$ and, if we denote

$$\psi_n := (\sqrt{2}B)^n\psi_0 = \sqrt{2}B\psi_{n-1},$$

from the above remarks we obtain

$$H\psi_n = \Big(n + \frac{1}{2}\Big)\psi_n \qquad (n = 0, 1, 2, \ldots)$$

and $\psi_n(x) = H_n(x)e^{-x^2/2}$. By induction over n, it follows that H_n is a polynomial with degree n. It is called a **Hermite polynomial**.

The functions ψ_n are mutually orthogonal, since they are eigenfunctions with different eigenvalues, and they generate $\mathcal{F}$.

To prove that $\mathcal{F}$ is a dense subspace of $L^2(\mathbf{R})$, let $f \in L^2(\mathbf{R})$ be such that $\int_{\mathbf{R}} f(x)x^ne^{-x^2/2}\,dx = (x^ne^{-x^2/2}, f(x))_2 = 0$ for all $n \in \mathbf{N}$. Then

$$F(z) := \int_{\mathbf{R}} f(x)e^{-x^2/2}e^{-2\pi ixz}\,dx$$

is defined and continuous on $\mathbf{C}$, and the Morera theorem shows that F is an entire function, with

$$F^{(n)}(z) = (-2\pi i)^n \int_{\mathbf{R}} x^nf(x)e^{-x^2/2}e^{-2\pi ixz}\,dx.$$

But $F^{(n)}(0) = (x^ne^{-x^2/2}, f(x))_2 = 0$ for all $n \in \mathbf{N}$, so that $F = 0$. From the Fourier inversion theorem we obtain $f(x)e^{-x^2/2} = 0$ and $f = 0$.

It follows that the eigenfunctions $\widetilde{\psi}_n := \|\psi_n\|_2^{-1}\psi_n$ of H are the elements of an orthonormal basis of $L^2(\mathbf{R})$, all of them contained in $\mathcal{S}(\mathbf{R})$, which is the domain of the essentially self-adjoint operator H. □

Remark 9.27. In the general setting, for any mh,

$$H = \frac{h\omega}{2}\left(x^2 - \frac{d^2}{dx^2}\right),$$

and we have $H\widetilde{\psi}_n = h\omega(n + \frac{1}{2})\widetilde{\psi}_n$. Thus

$$\sigma(H) = \{h\omega/2, h\omega(1 + 1/2), h\omega(2 + 1/2), \dots\}.$$

The wave functions $\widetilde{\psi}_n$ are known as the **bound states**, and the numbers are the **energy eigenvalues** of these bound states. The minimal energy is $h\omega/2$,[13] and $\widetilde{\psi}_0$ is the "**ground state**".

9.5. Appendix: Proof of the spectral theorem

The proof of Theorem 9.20 will be obtained in several steps. First, in Theorem 9.28, we define a functional calculus with bounded functions for spectral measures. Then this functional calculus will be extended to unbounded functions in Theorem 9.29. The final step will prove the spectral theorem for unbounded self-adjoint operators by the von Neuman method based on the use of the Cayley transform.

9.5.1. Functional calculus of a spectral measure. Our first step in the proof of the spectral theorem for unbounded self-adjoint operators will be to define a functional calculus associated to a general spectral measure

$$E : \mathcal{B}_K \to \mathcal{L}(H)$$

as the integral with respect to this operator-valued measure.

Denote by $L^\infty(E)$ the complex normed space of all E-essentially bounded complex functions (the functions coinciding E-a.e. being identified as usual) endowed with the natural operations and the norm

$$\|f\|_\infty = E\text{-}\sup|f|.$$

With the multiplication and complex conjugation, it becomes a commutative C^*-algebra, and the constant function 1 is the unit. Every $f \in L^\infty(E)$ has a bounded representative.

We always represent simple functions as

$$s = \sum_{n=1}^{N} \alpha_n \chi_{B_n} \in S(K),$$

where $\{B_1, \dots, B_N\}$ is a partition of K. Since every bounded measurable function is the uniform limit of simple functions, $S(K)$ is dense in $L^\infty(E)$, and we will start by defining the integral of simple functions:

---

[13] Max Planck first applied his quantum postulate to the harmonic oscillator, but he assumed that the lowest level energy was 0 instead of $h\omega/2$. See footnote 7 in this chapter.

As in the scalar case,

$$\int s\,dE := \sum_{n=1}^{N} \alpha_n E(B_n) \in \mathcal{L}(H)$$

is well-defined and uniquely determined, independently of the representation of s, by the relation

$$(\Big(\int s\,dE\Big)x, y)_H = \int_K s\,dE_{x,y} \qquad (x, y \in H),$$

since $\int_K s\,dE_{x,y} = \sum_{n=1}^N \alpha_n (E(B_n)x, y)_H = (\sum_{n=1}^N \alpha_n E(B_n)x, y)_H$.

It is readily checked that this integral is clearly linear, $\int 1\,dE = I$, and $(\int s\,dE)^* = \int \bar{s}\,dE$.

It is also multiplicative,

$$\int st\,dE = \int s\,dE \int t\,dE = \int t\,dE \int s\,dE, \tag{9.9}$$

since for a second simple function t we can suppose that $t = \sum_{n=1}^N \beta_n \chi_{B_n}$, with the same sets B_n as in s, and then

$$\begin{aligned}
\int s\,dE \int t\,dE &= \sum_{n=1}^{N} \beta_n \Big(\int s\,dE\Big) E(B_n) \\
&= \sum_{n=1}^{N} \beta_n \alpha_n E(B_n) E(B_n) = \sum_{n=1}^{N} \alpha_n \beta_n E(B_n) \\
&= \int st\,dE.
\end{aligned}$$

Also

$$\Big\|\Big(\int s\,dE\Big)x\Big\|_H^2 = \int_K |s|^2\,dE_{x,x} \qquad (x \in H,\ s \in S(K))$$

since

$$\begin{aligned}
(\Big(\int s\,dE\Big)x, \Big(\int s\,dE\Big)x)_H &= (\Big(\int s\,dE\Big)^* \Big(\int s\,dE\Big)x, x)_H \\
&= (\Big(\int |s|^2\,dE\Big)x, x)_H.
\end{aligned}$$

This yields

$$\Big\|\int s\,dE\Big\| \le \|s\|_\infty$$

and, in fact, the integral is isometric. Indeed, if we choose n so that $\|s\|_\infty = |\alpha_n|$ with $E(B_n) \neq 0$ and $x \in \operatorname{Im} E(B_n)$, then

$$\Big(\int s\,dE\Big)x = \alpha_n E(B_n)x = \alpha_n x$$

and necessarily

$$\Big\|\Big(\int s\,dE\Big)x\Big\|_H = \|s\|_\infty.$$

Now the integral can be extended over $L^\infty(E)$ by continuity, since it is a bounded linear map from the dense vector subspace $S(K)$ of $L^\infty(E)$ to the Banach space $\mathcal{L}(H)$.

We will denote

$$\Phi_E(f) := \int f\,dE = \lim_n \int s_n\,dE$$

if $s_k \to f$ in $L^\infty(E)$ ($s_k \in S(K)$).

The identities $(\Phi_E(s_k)x, y)_H = \int_K s_k\,dE_{x,y}$ extend to

$$(\Phi_E(f)x, y)_H = \int_K f\,dE_{x,y}$$

by taking limits. All the properties of Φ_E contained in the following theorem are now obvious:

Theorem 9.28. *If $E : \mathcal{B}_K \to \mathcal{L}(H)$ is a spectral measure, then there is a unique homomorphism of C^*-algebras $\Phi_E : L^\infty(K) \to \mathcal{L}(H)$ such that*

$$(\Phi_E(f)x, y)_H = \int_K f\,dE_{x,y} \qquad (x, y \in H,\ f \in L^\infty(K)).$$

This homomorphism also satisfies

$$\|\Phi_E(f)x\|_H^2 = \int_K |f|^2\,dE_{x,x} \qquad (x \in H,\ f \in L^\infty(K)). \tag{9.10}$$

9.5.2. Unbounded functions of bounded normal operators. To extend the functional calculus $f(T) = \Phi_f(T)$ of a bounded normal operator with bounded functions to unbounded measurable functions h, we start by extending to unbounded functions the functional calculus of Theorem 9.28 for any spectral measure E:

Theorem 9.29. *Suppose K a locally compact subset of $\mathbf{C}$, $E : \mathcal{B}_K \to \mathcal{L}(H)$ a spectral measure, h a Borel measurable function on $K \subset \mathbf{C}$, and*

$$\mathcal{D}(h) := \Big\{x \in H;\ \int_K |h(\lambda)|^2\,dE_{x,x} < \infty\Big\}.$$

Then there is a unique linear operator $\Phi_E(h)$ on H, represented as

$$\Phi_E(h) = \int_K h\,dE,$$

with domain $\mathcal{D}(\Phi_E(h)) = \mathcal{D}(h)$ and such that

$$(\Phi_E(h)x, y)_H = \int_K h(\lambda)\,dE_{x,y}(\lambda) \qquad (x \in \mathcal{D}(h),\ y \in H).$$

This operator is densely defined and, if f and h are Borel mesurable functions on K, the following properties hold:

(a) $\|\Phi_E(h)x\|_H^2 = \int_K |h|^2\, dE_{x,x}$, *if* $x \in \mathcal{D}(h)$.

(b) $\Phi_E(f)\Phi_E(h) \subset \Phi_E(fh)$ *and* $\mathcal{D}(\Phi_E(f)\Phi_E(h)) = \mathcal{D}(h) \cap \mathcal{D}(fh)$.

(c) $\Phi_E(h)^* = \Phi_E(\bar{h})$ *and* $\Phi_E(h)^*\Phi_E(h) = \Phi_E(|h|^2) = \Phi_E(h)\Phi_E(h)^*$.

Proof. It is easy to check that $\mathcal{D}(h)$ is a linear subspace of H. For instance, $\|E(B)(x+y)\|_H^2 \le 2\|E(B)x\|_H^2 + 2\|E(B)y\|_H^2$ so that

$$E_{x+y,x+y}(B) \le 2E_{x,x}(B) + 2E_{y+y}(B)$$

and $\mathcal{D}(h) + \mathcal{D}(h) \subset \mathcal{D}(h)$.

This subspace is dense. Indeed, if $y \in H$, we consider

$$B_n := \{|h| \le n\} \uparrow K,$$

so that, from the strong σ-additivity of E,

$$y = E(K)y = \lim_n E(B_n)y,$$

where $x_n := E(B_n)y \in \mathcal{D}(h)$ since

$$E(B)x_n = E(B)E(B_n)x_n = E(B \cap B_n)x_n \qquad (B \subset K)$$

and $E_{x_n,x_n}(B) = E_{x_n,x_n}(B \cap B_n)$, the restriction of E_{x_n,x_n} to B_n, so that

$$\int_K |h|^2\, dE_{x_n,x_n} = \int_{B_n} |h|^2\, dE_{x_n,x_n} \le n^2\|x_n\|_H^2 < \infty.$$

If h is bounded, then let us also prove the estimate

$$\Big|\int_K h\, dE_{x,y}\Big| \le \int_K |h|\, d|E_{x,y}| \le \Big(\int_K |h|^2\, dE_{x,x}\Big)^{1/2} \|y\|_H < \infty, \tag{9.11}$$

where $|E_{x,y}|$ is the total variation of the Borel complex measure $E_{x,y}$.

From the polar representation of a complex measure (see Lemma 4.12), we obtain a Borel measurable function ϱ such that $|\varrho| = 1$ and

$$\varrho h\, dE_{x,y} = |h|\, d|E_{x,y}|,$$

where $|E_{x,y}|$ denotes the total variation of $E_{x,y}$. Thus,

$$\begin{aligned}
\Big|\int_K h\, dE_{x,y}\Big| &\le \int_K |h|\, d|E_{x,y}| = \int_K \varrho h\, dE_{x,y} = (\Phi_E(\varrho h)x, y)_H \\
&\le \|\Phi_E(\varrho h)x\|_H \|y\|_H = \Big(\int_K |\varrho h|^2\, dE_{x,x}\Big)^{1/2} \|y\|_H \\
&= \Big(\int_K |h|^2\, dE_{x,x}\Big)^{1/2} \|y\|_H,
\end{aligned}$$

where in the second line we have used (9.10), and (9.11) holds.

When h is unbounded, to define $\Phi_E(h)x$ for every $x \in \mathcal{D}(h)$, we are going to show that $y \mapsto \int_K h\, dE_{x,y}$ is a bounded conjugate-linear form on H. Let us consider $h_n(z) = h(z)\chi_{B_n}(z) \to h(z)$ if $z \in K$, so that

$$\Big| \int_K h_n\, dE_{x,y} \Big| \leq \Big(\int_K |h_n|^2\, dE_{x,x} \Big)^{1/2} \|y\|_H,$$

and by letting $n \to \infty$, we also obtain (9.11) for h in this unbounded case if $x \in \mathcal{D}(h)$.

Then the conjugate-linear functional $y \mapsto \int_K h\, dE_{x,y}$ is bounded with norm $\leq (\int_K |h|^2\, dE_{x,x})^{1/2}$, and by the Riesz representation theorem there is a unique $\Phi_E(h)x \in H$ such that, for every $y \in H$,

$$(\Phi_E(h)x, y)_H = \int_K h(\lambda)\, dE_{x,y}(\lambda), \quad \|\Phi_E(h)x\|_H \leq \Big(\int_K |h|^2\, dE_{x,x} \Big)^{1/2}.$$

The operator $\Phi_E(h)$ is linear, since $E_{x,y}$ is linear in x, and densely defined.

We know that (a) holds if h is bounded. If it is unbounded, then let $h_k = h\chi_{B_k}$ and observe that $\mathcal{D}(h - h_k) = \mathcal{D}(h)$. By dominated convergence,

$$\|\Phi_E(h)x - \Phi_E(h_k)x\|_H^2 = \|\Phi_E(h - h_k)x\|_H^2 \leq \int_K |h - h_k|^2\, dE_{x,x} \to 0$$

as $k \to \infty$; according to Theorem 9.28, every h_k satisfies (a), which will follow for h by letting $k \to \infty$.

To prove (b) when f is bounded, we note that $\mathcal{D}(fh) \subset \mathcal{D}(h)$ and $dE_{x,\bar{\Phi}_E(f)z} = f\, dE_{x,z}$, since both complex measures coincide on every Borel set. It follows that, for every $z \in H$,

$$\begin{aligned} (\Phi_E(f)\Phi_E(h)x, z)_H &= (\Phi_E(h)x, \bar{\Phi}_E(f)z)_H = \int_K h\, dE_{x,\bar{\Phi}_E(f)z} \\ &= (\Phi_E(fh)x, z)_H \end{aligned}$$

and, if $x \in \mathcal{D}(h)$, we obtain from (a) that

$$\int_K |f|^2\, dE_{\Phi_E(h)x, \Phi_E(h)x} = \int_K |fh|^2\, dE_{x,x} \qquad (x \in \mathcal{D}(h)).$$

Hence, $\Phi_E(f)\Phi_E(h) \subset \Phi_E(fh)$.

If f is unbounded, then we take limits and

$$\int_K |f|^2\, dE_{\Phi_E(h)x, \Phi_E(h)x} = \int_K |fh|^2\, dE_{x,x} \qquad (x \in \mathcal{D}(h))$$

holds, so that $\Phi_E(h)x \in \mathcal{D}(f)$ if and only if $x \in \mathcal{D}(fh)$, and

$$\mathcal{D}(\Phi_E(f)\Phi_E(h)) = \{x \in \mathcal{D}(h);\ \Phi_E(h)x \in \mathcal{D}(f)\} = \mathcal{D}(h) \cap \mathcal{D}(fh),$$

as stated in (b).

Now let $x \in \mathcal{D}(h) \cap \mathcal{D}(fh)$ and consider the bounded functions $f_k = f\chi_{B_k}$, so that $f_k h \to fh$ in $L^2(E_{x,x})$. From (a) we know that $\Phi_E(f_k h)x \to \Phi_E(fh)x$,

$$\Phi_E(f)\Phi_E(h)x = \lim_k \Phi_E(f_k)\Phi_E(h)x = \lim_k \Phi_E(f_k h)x = \Phi_E(fh)x,$$

and (b) is true.

To prove (c), let $x, y \in \mathcal{D}(h) = \mathcal{D}(\bar{h})$. If $h_k = h\chi_{B_k}$, then

$$(\Phi_E(h)x, y)_H = \lim_k (\Phi_E(h_k)x, y)_H = \lim_k (x, \Phi_E(\bar{h}_k)y)_H = (x, \Phi_E(\bar{h})y)_H,$$

and it follows that $y \in \mathcal{D}(\Phi_E(h)^*)$ and $\bar{\Phi}_E(h) \subset \Phi_E(h)^*$. To finish the proof, let us show that $\mathcal{D}(\Phi_E(h)^*) \subset \mathcal{D}(h) = \mathcal{D}(\bar{h})$.

Let $z \in \mathcal{D}(\Phi_E(h)^*)$. We apply (b) to $h_k = h\chi_{B_k}$ and we have $\Phi_E(h_k) = \Phi_E(h)\Phi_E(\chi_{B_k})$ with $\Phi_E(\chi_{B_k})$ bounded and self-adjoint. Then

$$\begin{aligned} \Phi_E(\chi_{B_k})\Phi_E(h)^* &= \Phi_E(\chi_{B_k})^*\Phi_E(h)^* \subset (\Phi_E(h)\Phi_E(\chi_{B_k}))^* \\ &= \Phi_E(h_k)^* = \Phi_E(\bar{h}_k) \end{aligned}$$

and $\chi_{B_k}(\Phi_E(h)^*)z = \Phi_E(\bar{h}_k)z$. But $|\chi_k| \leq 1$, so that

$$\int_K |h_k|^2 dE_{z,z} = \int_K |\chi_{B_k}|^2 \, dE_{\Phi_E(h)^*z, \Phi_E(h)^*z} \leq E_{\Phi_E(h)^*z, \Phi_E(h)^*z}(K).$$

We obtain that $z \in \mathcal{D}(h)$ by letting $k \to \infty$.

The last part follows from (b), since $\mathcal{D}(\Phi_E(h\bar{h})) \subset \mathcal{D}(h)$. □

Remark 9.30. In Theorem 9.29, if $\Phi_E(B_0) = 0$, we can change K to $K \setminus B_0$:

$$(\Phi_E(h)x, y)_H := \int_{K\setminus B_0} h(\lambda)\, dE_{x,y}(\lambda) \qquad (x \in \mathcal{D}(\Phi_E(h)),\ y \in H)$$

if h is Borel measurable on $K \setminus B_0$.

If E is the spectral measure of a bounded normal operator T, then we write $h(T)$ for $\Phi_E(h)$, and then the results of Theorem 9.29 read

$$h(T) = \int_{\sigma(T)} h\, dE$$

on $\mathcal{D}(h) = \{x \in H;\ \|f\|^2_{E_{x,x}} < \infty\}$, in the sense that

$$(h(T)x, y)_H = \int_{\sigma(T)} h\, dE_{x,y} \qquad (x \in \mathcal{D}(h),\ y \in H).$$

Also

(a) $\|h(T)x\|^2_H = \int_{\sigma(T)} |h|^2\, dE_{x,x}$ if $x \in \mathcal{D}(h(T))$,

(b) $f(T)h(T) \subset (fh)(T)$, $\mathcal{D}(f(T)h(T)) = \mathcal{D}(h(T)) \cap \mathcal{D}((fh)(T))$ with $f(T)h(T) = (fh)(T)$ if and only if $\mathcal{D}((fh)(T)) \subset \mathcal{D}(h(T))$,

and

(c) $h(T)^* = \bar{h}(T)$ and $h(T)^*h(T) = |h|^2(T) = h(T)h(T)^*$.

9.5.3. The Cayley transform. We shall obtain a spectral representation theorem for self-adjoint operators using von Neumann's method of making a reduction to the case of unitary operators.

If T is a bounded self-adjoint operator on H, then the continuous functional calculus allows a direct definition of the Cayley transform of T as[14]

$$U = g(T) = (T - iI)(T + iI)^{-1},$$

where $g(t) = (t - i)/(t + i)$, a continuous bijection from $\mathbf{R}$ onto $\mathbf{S} \setminus \{1\}$, and it is a unitary operator (cf. Theorem 8.24).

Let us show that in fact this is also true for unbounded self-adjoint operators.

Let T be a self-adjoint operator on H. By the symmetry of T and from the identity $\|Ty \pm iy\|_H^2 = \|y\|_H^2 + \|Ty\|_H^2 \pm (iy, Ty)_H \pm (Ty, iy)_H$,

$$\|Ty \pm iy\|_H^2 = \|y\|_H^2 + \|Ty\|_H^2 \qquad \big(y \in \mathcal{D}(T)\big).$$

The operators $T \pm iI : \mathcal{D}(T) \to H$ are bijective and with continuous inverses, since $\pm i \in \sigma(T)^c$.

For every $x = Ty + iy \in \operatorname{Im}(T + iI) = H$ $(y \in \mathcal{D}(T))$, we define $Ux = U(Ty + iy) := Ty - iy$; that is,

$$Ux = (T - iI)(T + iI)^{-1}x \qquad (x \in H).$$

Then U is a bijective isometry of H, since $\|Ty + iy\|_H^2 = \|Ty - iy\|_H^2$ and $\operatorname{Im}(T \pm iI) = H$, and U is called the **Cayley transform** of T.

Lemma 9.31. *The Cayley transform*

$$U = (T - iI)(T + iI)^{-1}$$

of a self-adjoint operator T is unitary, $I - U$ is one-to-one, $\operatorname{Im}(I - U) = \mathcal{D}(T)$, *and*

$$T = i(I + U)(I - U)^{-1}$$

on $\mathcal{D}(T)$.

---

[14]Named after Arthur Cayley, this transform was originally described by Cayley (1846) as a mapping between skew-symmetric matrices and special orthogonal matrices. In complex analysis, the Cayley transform is the conformal mapping between the upper half-plane and the unit disc given by $g(z) = (z-i)/(z+i)$. It was J. von Neumann who, in 1929, first used it to map self-adjoint operators into unitary operators.

Proof. We have proved that U is unitary and, from the definition, $Ux = (T - iI)y$ if $x = (T + iI)y$ for every $y \in \mathcal{D}(T)$ and every $x \in H$. It follows that $(I + U)x = 2Ty$ and $(I - U)x = 2iy$, with $(I - U)(H) = \mathcal{D}(T)$. If $(I - U)x = 0$, then $y = 0$ and also $(I + U)x = 0$, so that a subtraction gives $2Ux = 0$, and $x = 0$. Finally, if $y \in \mathcal{D}(T)$, $2Ty = (I+U)(I-U)^{-1}(2iy)$. □

Remark 9.32. Since $I - U$ is one-to-one, 1 is not an eigenvalue of U.

9.5.4. Proof of Theorem 9.20: Let T be a self-adjoint operator on H. To construct the (unique) spectral measure E on $\sigma(T) \subset \mathbf{R}$ such that

$$T = \int_{\sigma(T)} t\, dE(t),$$

the Cayley transform U of T will help us to transfer the spectral representation of U to the spectral representation of T.

According to Theorem 8.24, the spectrum of U is a closed subset of the unit circle $\mathbf{S}$, and 1 is not an eigenvalue, so that the spectral measure E' of U satisfies $E'\{1\} = 0$, by Theorem 8.26. We can assume that it is defined on $\Omega = \mathbf{S} \setminus \{1\}$ and we have the functional calculus

$$f(U) = \int_{\sigma(U)} f(\lambda)\, dE'(\lambda) = \int_\Omega f(\lambda)\, dE'(\lambda) \qquad (f \in \mathbf{B}(\Omega)),$$

which was extended to unbounded functions in Subsection 9.5.2.

If $h(\lambda) := i(1 + \lambda)/(1 - \lambda)$ on Ω, then we also have

$$(h(U)x, y)_H = \int_\Omega h(\lambda)\, dE'_{x,y}(\lambda) \qquad (x \in \mathcal{D}(h(U)),\ y \in H),$$

with

$$\mathcal{D}(h(U)) = \{x \in H;\ \int_\Omega |h|^2\, dE'_{x,x} < \infty\}.$$

The operator $h(U)$ is self-adjoint, since h is real and $h(U)^* = \bar{h}(U) = h(U)$.

From the identity

$$h(\lambda)(1 - \lambda) = i(1 + \lambda),$$

an application of (b) in Theorem 9.29 gives

$$h(U)(I - U) = i(I + U),$$

since $\mathcal{D}(I - U) = H$. In particular, $\operatorname{Im}(I - U) \subset \mathcal{D}(h(U))$.

From the properties of the Cayley transform, $T = i(I + U)(I - U)^{-1}$, and then

$$T(I - U) = i(I + U), \quad \mathcal{D}(T) = \operatorname{Im}(I - U) \subset \mathcal{D}(h(U)),$$

so that $h(U)$ is a self-adjoint extension of the self-adjoint operator T. But, T being maximally symmetric, $T = h(U)$. This is,

$$(Tx, y)_H = \int_\Omega h(\lambda)\, dE'_{x,y}(\lambda) \qquad (x \in \mathcal{D}(T),\, y \in H).$$

The function $t = h(\lambda)$ is a homeomorphism between Ω and $\mathbf{R}$ that allows us to define $E(B) := E'(h^{-1}(B))$, and it is readily checked that E is a spectral measure on $\mathbf{R}$ such that

$$(Tx, y)_H = \int_{\mathbf{R}} t\, dE_{x,y}(t) \qquad (x \in \mathcal{D}(T),\, y \in H).$$

Conversely, if E is a spectral measure on $\mathbf{R}$ which satisfies

$$(Tx, y)_H = \int_{\mathbf{R}} t\, dE_{x,y}(t) \qquad (x \in \mathcal{D}(T),\, y \in H),$$

by defining $E'(B) := E(h(B))$, we obtain a spectral measure on Ω such that

$$(h(U)x, y)_H = \int_\Omega h(\lambda)\, dE'_{x,y}(\lambda) \qquad (x \in \mathcal{D}(h(U)),\, y \in H).$$

But $U = h^{-1}(h(U))$ and

$$(Ux, y)_H = \int_\Omega \lambda\, dE'_{x,y}(\lambda) \qquad (x, y \in H).$$

From the uniqueness of E' with this property, the uniqueness of E follows.

Of course, the functional calculus for the spectral measure E defines the functional calculus $f(T) = \int_{-\infty}^{+\infty} f\, dE$ for $T = \int_{-\infty}^{+\infty} \lambda\, dE(\lambda)$, and $f(T) = f(h(U))$.

9.6. Exercises

Exercise 9.1. Let $T : \mathcal{D}(T) \subset H \to H$ be a linear and bounded operator. Prove that T has a unique continuous extension on $\overline{\mathcal{D}(T)}$ and that it has a bounded linear extension to H. Show that this last extension is unique if and only if $\mathcal{D}(T)$ is dense in H.

Exercise 9.2. Prove that if T is a symmetric operator on a Hilbert space H and $\mathcal{D}(T) = H$, then T is bounded.

Exercise 9.3. Prove that the derivative operator D is unbounded on $L^2(\mathbf{R})$.

Exercise 9.4. If T is an unbounded densely defined linear operator on a Hilbert space, then prove that $(\operatorname{Im} T)^\perp = \operatorname{Ker} T^*$.

Exercise 9.5. If T is a linear operator on H and $\lambda \in \sigma(T)^c$, then prove that $\|R_T(\lambda)\| \geq 1/d(\lambda, \sigma(T))$.

Exercise 9.6. Show that, if T is a symmetric operator on H and $\operatorname{Im} T = H$, then T is self-adjoint.

Exercise 9.7. If T is an injective self-adjoint operator on $\mathcal{D}(T) \subset H$, then show that $\operatorname{Im} T = \mathcal{D}(T^{-1})$ is dense in H and that T^{-1} is also self-adjoint.

Exercise 9.8. Prove that the residual spectrum of a self-adjoint operator on a Hilbert space H is empty.

Exercise 9.9. Suppose A is a bounded self-adjoint operator on a Hilbert space H and let

$$A = \int_{\sigma(A)} \lambda \, dE(\lambda)$$

be the spectral representation of A. A vector $z \in H$ is said to be **cyclic** for A if the set $\{A^n z\}_{n=0}^{\infty}$ is total in H.

If A has a cyclic vector z and $\mu = E_{z,z}$, then prove that A is unitarily equivalent to the multiplication operator $M : f(t) \mapsto tf(t)$ of $L^2(\mu)$; that is, $M = U^{-1}AU$ where $U : L^2(\mu) \to H$ is unitary.

Exercise 9.10. Let

$$A = \int_{\sigma(A)} \lambda \, dE(\lambda)$$

be the spectral resolution of a bounded self-adjoint operator of H and denote

$$F(t) := E(-\infty, t] = E(\sigma(A) \cap (-\infty, t]).$$

Prove that the operator-valued function $F : \mathbf{R} \to \mathcal{L}(H)$ satisfies the following properties:

(a) If $s \leq t$, then $F(s) \leq F(t)$; that is, $(F(s)x, x)_H \leq (F(t)x, x)_H$ for every $x \in H$.

(b) $F(t) = 0$ if $t < m(A)$ and $F(t) = I$ if $t \geq M(A)$.

(c) $F(t+) = F(t)$; that is, $\lim_{s \downarrow t} F(s) = F(t)$ in $\mathcal{L}(H)$.

If $a < m(A)$ and $b > M(A)$, then show that with convergence in $\mathcal{L}(H)$

$$A = \int_a^b t \, dF(t) = \int_{m(A)+}^{M(A)} t \, dF(t) = \int_{\mathbf{R}} f \, dF(t)$$

as a Stieltjes integral.

Exercise 9.11. On $L^2(0,1)$, let $S = iD$ with domain $H^1(0,1)$. Prove the following facts:

(a) $\operatorname{Im} S = L^2(\mathbf{R})$.

(b) $S^* = iD$ with domain $H_0^1(0,1)$.

(c) S is a non-symmetric extension of iD with $\mathcal{D}(iD) = H^2(0,1)$.

Exercise 9.12. On $L^2(0,1)$, let $R = iD$ with domain $H_0^1(0,1)$ (i.e, S^* in Exercice 9.11). Prove the following facts:

(a) $\operatorname{Im} R = \{u \in L^2(\mathbf{R});\ \int_0^1 u(t)\,dt = 0\}$.

(b) $R^* = iD$ with domain $H^1(0,1)$ (i.e, $R^* = S$ of Exercice 9.11).

Exercise 9.13. As an application of Theorem 9.17, show that the operator $-D^2 = -d^2/dx^2$ in $L^2(0,1)$ with domain the C^∞ functions f on $[0,1]$ such that $f(0) = f(1) = 0$ is essentially self-adjoint.

Exercise 9.14. Show also that the operator $-D^2 = -d^2/dx^2$ in $L^2(0,1)$ with domain the C^∞ functions f on $[0,1]$ such that $f'(0) = f'(1) = 0$ is essentially self-adjoint.

Exercise 9.15. Prove that $-D^2 = -d^2/dx^2$ with domain $\mathcal{D}(0,1)$ is not an essentially self-adjoint operator in $L^2(0,1)$.

Exercise 9.16. Let V be a nonnegative continuous function on $[0,1]$. Then the differential operator $T = -d^2/dx^2 + V$ on $L^2(0,1)$ with domain $\mathcal{D}^2(0,1)$ has a self-adjoint Friedrichs extension.

Exercise 9.17. Let

$$Q_k\varphi(x) = x_k\varphi(x), \quad P_k\varphi = \frac{h}{2\pi i}\partial_k\varphi \qquad (1 \le k \le n)$$

represent the position and momentum operators on $L^2(\mathbf{R}^n)$.

Prove that they are unbounded self-adjoint operators whose commutators satisfy the relations

$$[Q_j, Q_k] = 0, \quad [P_j, P_k] = 0, \quad [P_j, Q_k] = \delta_{j,k}\frac{h}{2\pi i}I.$$

Note: These are called the **canonical commutation relations** satisfied by the system $\{Q_1, \ldots, Q_n; P_1, \ldots, P_n\}$ of $2n$ self-adjoint operators, and it is said that Q_k is **canonically conjugate** to P_k.

Exercise 9.18. Prove that, if A_1 and A_2 are two bounded self-adjoint operators in a Hilbert space, then $A_1A_2 = A_2A_1$ if and only if their spectral measures E^1 and E^2 commute as in (9.8): $E^1(B_1)E^2(B_2) = E^2(B_2)E^1(B_1)$ for all $B_1, B_2 \in \mathcal{B}_{\mathbf{R}}$.

Exercise 9.19. Let

$$A_1 = \int_{\mathbf{R}} \lambda\, dE^1(\lambda), \quad A_2 = \int_{\mathbf{R}} \lambda\, dE^1(\lambda)$$

be two self-adjoint operators in a Hilbert space H. If they commute (in the sense that their spectral measures commute), prove that there exists a unique spectral measure E on $\mathbf{R}^2$ such that

$$E(B_1 \times B_2) = E(B_1)E(B_2) \qquad (B_1, B_2 \in \mathcal{B}_{\mathbf{R}}).$$

In the case of the position operators $A_1 = Q_1$ and $A_2 = Q_2$ on $L^2(\mathbf{R}^2)$, show that $E(B) = \chi_B \cdot$ $(B \subset \mathcal{B}_{\mathbf{R}^2})$.

Exercise 9.20. Find the infinitesimal generator of the one-parameter group of unitary operators $U_t f(x) := f(x+t)$ on $L^2(\mathbf{R})$.

Exercise 9.21. Suppose that $g : \mathbf{R} \to \mathbf{R}$ is a continuous function. Describe the multiplication $g\cdot$ as a self-adjoint operator in $L^2(\mathbf{R})$ and $U_t f := e^{itgf}$ as a one-parameter group of unitary operators. Find the infinitesimal generator A of U_t $(t \in \mathbf{R})$.

References for further reading

E. Hille and R. S. Phillips, *Functional Analysis and Semigroups.*

T. Kato, *Perturbation Theory for Linear Operators.*

P. D. Lax, *Functional Analysis.*

E. H. Lieb and M. Loss, *Analysis.*

E. Prugovečki, *Quantum Mechanics in Hilbert Space.*

M. Reed and B. Simon, *Methods of Modern Mathematical Physics.*

F. Riesz and B. Sz. Nagy, *Leçons d'analyse fonctionelle.*

W. Rudin, *Functional Analysis.*

A. E. Taylor and D. C. Lay, *Introduction to Functional Analysis.*

J. von Neumann, *Mathematical Foundations of Quantum Mechanics.*

K. Yosida, *Functional Analysis.*

Hints to exercises

Chapter 1

Exercise 1.1. The identity is a homeomorphism between the two metric spaces, and $\{1, 2, 3, \ldots\}$ is a Cauchy sequence in $(\mathbf{R}, d)$ but not in $(\mathbf{R}, |\cdot|)$.

Exercise 1.2. A finite number of balls $B(a, m)$ $(m \in \mathbf{N})$ cover the compact set.

Exercise 1.3. If every sequence $\{a_k\} \subset A$ has a convergent subsequence in M, for every sequence $\{x_k\} \subset \bar{A}$, choose $\{a_k\} \subset A$ with $d(x_k, a_k) < 1/k$. If $a_{k_m} \to x$, also $x_{k_m} \to x$.

Exercise 1.4. If U is an open neighborhood of $x \in K$, using the fact that $\bar{U}$ is compact and that K is Hausdorff, show that U contains a compact neighborhood W of x in $\bar{U}$.

Exercise 1.5. If $X = F \cap G$ (F closed and G open) and $a \in X$, then $\bar{B}(a, R) \subset G$ for some $R > 0$ and $\bar{B}(a, r) \cap F$ $(r < R)$ is a neighborhood basis of a in X.

If $X \subset \mathbf{R}^n$ is locally compact, for every $a \in X$ consider a compact neighborhood $W(a)$; then $\operatorname{Int} W(a)$ is an open neighborhood of a and $\operatorname{Int} W(a) = U(a) \cap X$ for some open set $U(a)$ in $\mathbf{R}^n$, so that $G = \bigcup_{a \in X} U(a)$ is also open. Show that $X = G \setminus (G \setminus X) = G \cap (G \setminus A)^c$.

Exercise 1.6. If $\{f_n\}_{n=1}^{\infty} \subset M$ and $A := \bigcup_{n=1}^{\infty} \{f_n \neq 0\} = \{x_1, x_2, \ldots\}$, I being compact, by a diagonal argument select f_{n_k} so that $f_{n_k}(x_j) \to \ell_j \in I$. If $f(x_j) := \ell_j$ and $f(x) = 0$ if $x \notin A$, then $f \in M$ and $f_{n_k} \to f$ in M.

Exercise 1.7. See J. L. Kelley [**24**, Chapter 3].

Exercise 1.8. Consider $I : (K, \mathcal{T}') \to (K, \mathcal{T})$ and use that $I(F)$ is a compact subset in $(K, \mathcal{T})$ to show that I^{-1} is continuous.

Exercise 1.9. (b) Show that $x_n \to x_0$ in X if and only if $x_n = x_0$ for every $n \geq n_0$.

Exercise 1.10. See W. Rudin [**39**, Chapter 1].

Exercise 1.11. Every function is measurable and, according to the definition (1.3) of the Lebesgue integral, the integral of a nonnegative function $f = \{f(j)\}_{j \in J}$ is the sum

$$\sum_{j \in J} f(j) := \sup \Big\{ \sum_{k \in F} f(k);\ k \in K,\ K \subset \ \text{finite} \Big\}.$$

If $\sum_{j \in J} |f(j)| < \infty$, then $N := \{j \in J;\ f(j) \neq 0\} = \bigcup_{n=1}^{\infty} \{j \in J;\ |f(j)| \geq 1/n\}$ is at most countable, since every $\{j \in J;\ |f(j)| \geq 1/n\}$ is finite. Hence $\sum_{j \in J} |f(j)| = \sum_{n \in N} |f(n)|$ and

$$\int f \, d\nu = \int f^{+} \, d\nu - \int f^{-} \, d\nu = \sum_{k \in N} f^{+}(k) - \sum_{k \in N} f^{-}(k) = \sum_{k \in N} f(k).$$

Exercise 1.12. Use monotone convergence (first prove that $(1 + x/n)^n$ is increasing), dominated convergence, and Fatou's lemma, respectively.

Exercise 1.13. Show that $I = 2J$ with

$$J = \int_0^1 dx \int_0^x \frac{dy}{(x-y)^{\alpha}}, \qquad \int_0^x \frac{dy}{(x-y)^{\alpha}} = \lim_{\varepsilon \downarrow 0} \int_0^{x-\varepsilon} \frac{dy}{(x-y)^{\alpha}}.$$

Exercise 1.14. If $E = \biguplus_k (a_k, b_k)$, then $\sum_k |F(b_k) - F(a_k)| \leq \int_E |f(t)| \, dt = \mu(E)$, where μ is a finite measure such that $\mu(A) = 0$ if $|A| = 0$. It is shown that $\mu(E) \to 0$ as $|E| \to 0$ by supposing that, for some $\varepsilon > 0$ there exists E_n such that $\mu(E_n) \geq \varepsilon$ and $|E_n| \leq 1/2^n$ for every $n \in \mathbf{N}$, since then $E := \bigcap_{k \geq 1} \bigcup_{n \geq k} E_n$ would satisfy $|E| = 0$ but $\mu(E) \geq n$.

Exercise 1.15. See Royden [**37**, III.3].

Exercise 1.16. μ is the linear combination of four finite measures.

Exercise 1.17. Using the representation $\int_B f \, d\mu = \int_B fh \, d|\mu|$ with h as in (1.10), we have both $\left| \int_B h \, d\mu \right| = |\mu|(B)$ with $|h| = 1$ and also $\left| \int_B f \, d\mu \right| \leq \int_B |f| \, d|\mu| \leq |\mu|(B)$ if $|f| \leq 1$.

Exercise 1.18. The Riesz-Markov theorem with $u(g) = \sum_{k=1}^{\infty} \lambda_k g(a_k)$ yields the first part.

Obviously $|\int g\,d\mu| \le \sum_{k=1}^{\infty} |\lambda_n|$ if $g \in \mathcal{C}_c(\mathbf{R}^n)$ and $|g| \le 1$. For the opposite estimate, if $\sum_{k>n_0} |\lambda_n| \le \varepsilon$, construct a convenient function g such that $g(a_k) = \operatorname{sgn} \lambda_k$ if $k \le n_0$. Then $\sum_{k\le n_0} |\lambda_n| \le |u(g)| + \varepsilon$ and

$$\sum_{k=1}^{\infty} |\lambda_n| = \sup_{n_0} \sum_{k\le n_0} |\lambda_n| \le \sup_{|g|\le 1} |u(g)|.$$

Similarly $\sum_{a_k\in G} |\lambda_n| = |\mu|(G)$ for any open set $G \subset \mathbf{R}^n$ and $|\mu|$ is associated to $\{|\lambda_k|\}$ in the same way that μ is associated to $\{\lambda_k\}$. Clearly $|\mu|(G) = 0$ if $G = F^c$.

Chapter 2

Exercise 2.1. If $n = 1$, use induction and $\binom{m+1}{k} = \binom{m}{k} + \binom{m}{k-1}$.

If $n > 1$ and $\alpha = (\alpha_1, \dots, \alpha_{n+1})$, write $\bar\alpha := (\alpha_1, \dots, \alpha_n, 0)$ and

$$D^{\bar\alpha}(\partial_{n+1}^{\alpha_{n+1}}(fg)) = \sum_{k=0}^{\alpha_{n+1}} \binom{\alpha_{n+1}}{k} D^{\bar\alpha}\Big(\partial_{n+1}^k f \partial_{n+1}^{\alpha_{n+1}-k} g\Big).$$

Then

$$D^{\bar\alpha}\Big(\partial_{n+1}^k f \partial_{n+1}^{\alpha_{n+1}-k} g\Big) = \sum_{\beta\le\bar\alpha} \binom{\bar\alpha}{\beta} D^\beta(\partial_{n+1}^k f) D^{\bar\alpha-\beta}(\partial_{n+1}^{\alpha_{n+1}-k} g).$$

Finally,

$$\sum_{k=0}^{\alpha_{n+1}} \binom{\alpha_{n+1}}{k} \sum_{\beta\le\bar\alpha} \binom{\bar\alpha}{\beta} D^\beta(\partial_{n+1}^k f) D^{\bar\alpha-\beta}(\partial_{n+1}^{\alpha_{n+1}-k} g) = \sum_{\beta\le\alpha} \binom{\alpha}{\beta} (D^\beta f) D^{\alpha-\beta} g.$$

Exercise 2.2. If $U(x) = x + U \subset F$, a neighborhood of x, then also $U \subset F$ and $\bigcup_{m=1}^{\infty} mB_E(0,r) = E$.

Exercise 2.3. Suppose $x \in E \setminus \{0\}$ and let $U = \{0\}$. The product λx should be continuous in λ, so that we should have $\lambda x \in U$ if $|\lambda| \le \varepsilon$ for some $\varepsilon > 0$.

Exercise 2.4. If $0 < t < 1$, then $t \operatorname{Int} K + (1-t) \operatorname{Int} K \subset K$ and every $t \operatorname{Int} K + (1-t) \operatorname{Int} K \subset K$ is open, so that $t \operatorname{Int} K + (1-t) \operatorname{Int} K \subset \operatorname{Int} K$.

Exercise 2.5. Define $u(e_1) = 1$, $u(e_2) = 2$, ... if $\{e_n\}$ is the canonical basis sequence.

Exercise 2.6. If g and h are two different bounded continuous functions, then $G = \{f \ne g\}$ is a nonempty open set and $|G| > 0$, so that $g \ne h$ in $L^\infty(\mathbf{R}^n)$. Choose an uncountable family of intervals I_α, so that $\|\chi_{I_\alpha} - \chi_{I_\beta}\|_\infty \ge 1$ if $\alpha \ne \beta$ to show that $L^\infty(\mathbf{R}^n)$ is not separable. See Remark 2.18.

Exercise 2.7. Use the fact that, as a subspace of $\mathcal{C}(\bar{B}(0,m))$, $\mathcal{C}_c(B(0,m))$ is separable and it is dense in $L^p(B(0,m))$. Then approximate every $f \in L^p(\mathbf{R}^n)$ by $f\chi_{B(0,m)}$.

Exercise 2.8. $\mathcal{C}_0(\mathbf{R}^n)$ is complete and $\varrho(x/k)g(x) \to g(x)$ uniformly if $\bar{B}(0,1) \prec \varrho \prec B(0,2)$. Every $\{f \in \mathcal{C}_c(\mathbf{R}^n);\ \operatorname{supp} f \subset \bar{B}(0,m)\}$ is separable.

Exercise 2.9. (a) For every $n \in \mathbf{N}$, $K = \bigcup_{j=1}^{N(j)} B(c_{n,j}, 1/n)$, and the set $\{c_{n,j}\}_{n,j}$ is dense in K. (b) Apply the Stone-Weierstrass theorem to the subalgebra of $\mathcal{C}(K)$ generated by the functions $\varphi_{m,n}$.

Exercise 2.10. Since $\mathcal{C}(0,T)$ is dense in $L^2(0,T)$, $\mathcal{C}(\mathbf{T})$ is also dense in $L^2(\mathbf{T})$. The Stone-Weierstrass theorem shows that the algebra of trigonometric polynomials is dense in $\mathcal{C}(\mathbf{T})$, and the uniform convergence implies the convergence in $L^2(\mathbf{T})$.

Exercise 2.11. Describe $\mathcal{C}^1[a,b]$ as a closed subspace of $\mathcal{C}[a,b] \times \mathcal{C}[a,b]$.

Exercise 2.12. The constants are 1 and $n^{1/p}$.

Exercise 2.13. Show that $|u(f)| = 1$ with $|f| \leq 1$ implies $f = \operatorname{sgn} g$ on every interval $(1/(n+1), 1/n)$ and that f cannot be continuous, so that $u(B_{\mathcal{C}}) \subset (-1,1)$. If $-1 < r < 1$, we define f such that $u(f) = r$ as $f = \sum_{n=1}^{\infty} (-1)^n \varrho_n(t)$ with $(-1)^n \varrho_n \prec (1/(n+1), 1/n)$.

Exercise 2.14. $T_K 1 = \|T_K\|$.

Exercise 2.15. If $M = \max_{0 \leq y \leq 1} \int_0^1 |K(x,y)|\, dx$, then $\|Tf\|_1 \leq M\|f\|_1$ by Fubini-Tonelli, and $\|T\| \leq M$.

To prove that $\|Th\|_1 \geq M - \varepsilon$ with $\|h\|_{[0,1]} \leq 1$, note that $M = \int_0^1 |K(x,y_0)|\, dx$ for some $y_0 \in [0,1]$, use the uniform continuity of K to choose $y_1 \leq y_0 \leq y_2$ in $[0,1]$ so that $|K(x,y) - K(x,y_0)| \leq \varepsilon$ if $y \in [y_1, y_2]$, and define $h = (y_2 - y_1)^{-1}\chi_{[y_1,y_2]}$.

Exercise 2.16. $\|T_K\| = \|T_K 1\|_p$ ($p = \infty$ or 1).

Exercise 2.17. Show that $T^n = 0$, since its kernel is $K_n = 0$ if $n > 1$, and then $(I-T)^{-1} = I + T$. Note that $v_0(y)\sin(ny)$ is an odd function when v_0 is even.

Exercise 2.18. Write the equations in the form $(pu')' - qu = f$ and the Cauchy problems as $\int_0^1 K(x,y)u(y)\, dy - u(x) = g(x)$. Then (a) $f = 0$, $K(x,y) = y - x$, $g(x) = -x$; (b) $f(x) = \cos(x)$, $K(x,y) = y - x$, $g(x) = -x + \cos x$; (c) $f = 0$, $K(x,y) = (a_0/a_1)(\exp(a_1 y - a_1 x) - 1)$, $g(x) =$

$-\alpha + (b/\alpha_1)(\exp(-a_1 x) - 1)$; (d) $f = 0$, $K(x, y) = x(1 - \exp(x^2/2 - y^2/2))$, $g(x) = -x\exp(x^2/2)$.

Exercise 2.19. By induction and Fubini's theorem,

$$\int_a^x dx_{n-1} \int_a^{x_{n-1}} dx_{n-2} \cdots \int_a^{x_1} f(t)\,dt = \frac{1}{(n-1)!}\int_a^x (x-t)^{n-1} f(t)\,dt.$$

Exercise 2.20. Consider $v = u^{(n)}$, integrate on $[a, x]$, use the initial conditions $u^{(j)}(a) = c_j$, and apply the result of Exercise 2.19.

Exercise 2.21. Show that $\operatorname{supp} f * g$ contains $[0, \infty) \times \{0\}$ but $\operatorname{supp} f + \operatorname{supp} g = (0, \infty) \times (0, \infty)$, a sum of noncompact sets.

Exercise 2.22. If $s(f)(x) = f(-x)$, show that $|(f * g)(b) - (f * g)(a)| \leq C\|\tau_b s(f) - \tau_a s(f)\|$ and apply Theorem 2.14.

Exercise 2.23. Just compute the convolutions.

Exercise 2.24. See the proof of Theorem 2.41(a).

Exercise 2.25. Write

$$2F_{2N} - F_N = \frac{1}{N}\sum_{n=0}^{2N-1} D_n - \frac{1}{N}\sum_{n=0}^{N-1} D_n = \frac{1}{N}\sum_{n=N}^{2N-1} D_n.$$

Exercise 2.26. $\|c\| = 1$.

Exercise 2.27. $W_t(x) = (1/t^n)W(x/t)$.

Exercise 2.28. If $P_F(x) = y$, then $y = \sum_{n\geq 1}(y, e_n)_H e_n$ (Fischer-Riesz). If $x = y + z$, then $(y, e_n)_H = (x, e_n)_H$.

Exercise 2.29. Check that

$$\int_Y |K(x,y)||f(y)|\,dy \leq C^{1/p'}\Big(\int_Y |K(x,y)||f(y)|^p\,dy\Big)^{1/p}$$

and then integrate.

Exercise 2.30. Note that T is $(1, 1)$ and (∞, ∞) with $M_0 = M_1 = C$ and $1/p = 1/q = 1 - \vartheta$.

Exercise 2.31. $\|\tilde{T}\| \leq 2\|T\|$.

Exercise 2.32. Let $\vartheta = 1/2$ and note that T is of type $(4, 4/3)$ with constant $22^{1/2}(2^{1/2})^{1/2}$ (Exercise 2.31). Show that $\|T(1,2)\|_{4/3} > 2^{3/4}\|(1,2)\|_4$ if $(x, y) = (1, 2)$.

Exercise 2.33. Obviously $|c_k(f)| \le \|f\|_1$, and also $\|c(f)\|_2 = \|f\|_2$ (Parseval). If $1 < p < 2$, apply the Riesz-Thorin theorem to $c : L^p(\mathbf{T}) \to \ell^{p'}$.

Chapter 3

Exercise 3.1. (a) See (4.11).

Exercise 3.2. If $A \subset rU$ and U is convex, also $\mathrm{co}(A) \subset rU$. If A is open, then

$$\mathrm{co}(A) = \bigcup\{t_1 A + \cdots + t_n A; n \in \mathbf{N},\, t_j \ge 0,\, t_1 + \cdots + t_n = 1\}$$

is also open.

Exercise 3.3 Let U be a neighborhood of zero in E. Then $\lambda V \subset U$ if $|\lambda| \le \delta$, for some $\delta > 0$ and some neighborhood of zero, V. It follows that $\bigcup\{\lambda V; |\lambda| \le \delta\} \subset U$ is balanced.

Exercise 3.4. If $f_n \to f$ uniformly on compact sets, then $0 = \int_\gamma f_n(z)\,dz \to \int_\gamma f(z)\,dz = 0$, and f is analytic. Hence, $\mathcal{H}(\Omega)$ is closed in $\mathcal{C}(\Omega)$.

Exercise 3.5. Use the open mapping theorem.

Exercise 3.6. (d) $\|\lambda x\| \le \sum_{n=1}^N 2^{-n} \frac{p_n(\lambda x)}{1+p_n(\lambda x)} + 2^{-N} \le \varepsilon$ if N is such that $2^{-N} \le \varepsilon/2$, and allow $\lambda \to 0$, so that $\sum_{n=1}^N 2^{-n} \frac{p_n(\lambda x)}{1+p_n(\lambda x)} \le \varepsilon$. (f) $\|nx\| \le 1$ and $\|nx\| = n\|x\| \not\to \infty$.

Exercise 3.7. It should be $\|h\| \le 1/2$.

Exercise 3.8. If the semi-norms p_n^j define the topology of E^j, then the semi-norms $q_n(x) := \max(p_n^1(x^1), \ldots, p_n^m(x^m))$ ($x = (x^1, \ldots, x^m)$) define the product topology on E.

If $E = E^1 \times \cdots \times E^m \times \cdots$, a countable product, the topology is defined by the sequence of semi-norms q_m ($x = (x^1, \ldots, x^m, \ldots)$).

Embed $\mathcal{E}(\Omega) \hookrightarrow \prod_{k=0}^\infty \mathcal{C}(\Omega)$ by $f \mapsto (f^{(k)})$.

Exercise 3.9. (a) If $\{f_{n_k}\}$ is a subsequence, then $f_{n_k}(t) \to 0$ if $t \ne 0$, and $f_{n_k}(0) \to 1$. (b) Choose $a \in \Omega$ and define

$$f_n(t) := \frac{d(t, B(a, 1/n)^c)}{d(x,a) + d(t, B(a,1/n)^c)}$$

with n large enough.

Exercise 3.10. Consider $M \ni x_n \to x \in E_1$ and prove that Tx_n is a Cauchy sequence with a limit $y \in E_2$ such that, if also $M \ni x_n' \to x \in E_1$, then necessarily $Tx_1, Tx_1', Tx_2, Tx_2', \ldots$ is a Cauchy sequence with the same limit y. Define $\tilde{T}\tilde{x} := y$.

Exercise 3.11. (a) First show that the topology in E/F is the collection of all sets $G \subset E/F$ such that $\pi^{-1}(G)$ is an open set in E, so that π is continuous, and if a set $M \subset E/F$ is closed, then $\pi^{-1}(M)$ is closed in E.

(b) If $\|\cdot\|_E$ is an F-norm associated to the topology on E, then check that $\|\pi(x)\| = \inf_{z\in F} \|x-z\|_E$ is a corresponding F-norm for E/F. To prove that every Cauchy sequence $\{u_n\}$ in E/F must converge, choose $\|u_{n_k} - u_{n_{k+1}}\| < 1/2^k$ and $\pi(x_k) = u_{n_k}$ so that $\|x_k - x_{k+1}\|_E < 1/2^k$ still. If $x_k \to x$ in E, then $u_{n_k} \to \pi(x)$ in E/F and also $u_n \to \pi(x)$ in E/F.

(c) $F + M = \pi^{-1}(\pi(M))$ is closed, since $\pi(M)$ is finite dimensional in E/F and every finite dimensional subspace is closed by Theorem 2.25 (in the case of a Fréchet space, cf. Köthe [**26**, 15.5] or Rudin [**38**, 1.21]).

Exercise 3.12. Show that $\{e_n\} \subset F + M$ and $F + M$ is dense. The element $x_0 = \sum_{n=1}^{\infty} \frac{1}{n} e_{-n}$ is well-defined. Suppose $x_0 = x_1 + x_2$ with $x_2 = \lim b_m$ ($b_m \in [u_n; n \geq 1]$) and $x_1 \in F$ and show that, if $n \geq 1$, then $(-ne_{-n} + e_n, x_2)_H = 0$ and also $(x_1, e_{-n})_H = 0$. It follows that $(x_0, -ne_{-n} + e_n)_H = -1$ so that $(x_1, e_n)_H = -1$, and then $\sum_n |(x_1, e_n)_H|^2 = \infty$, a contradiction.

Exercise 3.13. Write $[a, b] = \bigcup_\alpha \{x_\alpha\}$, where $\operatorname{Int}(\{x_\alpha\}) = \emptyset$.

Exercise 3.14. If $F = \bigcup_{n=1}^{\infty} [e_1, \ldots, e_n]$ and $\operatorname{Int}([e_1, \ldots, e_m]) \neq \emptyset$, then $F = [e_1, \ldots, e_m]$.

Exercise 3.15. Consider $f(x) = 1/x$ ($f(0) := 0$).

Exercise 3.16. Use the closed graph theorem.

Exercise 3.17. If $f_n(t) = t^n/n$, then $f_n \to 0$ in $(\mathcal{C}^1[0,1], \|\cdot\|_{[0,1]})$ but $f_n'(t) = t^{n-1}$ does not converge in $(\mathcal{C}^1[0,1], \|\cdot\|_{[0,1]})$.

Exercise 3.18. See Remark 3.12.

Exercise 3.19. The set $F_n := \{\psi(x) \leq n\}$ is closed and $M = \bigcup_n F_n$. By Corollary 3.9 of Baire's theorem, $\operatorname{Int} F_m \neq \emptyset$ for some m.

Exercise 3.20. Apply the uniform boundedness principle (Theorem 3.14) to the sequence $\{J_n\}$.

Exercise 3.21. (a) If $(x_n, y_n) \to (x_0, y_0)$ in $E_1 \times E_2$, show that the linear mappings $T_n = B(\cdot, y_n)$ are uniformly bounded and write $B(x_n, y_n) - B(x_0, y_0) = T_n(x_n - x_0) + B(x_0, y_n - y_0)$. (b) Choose P_n polynomials such that $P_n(x) \to x^{-1/2}$ in $L^1(0,1)$ (polynomials are dense in $L^1(0,1)$, since $C_c(0,1)$ is dense by Corollary 2.13 and polynomials are uniformly dense in $C_c(0,1)$) and observe that $\{P_n\}$ is a bounded sequence such that $\{B(P_n, P_n)\}$ is unbounded.

Exercise 3.22. See Exercise 3.21.

Exercise 3.23. Use the closed graph theorem.

Chapter 4

Exercise 4.1. See the proof of Theorem 4.14, now for the vector subspace

$$F = \{x \in \ell^\infty;\ \exists \lim_n \Lambda_n x \in \mathbf{R}\}.$$

Exercise 4.2. See Theorem 4.17.

Exercise 4.3. See, for instance, Köthe [**26**, 15.8].

Exercise 4.4. If π_n is the n-projection of ℓ^∞, apply the Hahn-Banach theorem to extend $\pi_n \circ T$ to $T_n \in E'$ and define $\widetilde{T}(x) = (T_n(x)) \in \ell^\infty$.

Exercise 4.5. If $u \neq 0$ and $\operatorname{Ker} u$ is closed, then $x + U \subset E \setminus \operatorname{Ker} u$ for some neighborhood of zero, U, which can be supposed balanced, and then $u(U) \subset \mathbf{K}$ is also balanced. Show that either $u(U)$ is bounded, and then u is continuous, or $u(U) = \mathbf{K}$, in which case $u(x) = -u(y)$ for some $y \in U$ and then $x + y \in (x + U) \cap \operatorname{Ker} u$, a contradiction.

Exercise 4.6. Note that $\{e_n\}$ is a linearly independent system. Check that $\operatorname{Ker} \pi_n = [e_j;\ j \neq n] = \overline{[e_j;\ j \neq n]}$ (show that $x \in \overline{[e_j;\ j \neq n]} \setminus [e_j;\ j \neq n]$ would imply $\pi_n(x)e_n \in \overline{[e_j;\ j \neq n]}$), so that π_n is continuous (cf. Exercise 4.5). Observe that $x \in \operatorname{co}(K)$ if and only if $x = \sum_{n=1}^N \pi_n(x)e_n$ with $\pi_n(x) \geq 1$ and $\sum_{n=1}^N \pi_n(x) \leq 1$. To prove that $\operatorname{co}(K)$ is closed, if $x = \sum_{n=1}^N \pi_n(x)e_n \notin H$, consider the cases (a) $\pi_{n_0}(x) \notin [0,1]$ and (b) $\sum_{n=1}^N \pi_n(x) \notin [0,1]$. Check that $x \in V \subset \operatorname{co}(K)^c$ if $V = \pi_{n_0}^{-1}([0,1]^c)$ in case (a), and $V = \left(\sum_{n=1}^N \pi_n\right)^{-1}([0,1]^c)$ in case (b).

Example: The linear hull of an orthonormal sequence u_n in a Hilbert space and $e_n = (1/n)u_n$.

Exercise 4.7. If $\tilde{x} = \lim_n x_n$ and $\tilde{y} = \lim_n y_n$ in $\tilde{H}$ with $x_n, y_n \in H$, show that $\{(x_n, y_n)_H\}$ is a Cauchy sequence of numbers and that $(\tilde{x}, \tilde{y}) := \lim (x_n, y_n)_H$ is a well-defined inner product.

Exercise 4.8. If $u_x = (\cdot, x)$, then $(a, x)_H = 0$ if and only if $u_x(a) = 0$. Thus, $x \in A^\perp$ if and only if $u_x \in A^o$.

Exercise 4.9. $\|x\|_2 \leq \|x\|_V$ and V is dense in ℓ^2.

Exercise 4.10. If B is continuous at $(0,0)$, then $\|B(x,y)\|_G \le 1$ if $\|x\|_E < \varepsilon$ and $\|y\|_F \le \varepsilon$, for some $\varepsilon > 0$. Then $\|B((\varepsilon/\|x\|_E)x, (\varepsilon/\|y\|_F)y)\|_G \le 1$ (if $x, y \ne 0$), and $\|B(x,y)\|_G \le \varepsilon^{-2}\|x\|_E\|y\|_F$.

Conversely, the condition $\|B(x,y)\|_G \le C\|x\|_E\|y\|_F$ yields

$$\|B(x,y) - B(a,b)\| \le C\|x-a\|_E\|y\|_F + C\|a\|_E\|y-b\|_F.$$

Exercise 4.11. Suppose $\|f_k^{(j)}\|_{[a,b]} \le 1$ ($k \in \mathbf{N}$ and $0 \le j \le m+1$) and use the mean value theorem to apply Theorem 4.28.

Exercise 4.12. Similar to Exercise 4.11.

Exercise 4.13. If $m(E) = 0$ and $\mu(E) > 0$, then $m([a,c] \setminus E) = c$; if also $\mu(A) = c$, then $m(A \cup E) = c$ but $\mu(B) > c$.

To prove that $\mu = m$, show first that also $m(A) = c/2 \Rightarrow \mu(A) = c/2$ when $c \le 1/2$. If, for instance, $m(A_1) = c/2$ and $\mu(A_1) < c/2$, choose A_2 and A_3, all of them disjoint and such that $m(A_2) = m(A_3) = c/2$; since $m(A_1 \cup A_2) = m(A_2 \cup A_3) = m(A_1 \cup A_3) = c$, it follows that $\mu(A_2) > c/2$, $\mu(A_3) < c/2$, and $\mu(A_3) > c/2$, a contradiction. If $c > 1/2$, extend μ to $[0,2]$ by defining $\mu(A) = \mu(A-1)$ if $A \subset (1,2]$.

Exercise 4.14. The condition is $m(E \setminus F) = 0$, and then $d\mu_E = \chi_E d\mu_F$.

Exercise 4.15. If $\|x\|_p = 1$, define $y_k = \operatorname{sgn} x_k |x_k|^{p-1}$.

Exercise 4.16. Consider $x_k = 1 - 1/k$.

Exercise 4.17. $|v(x)| \le \|x\|_\infty$ and v extends to $v \in (\ell^\infty)'$. From $v(e_k) = \langle e_k, y\rangle = y^k$ it would follow that $y = 0$.

Exercise 4.18. See Exercise 4.17, with $\mathcal{C}[a,b]$ as a substitute of c.

Exercise 4.19. If $v \in c_0'$, define $x^k = v(e_k)$. If $x_N = \sum_{k=1}^N x^k e_k \to x$ in c_0, so that $v(y) = \lim_{k=1}^N y^k x^k = \langle x, y\rangle$, choose $z^k = \operatorname{sgn} y^k$ ($z^k = 0$ if $y^k = 0$), so that $v(z) \le \|v\|$, and $|v(y)| \le \|y\|_1\|x\|_\infty$.

Exercise 4.20. Represent the transpose of T as $T' : (y_n) \in \ell^\infty \mapsto (y_n/n) \in \ell^\infty$, with $\operatorname{Im} T' \subset c_o$ and $\operatorname{Ker} T = \{0\}$.

Exercise 4.21. Approximate uniformly every $g \in \mathcal{C}_c(a,b)$ by step functions.

Exercise 4.22. If $K \ge 0$, Hölder's inequality yields

$$\int_{-\infty}^{+\infty} \Big(\int_{-\infty}^{+\infty} K(x,y)\,dy\Big)|g(x)|\,dx \le \Big(\int_{-\infty}^{+\infty} K(x,y)^p\,dx\Big)^{1/p}\|g\|_{p'}.$$

If $K(\cdot,y) \in L^p(\mathbf{R})$, consider $|K(x,y)|$.

Exercise 4.23. Note that $\tilde{\mu}$ has to be finite.

Exercise 4.24. Let $u(1) = \sup_{|g|\le 1} |u(g)|$ and assume $u(1) = 1$. If $0 \le f \le 1$, define $g = 2f - 1$ and then $|u(g)| \le 1$; thus $u(f) = (1 + u(g))/2 \ge 0$.

Exercise 4.25. (a) $|u_g(f)| \le \frac{1}{2\pi} |\int_{\gamma_r} f(z)g(z)\,dz| \le \|f\|_{\bar{D}(0,r)} \|g\|_{D(0,r)^c}$. Moreover, $\int_{\gamma_r} f(z)g(z)\,dz = \int_{\gamma_s} f(z)g(z)\,dz$ if $\varrho < r < s < 1$.

(b) $|\int_{\varrho\bar{D}} f\,d\mu| \le \int_{\varrho\bar{D}} |f|\,d\mu \le \mu(D)\|f\|_{\varrho\bar{D}}$.

(c) Derivation under the integral yields that g_μ is holomorphic.

(d) Fubini's theorem combined with the Cauchy integral formula shows that

$$\frac{1}{2\pi i} \int_{\gamma_r} f(z) g_\mu(z)\,dz = \int_{\varrho\bar{D}} \frac{1}{z - \omega}\,d\mu(\omega).$$

Exercise 4.26. Consider T as a restriction of T''.

Exercise 4.27. If $Tf(x) := \int_0^x f(t)\,dt$ $(0 \le x \le 1)$ and $Tf = 0$, then differentiation shows that $f = 0$.

Exercise 4.28. If $K = \overline{\{\lambda_1, \lambda_2, \ldots\}}$, define $T \in \mathcal{L}(\ell^2)$ such that $Te_k = \lambda_k e_k$.

Exercise 4.29. See Exercise 4.28.

Exercise 4.30. If $K = \overline{T(B_E)}$, consider $K \subset \bigcup_{k=1}^{N(n)} B(c_k^n, 1/n)$ and $F_n = [c_1^n, \ldots, c_k^{N(n)}]$. Show that $\lim_{n\to\infty} \|T - P_{F_n} T\| = 0$.

Chapter 5

Exercise 5.1. Not every $x \in (\ell^1)'$ attains its norm on the unit sphere of ℓ^1 (Exercise 4.16).

Exercise 5.2. Since $L^2(\mathbf{R}) = L^2(\mathbf{R})'$ and $L^2(\mathbf{R})$ is separable, $B_{L^2(\mathbf{R})}$ is w^*-compact and metrizable. For the last part, see Exercise 5.7.

Exercise 5.3. Suppose $u(f_n) \not\to 0$, so that there exist $\delta > 0$ and $\{f_{n_k}\}$ so that $n_{k+1} > 2n_k$ and $u(f_{n_k}) > \delta$ (or $< -\delta$). Define $y_N := \sum_{k=1}^N f_{n_k}$, which satisfies $y_N(x) < 4$ everywhere. It follows from $u(f_{n_k}) > \delta$ that $u(y_N) > N\delta$, and u would be unbounded. Hence, $f_n \to 0$ weakly, but $\|f_n\|_{[0,1]} = 1$.

Exercise 5.4. Suppose that $g(0) = 0$ and choose φ to be zero near 0 and such that $\|g - \varphi\|_{[-1,1]} \le \varepsilon$. Show that (3) implies $\int_{-1}^1 (g - \varphi)h_n \le c\varepsilon$ and $\limsup_n \int_{-1}^1 gh_n \le \varepsilon$, so that $\langle g, h_n\rangle \to g(0)$. If $g(0) = C$, write $g = g_0 + C$.

For the converse, choose $g = 1$ to prove (1). To prove (3), apply the uniform boundedness principle to the sequence $\langle \cdot, h_n\rangle$.

Exercise 5.5. Let $\|x\|_E = u(x)$, $\|u\|_{E'} = 1$. Then $\|x\|_E = \lim_n |u(x_n)| \le \liminf_n \|x_n\|_E$.

Exercise 5.6. $K = B_{E'}$ is w^*-compact, and $x \mapsto \widehat{x}$ ($\widehat{x}(u) = u(x)$ if $\|u\|_{E'} \leq 1$) is a linear isometry of E into $\mathcal{C}(K)$.

Exercise 5.7. If $x_n \to 0$ in E, then $p_u(x_n) = |u(x_n)| \to 0$ with $u \in E'$. Let $\{e_n\}_{n=1}^{\infty}$ be an orthonormal system in a Hilbert space. Then $e_n \not\to 0$ in H, but $e_n \to 0$ weakly, since $\sum_{n=1}^{\infty} |(e_n, x)_H|^2 \leq \|x\|_H^2 < \infty$ and $(e_n, x)_H \to 0$.

Exercise 5.8. By Theorem 5.3, the weak closure K of $\mathrm{co}(\{x_n\})$ is also its closure, $x \in K$, and x is the limit of points in $\mathrm{co}(\{x_n\})$.

Exercise 5.9. Consider $A \subset E$ as a subset of $E'' = \mathcal{L}(E', \mathbf{K})$.

Exercise 5.10. Assume (b): the graph of T is closed and (a) holds, since if $x_n \to x$ in E and $Tx_n \to y$ in F, then it follows from $u(x_n) \to u(x)$ for all $u \in E'$ that $v(Tx_n) \to v(Tx)$ ($u \in v \circ T \in E'$) and also $v(Tx_n) \to v(y)$. Thus, $v(Tx) = v(y)$ for all $v \in F'$, and $Tx = y$. To show that (b) and (c) follow from (a), note that $p_x(T'u) = |u \circ T(x)| = p_{Tx}(u)$.

Exercise 5.11. Let $U_n = \{\max\{p_{x_1}, \ldots, p_{x_N}\} < 1/M\}$. If $x \in E$, every ball $\{p_x < \varepsilon\}$ contains some U_n; that is, $|u(x_j)| < 1/n \, \forall j \Rightarrow |u(x)| < 1$. Hence $u(x_1) = \cdots = u(x_N) = 0 \Rightarrow u(x) = 0$ and $x \in [x_1, \ldots, x_n]$.

Exercise 5.12. None of the semi-norms $\sup_{1 \leq j \leq N} |\langle x, u_j \rangle|$ defining the weak topology can be a norm.

Exercise 5.13. If T is compact and $x_n \to 0$ weakly, every subsequence of $\{Tx_n\}$ has a subsequence which is norm-convergent to 0, so $\|Tx_n\|_H \to 0$ in H.

For the converse note that B_H is weakly compact and metrizable. Then, for every $\{x_k\} \subset B_H$, there is a weakly convergent subsequence, $x_{k_n} \to x$ in B_H, so that $x_{k_n} - x \to 0$ weakly and $\|Tx_{k_n} - Tx\|_H \to 0$.

Exercise 5.14. There exists f in E' which strictly separates K_1 and K_2.

Exercise 5.15. Denote by $\bar{B}_E$ the weak* closure in E'' of the closed unit ball B_E and suppose that $\mathbf{K} = \mathbf{R}$. To prove that $\bar{B}_E = B_{E''}$, let $w_0 \in E''$ be any point not in $\bar{B}_E$, a weak* closed convex set. Then there is a $u \in E'$, $\|u\|_{E'} = 1$, such that $\sup_{w \in \bar{B}_E} \langle u, w \rangle < \langle u, w_0 \rangle$, and it follows that $\|w_0\|_{E''} > 1$. Therefore $B_{E''} \subset \bar{B}_E$.

Exercise 5.16. (a) If B_E is weakly compact, use Goldstine's theorem to show that $B_E = B_{E''}$. The converse follows from Alaoglu's theorem.

(b) The weak topology of F is the restriction of the weak topology of E.

(c) If E is reflexive, then $\sigma(E', E) = \sigma(E', E'')$ and $B_{E'}$ is $\sigma(E', E)$-compact and $\sigma(E', E'')$-compact, so that E' is reflexive. If E' is reflexive, then E is a closed subspace of the reflexive space E''.

Exercise 5.17. To show that $w = \langle x_0, \cdot \rangle$ for every $w \in H''$, write $\tau u = z$ if $u = (\cdot, z)_H \in H'$. Then $\tau : H' \to H$ is a bijective skewlinear isometry, and H' is a Hilbert space with the inner product defined by $(u_1, u_2)_{H'} = (\tau u_2, \tau u_1)_H$. Finally, $w(u) = (u, u_0)_{H'} = (\tau u_0, \tau u)_H$ for some $u_0 \in H'$. Choose $x_0 = \tau u_0$.

The mapping $f \in L^1(0,1) \mapsto \langle \cdot, f \rangle \in L^1(0,1)''$ is not exhaustive: if $w = \langle \cdot, f_0 \rangle$ for some $f_0 \in L^1(0,1)$, since $L^1(0,1)' = \{u_g;\, g \in L^\infty(0,1)\}$ with $u_g(f) = \int_0^1 f(t)g(t)\,dt$, necessarily $w(u_g) = u_g(f_0)$ $\forall g \in L^\infty(0,1)$, and the mapping J of Exercise 4.18 is not exhaustive.

As the dual of $L^1(0,1)$, $L^\infty(0,1)$ is not reflexive (Exercise 5.16(c)).

Exercise 5.18. For the last part, see Köthe [**26**, 22.4(2)].

Exercise 5.19. Choose $f_n(t) = e^{int}$ and apply the Riemann-Lebesgue lemma.

Exercise 5.20. If $\int_0^1 fg = 0$ for all $g \in \mathcal{C}[0,1]$, with $f \in L^1(0,1)$, then also $\int_a^b f = 0$ for every $(a,b) \subset (0,1)$ and it follows that $f = 0$. In $L^\infty(0,1)$ a limit of continuous functions is also continuous.

See the proof of Theorem 5.11 to solve Exercises 5.21–5.24

Chapter 6

Exercise 6.1. $\psi'(x) = g(x)g(1-r-x) \geq 0$, $\psi(x) = \int_0^{1-r} g(t)g(1-r-t)\,dt = C$, a constant, if $x \geq 1-r$, and $\psi(x) = 0$ if and only if $x \leq 0$. Hence $\varrho(x) = 0$ if and only if $x \leq -1$ or $x \geq 1$, and $\varrho(x) = C^2$, a constant, if $-r \leq x < r$. Choose $c = 1/C^2$.

Exercise 6.2. Since $\mathcal{C}_c(\Omega)$ is dense in $L^p(\Omega)$ (Corollary 2.13), approximate every $f \in \mathcal{C}_K(\Omega) \subset \mathcal{C}_c(\mathbf{R}^n)$ by $f * \varrho_\varepsilon$ as in Theorem 6.2, with $\operatorname{supp} f * \varrho_\varepsilon \subset \Omega$.

Exercise 6.3. The distance $d(\varphi, \psi) := \sum_{N=0}^\infty 2^{-N} \|\varphi - \psi\|_N / (1 + \|\varphi - \psi\|_N)$ defines the topology. Choose $\varrho \in \mathcal{D}_{[0,1]}(\mathbf{R})$ and show that the test functions $\varphi_N = \sum_{k=0}^N \tau_k \varrho$ form a Cauchy sequence in the new topology $\mathcal{T}$ but $\{\varphi_N\}$ it is not convergent in the convergence we are considering for test functions.

Exercise 6.4. Note that $|\varphi_n^{(m)}| \leq n^{-m-1} \sup_{\mathbf{R}} |\varphi^{(m)}|$ and that $\bigcup_n \operatorname{supp} \varphi_n$ is unbounded.

Exercise 6.5. The family $\mathcal{T}$ of all unions of sets of the form $\varphi + U$ ($\varphi \in \mathcal{D}(\Omega)$, $U \in \mathcal{U}$) is a topology on $\mathcal{D}(\Omega)$. To check (a), first show that if $\varphi \in G_1 \cap G_2$ with $G_1, G_2 \in \mathcal{T}$, then $\varphi + U \subset G_1 \cap G_2$ for some $U \in \mathcal{U}$.

To prove that every Cauchy sequence is contained in some $\mathcal{D}_K(\Omega)$, suppose that this is not the case, so that there are terms φ_k of the sequence and distinct points $x_k \in \Omega$ ($k \in \mathbf{N}$) such that $\varphi_k(x_k) \neq 0$ and with no limit points in Ω. Then $U := \{\varphi;\, k|\varphi(x_k)| < |\varphi_k(x_k)|\} \in \mathcal{U}$, since $K \in \mathcal{K}(\Omega)$ contains finitely many points x_k and then $U \cap \mathcal{D}_K(\Omega) \in \mathcal{T}$.

Exercise 6.6. (a) By the Leibniz formula,

$$f_n^{(m)}(t) = \frac{c_n}{n!}\sum_{j=0}^{m}\binom{m}{j}\frac{n!}{(n-j)!}t^{n-j}r_n^{m-j}\varphi^{(k-j)}(r_n t)$$

and $\operatorname{supp}\varphi(r_n t) \subset [-R/r_n, R/r_n]$.

(b) $r_n^{m-n} \le 1/r_n$ when $m < n$, and choose r_n large.

From (b), $\|f_n^{(p)}\|_{\mathbf{R}} \le q_{n-1}(f_n) \le 1/2^n$ whenever $n > p$, so that $\sum_{n=1}^{\infty} f_n^{(p)}$ is uniformly convergent on compact subsets and $f \in \mathcal{E}(\mathbf{R})$. Also $f_n(t) = c_n t^n/n!$ if $|t| \le 1/r_n$, so that $f_n^{(p)}(0) = 0$ when $p \neq n$ and $f_n^{(n)}(0) = c_n$.

Exercise 6.7 Note that $|\langle\varphi, f\rangle| \le (\int_K |f|)q_0(\varphi)$ and $|\langle\varphi, \mu\rangle| \le |\mu|(K)q_0(\varphi)$ for every $\varphi \in \mathcal{D}_K(\Omega)$.

Exercise 6.8. We would have $\int \varphi f = 0$ if $\varphi(a) = 0$ and then, as in Theorem 6.5, $f = 0$ a.e. on $\mathbf{R}^n \setminus \{a\}$, that is, $f = 0$ a.e. but $\delta_a \neq 0$.

Exercise 6.9. Apply δ' to the test functions $\psi_n(t) = -\sin(nt)\varphi(t)$, with $[-1/2, 1/2] \prec \varphi \prec (-1, 1)$. Then $\delta'(\psi_n) = n$ and, if $\delta' = \mu$,

$$n \le \left|\int_{[-1,1]} \varphi(t)\sin(nt)\, d\mu(t)\right| \le |\mu|([-1,1]).$$

Exercise 6.10. Write $\int K_\lambda(x)\varphi(x)\,dx - \varphi(0) = \int K_\lambda(x)(\varphi(x) - \varphi(0))\,dx$ and see the proof of Theorem 2.41.

Exercise 6.11. If $\operatorname{supp}\varphi \subset [-n, n]$ and $N \ge n$, then $\sum_{k=-N}^{N}\delta_k(\varphi) = \sum_{k=-n}^{n}\varphi(k)$. We can define $Ш(\varphi) := \sum_{k=-\infty}^{+\infty}\varphi(k)$, $Ш = \sum_{k=-n}^{+n}\delta_k$ on $\mathcal{D}_{[-n,n]}(\mathbf{R})$.

Exercise 6.12. The order is $\le m$, since $|\langle\varphi, \delta^{(m)}\rangle| \le \sum_{j=0}^{m}|\varphi(0)|$. To show that it is $> m-1$, apply $\delta^{(m)}$ to the functions $\psi_k\varrho$, where $[-1,1] \prec \varrho \prec (-2,2)$ and $\psi_k(x) = k^{1-m}\cos(kx)$ if m is even, else $\psi_k(x) = k^{1-m}\sin(kx)$.

Exercise 6.13. The distribution u satisfies $|u(\varphi)| \le C_K\sum_{n=0}^{N}\|\varphi^{(n)}\|_K = C_K q_N(\varphi)$ if $\varphi \in \mathcal{D}_K$ and $K \subset (1/N, \infty)$. To prove that there is no $N \in \mathbf{N}$ such that the above estimate holds for every compact set $K \subset (0, \infty)$, apply Exersice 6.12. There is no extension v of u to $\mathbf{R}$, since the continuity of v on $\mathcal{D}_{[-2,2]}$ would imply that the order should be finite.

Exercise 6.14. Let $K \prec \varrho \prec \Omega$, $|\varphi| \leq \|\varphi\|_K \varrho$ for every $\varphi \in \mathcal{D}_K(\Omega)$. If φ is real, then $-\|\varphi\|_K \varrho \leq \varphi \leq \|\varphi\|_K \varrho$ and $|u(\varphi)| \leq u(\varrho)\|\varphi\|_K$. If φ is not real, then $|u(\varphi)| \leq 2u(\varrho)\|\varphi\|_K$.

Exercise 6.15. Let $\psi \in \mathcal{C}[-a,b]$ be such that $\psi(t) = (\varphi(t) - \varphi(0))/t$ if $t \neq 0$. Then $u_2(\varphi) := \int_{-a}^{b} \frac{\varphi(t)-\varphi(0)}{t}\, dt = \int_{-a}^{b} \psi(t)\, dt$ defines a distribution on $\mathbf{R}$, and $u_f \in \mathcal{D}'(\mathbf{R})$ as the sum of three distributions. If $0 \notin \operatorname{supp}\varphi$, then $\psi(t) = \varphi(t)/t$ and $u_f(\varphi) = \int_{\mathbf{R}} \varphi(t)/t\, dt$.

Exercise 6.16. (a) If $\varphi_k \to 0$ in $\mathcal{D}(\mathbf{R})$, define $\psi_k(t) := (\varphi_k(t) - \varphi_k(0))/t$ ($\varphi_k'(0)$ if $t = 0$), continuous and such that $|\psi_k(t)| \leq \|\varphi_k'\|_{[-r,r]}$, to prove that $u_r(\varphi_k) \to 0$. (b) Observe that $\int_{\varepsilon \leq |t| \leq r} \frac{\varphi(t)-\varphi(0)}{t}\, dt \to \int_{-r}^{r} \frac{\varphi(t)-\varphi(0)}{t}\, dt$ if $\varepsilon \downarrow 0$.

Exercise 6.17. Write

$$u_f(\varphi) = \int_{\{|x|\geq 1\}} f(x)\varphi_k(x)dx + \int_{\{|x|<1\}} f(x)(\varphi_k(x) - T_m\varphi_k(x))dx$$

to prove that $u_f(\varphi_k) \to 0$ as $\varphi_k \to 0$ in $\mathcal{D}(\mathbf{R}^n)$. Note that if $0 \notin \operatorname{supp}\varphi$, then

$$u_f(\varphi) = \int_{\mathbf{R}^n} f(x)\varphi(x)dx = \langle \varphi, f \rangle,$$

since $\varphi(x) - T_m\varphi(x) = \varphi(x)$.

Exercise 6.18. If $\operatorname{supp}\varphi \subset [-N, N]$, then $-\int_0^\infty \varphi' = -\varphi(N) + \varphi(0) = \varphi(0) = \delta(\varphi)$, so that $Y' = \delta$. By induction, $Y^{(n+1)}(\varphi) = \pm\varphi^{(n)}(0)$.

Exercise 6.19. If $\operatorname{supp}\varphi \subset [-n,n]$, then $\int_{[-n,n]} f(t)\, dt = 2\int_0^n \log t\, dt = n\log n - n$ and $f \in L^1_{\mathrm{loc}}(\mathbf{R})$. Moreover $-\langle \varphi', f \rangle = -\lim_{\varepsilon\downarrow 0} \int_{\varepsilon\leq|t|\leq r} \varphi'(t)f(t)\, dt$ and

$$\int_{\varepsilon\leq|t|\leq r} \varphi'(t)f(t)\, dt = -\varphi(\varepsilon)\log\varepsilon + \varphi(\varepsilon)\log\varepsilon - \int_{\varepsilon\leq|t|\leq r} \frac{\varphi(t)}{t} dt \to -\langle \varphi', vp\frac{1}{t} \rangle$$

since $|\log\varepsilon(\varphi(-\varepsilon) - \varphi(\varepsilon))| \leq \varepsilon|\log\varepsilon|\|\varphi'\|_{[-n,n]} \to 0$ as $\varepsilon \to 0$.

Exercise 6.20. If $\operatorname{supp}\varphi \subset [-a,a]$, $\Delta = \{-a \leq x \leq y \leq a\}$, and $I = \int_\Delta \varphi'(x)f'(y)\, dx\, dy$, then by Fubini's theorem

$$I = \int_{\mathbf{R}} f'(y)\varphi(y)\, dy = -\int_{\mathbf{R}} \varphi'(x)f(x)\, dx.$$

Exercise 6.21. $\langle \varphi, D^\alpha u \rangle := (-1)^{|\alpha|}\langle D^\alpha\varphi, u \rangle$.

Exercise 6.22. If $P(D) = cD^\alpha$, note that $\langle \varphi, cD^\alpha u \rangle = (-1)^\alpha \langle D^\alpha(c\varphi), u \rangle$.

Exercise 6.23. See Theorems 6.17 and 6.18.

Exercise 6.24. If $u(\varphi)=0$ when $\operatorname{supp}\varphi\subset G$, then $D^\alpha u(\varphi)=\pm u(D^\alpha u)=0$, since $\operatorname{supp} D^\alpha\varphi\subset G$.

Exercise 6.25. $\operatorname{supp} Y=[0,\infty)$ and $\operatorname{supp}\delta'=\{0\}$.

Exercise 6.26. If $v\in\mathcal{E}(\Omega)$ and $\operatorname{supp} v\prec \operatorname{Int} K$, then $v=\varrho v$. This yields $|v(\varphi)|\le C_K q_N(\varphi)$ for every $\varphi\in\mathcal{D}_K(\Omega)$ and the Leibniz formula shows that $q_N(\varrho\varphi)\le Mq_N(\varphi)$. Hence $|v(\varphi)|=|v(\varrho\varphi)|\le C_K M q_N(\varphi)$.

Exercise 6.27. $1*(\delta'*Y)=1$ and $(1*\delta')*Y=0$. Two of the supports are not compact.

Exercise 6.28. See Theorem 6.18.

Exercise 6.29. $Y_m(x)\to Y(x)$ if $x\ne 0$ and $|Y_m(x)|\le\int_{-N}^N|d_m(t)|<\infty$ on $[-N,N]$ $(N\ge 1)$. Then $Y_m\to Y$ in $\mathcal{D}'(\mathbf{R})$ by dominated convergence. Moreover $Y_m'\to\delta$ in $\mathcal{D}'(\mathbf{R})$ and $Y_m'=d_m$ (integrate by parts in $\int_{[a,b]}\varphi(t)Y_m'(t)\,dt$, when Y_m is C^1 or absolutely continuous and $Y'=d_m$ in the distributional sense).

Finally, $d_m(t):=\frac{1}{\pi}\frac{m}{m^2t^2+1}$ is an approximation of δ.

Exercise 6.30. Two different solutions are $u=1$ and $u=Y$.

Exercise 6.31. The general solution of $P(D)=0$ is $F(t)=Ae^t+B\cos t+C\sin t$. From $f''(0)=1$ and $f(0)=f'(0)=0$ we obtain $A=1/2e$, $B=C=-1/2$. Hence, $E(t)=F(t)+\big((e^t)/(2e)-(1/2)\cos t-(1/2)\sin t\big)Y(t)$.

Exercise 6.32. The general solution of $P(D)=0$ is $F(t)=Ae^{-t}+Bte^{-t}+Ct^2e^{-t}$. From $f''(0)=1$ and $f(0)=f'(0)=0$ we obtain $A=B=0$ and $C=1/2$. Hence, $E(t)=F(t)+((1/2)t^2e^{-t})Y(t)$.

Exercise 6.33. Take derivatives under the integral.

Exercise 6.34. Check that $\partial_{\bar z}f=\pi\delta$ as follows: Define the continuous functions $f_n(z)=1/z$ if $|z|>1/n$ and $f_n(z)=n^2\bar z$ if $|z|\le 1/n$. Then show that $\partial_{\bar z}f_n=n^2\chi_{\bar D(0,1/n)}$, which tend to $\pi\delta$ as $n\to\infty$. Finally prove that $f_n\to 1/z$ in $\mathcal{D}'(\mathbf{R}^2)$ (use dominated convergence).

Exercise 6.35. It is easy to see that $F(y):=\langle\varphi(x,y),u(x-cy)\rangle$ defines a continuous function and that $\langle\varphi_k,v\rangle\to 0$ as $\varphi_k\to 0$ in $\mathcal{D}(\mathbf{R}^2)$ (use dominated convergence). If $u\in\mathcal{E}(\mathbf{R})$, a direct computation shows that $\partial_t^2v-c^2\partial_x^2v=0$. When $u\in L^1_{\mathrm{loc}}(\mathbf{R})$ and L is the wave operator, use the fact that $\langle\varphi,Lv\rangle=\langle L\varphi,v\rangle$.

Exercise 6.36. Apply Theorem 6.31.

Exercise 6.37. Write $E_2 = E_2^0 + E_2^\infty$, $E_2^0 = \chi_{\{|x|\le 1\}}$. To check that $u = f * E_2^0 + f * E_2^\infty$ is defined and locally integrable, note that $f * E_2^\infty$ exists everywhere and is locally bounded, since $\log|x-y| - \log|y| \to 0$ as $y \to \infty$. Every function $f_n := f\chi_{\{|x|\le n\}}$ satisfies the same conditions as f and $u_n := f_n * E_2 \to u$ in the distributional sense, by dominated convergence, so that also $\Delta u_n \to \Delta u$ in $\mathcal{D}'(\mathbf{R}^2)$. Moreover $\langle u_n, \Delta\varphi\rangle = \langle E_2, \Delta(\tilde{f}_n * \varphi)\rangle = \langle f_n, \varphi\rangle$. Therefore $\Delta u = \lim_n \Delta u_n = \lim_n f_n = f$.

Exercise 6.38. Write $u_r(y) = u(ry)$, choose $g \in \mathcal{C}(S)$ so that $\|f-g\|_p \le \varepsilon$, and let $v(x) = \int_S P(x,y)g(y)\,d\sigma(y)$. Then

$$\|f-u_r\|_p \le \|f-g\|_p + \|g-v_r\|_p + \|v_r-u_r\|_p \le 2\varepsilon + \|v_r-u_r\|_p$$

when r is close to 1. Prove that $\int_S P(rx,y)\,d\sigma(y) = \int_S P(rx,y)\,d\sigma(x) = 1$ (use the mean value theorem) to show that $f \in L^p(S) \mapsto u_r \in L^p(S)$ has norm 1, so that $\|v_r - u_r\|_p \to 0$ as $g \to f$ in $L^p(S)$.

Chapter 7

Exercise 7.1. Note that f_1 is the derivative of $-e^{-t^2}/2$, f_2 is the translation of a dilation of $\chi_{(-1/2,1/2)}$, and f_3 and f_4 are related to that Poisson kernel.

Exercise 7.2. Take derivatives under the integral, and use the properties of summability kernels.

Exercise 7.3. Compare K_t with the Gauss-Weierstrass kernel W_h.

Exercise 7.4. $\|\widehat{f}\|_\infty \le \|f\|_1$ and $\widehat{W} = W$, so that $\|\mathcal{F}\| = 1$.

Exercise 7.5. If $f = \widehat{\varphi}$, $g = \widehat{\psi}$, and $\operatorname{supp}\varphi \cap \operatorname{supp}\psi = \emptyset$, then $f * g = 0$.

Exercise 7.6. If $g(x) = x^\beta D^\alpha \varphi(x)$ is in $\mathcal{S}(\mathbf{R}^n)$, then also $g \in \mathcal{S}(\mathbf{R}^n) \subset L^p(\mathbf{R}^n)$. Moreover, $\|\varphi\|_p = \|\omega_{-2N}\omega_{2N}\varphi\|_p \le Cq_N(\varphi)$ if $Np > n/2$.

If $\omega_{2N}\varphi \in L^p(\mathbf{R}^n)$ for all N, then $\varphi \in L^1(\mathbf{R}^n)$. If the functions $x^\beta D^\alpha\varphi(x)$ are in $L^1(\mathbf{R}^n) \cap \mathcal{E}(\mathbf{R}^n)$, then φ is bounded, and so is every $x^\beta D^\alpha \varphi(x)$.

Exercise 7.7. From $Qf = P \in \mathcal{S}(\mathbf{R}^n)$, P a polynomial, it follows that $P = 0$.

Exercise 7.8. Apply the closed graph theorem.

Exercise 7.9. Write $\widehat{f} = \varrho_R \widehat{f}$ with $[-R,R] \prec \varrho_R \prec (-R-1, R+1)$ and apply the Fourier transform to $f^{(n)}$.

Exercise 7.10. Consider $f = f\chi_{\{|f|>1\}} + f\chi_{\{|f|\le 1\}} \in L^1(\mathbf{R}^n) + L^2(\mathbf{R}^n)$. Since $\|\widehat{f}\|_\infty \le \|f\|_1$ and $\|\widehat{f}\|_2 = \|f\|_2$, the result follows from the Riesz-Thorin theorem by choosing $\theta = 2/p'$.

Exercise 7.11. $\mathcal{F}(\operatorname{sinc}) = \widetilde{\mathcal{F}}(\operatorname{sinc}) = \chi_{(-1/2,1/2)}$ is not continuous, and $\|\operatorname{sinc}\|_2 = \|\chi_{(-1/2,1/2)}\|_2 = 1$.

Exercise 7.12. Define $v(x,y) = u(x,y) + y$.

Exercise 7.13. If $x[k] \in \ell^p \subset \ell^\infty$, then $|x[k]| \leq |x[k]| \leq C|k|$. If $\varphi \in \mathcal{D}_{[-N,N]}(\mathbf{R})$, then $\langle \varphi, u_x \rangle = \sum_{k=-N}^{N} x[k]\varphi(k) \to 0$ as $\varphi \to 0$ in $\mathcal{D}_{[-N,N]}(\mathbf{R})$.

Exercise 7.14. See Exercise 7.10.

Exercise 7.15. Since

$$\int_{-1/2}^{1/2} \sum_{k=-\infty}^{k=\infty} |\varphi(t-k)|\, dt = \|\varphi\|_1 < \infty,$$

$\varphi_1(t) = \sum_{k=-\infty}^{+\infty} \varphi(t-k)$ is well-defined on interval $(-1/2, 1/2]$ and the series converges a.e. and in L^1. The Fourier coefficients are

$$c_k(\varphi_1) = \sum_{k=-\infty}^{+\infty} \int_{-1/2}^{1/2} \varphi(t-k) e^{-2k\pi it}\, dt = \widehat{\varphi}(k)$$

and $\sum_{k=-\infty}^{+\infty} |\widehat{\varphi}(k)| < \infty$ since $\widehat{\varphi} \in \mathcal{S}(\mathbf{R})$. Therefore the Fourier series is uniformly convergent and $\sum_{k=-\infty}^{+\infty} \varphi(k) = \varphi_1(0) = \sum_{k=-\infty}^{+\infty} \widehat{\varphi}(k)$, the sum of the Fourier series at the origin. Note that (d) follows from (c).

Exercise 7.16. In $\mathbf{R}^N$ the norms $\sum_{k=1}^{N} |x_k|$, $\max_{k=1}^{N} |x|_k$, and $(\sum_{k=1}^{N} |x_k|^p)^{1/p}$ are equivalent.

Exercise 7.17. $|u(x) - u(y)| \leq |x-y|^{1/p'} \|u'\|_p$.

Exercise 7.18. As in Theorem 7.25, $v(t) = \int_c^t u'(s)\, ds + C$ is continuous and $|v(t)| \leq \|u'\|_1 + |C|$. Also $W^{1,p}(a,\infty) \hookrightarrow \mathcal{C}[a,\infty) \cap L^\infty(a,\infty)$ is continuous, by the closed graph theorem.

Exercise 7.19. By Theorem 7.25, $Ru \in \mathcal{C}(\mathbf{R})$, and $\|Ru\|_p^p = 2\|u\|_p^p$. If $v = u'$, the distributional derivative of u on $(0,\infty)$, define $v(-t) := -v(t)$ if $t < 0$. Then

$$Ru(x) - u(0) = u(-x) - u(0) = \int_x^0 v(-t)\, dt = \int_0^x v(t)\, dt \quad (x < 0)$$

and $Ru(x) - u(0) = u(x) - u(0) = \int_0^x v(t)\, dt$ also holds if $x \geq 0$. Then $(Ru)' = v$, $\|(Ru)'\|_p^p = 2\|u'\|_p^p$, and $Ru \in W^{1,p}(\mathbf{R})$ with $\|Ru\|_{1,p} = 2^{1/p}\|u\|_{1,p}$.

Exercise 7.20. If $u_k \to 0$ in $H_0^1(\Omega)$ and $\bar{u}_k \to v$ in $H^1(\mathbf{R}^n)$, then $v = 0$.

Exercise 7.21. $(\partial_j Pu)_{|\Omega} = (\partial_j u)_{|\Omega}$ and $(\partial_j Pu)_{|\bar{\Omega}^c} = (\partial_j u)_{|\bar{\Omega}^c} = 0$, so that $(\partial_j Pu)_{|\partial\Omega^c} = (\partial_j u)_{|\partial\Omega^c}$ and $\partial\Omega$ is a null set. Hence $\partial_j Pu = P\partial_j u$ in $L^2(\mathbf{R}^n)$ and as distributions.

Exercise 7.22. Choose $u = 1$, a constant on $(-1,1)$. Then $u^o \in L^2(\mathbf{R})$, but $(u^o)' = \delta_{-1} - \delta_1 \notin L^2(\mathbf{R})$, since it is not a function.

Exercise 7.23. Use the closed graph theorem.

Exercise 7.24. The Fourier transform of u is of type $(1+|\xi|^2)^{-1}$ and

$$\int_0^\infty [(1+\xi^2)^{-1}\omega_s(\xi)]^2\,d\xi = \int_0^\infty (1+\xi^2)^{s-2}\,d\xi < \infty$$

if $s < 3/2$.

Exercise 7.25. $B(u,u) = \int_\Omega \left(|\nabla u(x)|^2 + |u(x)|^2\right) dx = \|u\|^2_{H_0^1}$, coercive.

Chapter 8

Exercise 8.1. If $e = (e,0)$ is the unit in $A \subset A_1$, then $\delta e = e \neq \delta$.

Exercise 8.2. Note that $\chi(1)\chi(1) = \chi(1)$ for every $\chi \in \Delta(\mathbf{C}_1)$, and $\chi(\delta) = 1$. Here $1 = (1,0)$ and $\delta = (0,1)$.

Exercise 8.3. (a) $z \in A(D)$ but $\bar{z} \notin A(D)$. (b) Suppose $\chi \in \Delta$ and let $g(z) = z$. Then $\sigma(g) = \bar{D}$, $\chi(P) = P(\alpha)$, and, by continuity, $\chi(f) = f(\alpha) = \delta_\alpha(f)$. (c) If $\alpha \in \bar{D}$, then $|f_j(\alpha)| > 0$ for at least one j and $\chi(f_j) \neq 0$. Then $J = f_1 A(D) + \cdots + f_n A(D)$ is not contained in a maximal ideal, since there is no δ_α such that $\delta_\alpha(f_j) = 0$ for every j $(1 \le j \le n)$.

Exercise 8.4. Suppose $e * f = f$ for every $f \in L^1(\mathbf{R})$. From the properties of the Fourier transform, $\widehat{e}\varphi = \varphi$ for every $\varphi \in \mathcal{S}(\mathbf{R})$, which implies $\widehat{e} = 1$.

Exercise 8.5. Similar to Exercise 8.4, with Fourier coefficients.

Exercise 8.6. The unit is $\delta = \{\delta[k]\}$ such that $\delta[0] = 1$ and $d[k] = 0$ if $k \neq 0$.

Exercise 8.7. Note that $L_{ab} = L_a L_b$, $L_e = I$, and $\|L_a\| = \sup_{\|x\|\le 1} \|ax\| \le \|a\|$ and $\|L_a e\| = \|a\|$.

Exercise 8.8. (a) $|\gamma_\chi(\alpha+\beta) - \gamma_\chi(\alpha)| = |\chi(\tau_{-\alpha}(\tau_{-\beta}u - u))| \le \|\tau_{-\beta}u - u\|_1 \to 0$ as $\beta \to 0$ and $\gamma_\chi(\alpha+\beta) = \chi(\tau_{-\alpha}u * \tau_{-\alpha}u) = \gamma_\chi(\alpha)\gamma_\chi(\beta)$. It follows from $\gamma_\chi(\alpha)\gamma_\chi(-\alpha) = \gamma_\chi(0) = 1$ and from $\gamma_\chi(n\alpha) = \gamma_\chi(\alpha)^n$ that $|\gamma_\chi(\alpha)| = 1$ (γ_χ is bounded).

(b) A continuous solution $\gamma : \mathbf{R} \to \mathbf{T}$ of $\gamma(\alpha+\beta) = \gamma(\alpha)\gamma(\beta)$ has the form $\gamma_\chi(\alpha) = e^{i\xi\alpha}$ for some $\xi \in \mathbf{R}$. The following steps lead to a proof: (1) $\gamma(n\alpha) = \gamma(\alpha)^n$. (2) $\omega_n := \operatorname{Arg}(\gamma(2^{-n})) \to 0$ as $n \to \infty$. (3) $2\omega_{n+1} - \omega_n \in 2\pi\mathbf{Z}$. (4) $2\omega_{n+1} = \omega_n$ for all $n \ge p$ for some $p \in \mathbf{N}$. (5) $\omega_n = 2^{p-n}\omega_p$ for all $n \ge p$. (6) If $\xi = 2^p\omega_p$, then $\gamma(\alpha) = e^{i\xi\alpha}$ for all $\alpha \in \{m2^n;\, m,n \in \mathbf{Z},\, n \ge p\}$, which is a dense subset of $\mathbf{R}$.

(c) From $u*f(\xi)=\int_{\mathbf{R}} f(\alpha)\tau_{-\alpha}u(\xi)\,d\alpha$, $\chi(f)=\chi(u*f)=\int_{\mathbf{R}} f(\alpha)\gamma_\chi(-a)\,d\alpha$ and χ is continuous on $L^1(\mathbf{R})$.

Exercise 8.9. If $F_\alpha = G_\alpha^c$, then $\left(\bigcup G_\alpha\right)^c = \bigcap f^{-1}(F_\alpha^c)$. Also $\mu(f^{-1}(\{|z| > \|f\|_\infty\})) = 0$ and $\mu(f^{-1}(\{|z| > |\lambda|\})) \neq 0$ if $|\lambda| < \|f\|_\infty$. Finally, $\lambda \notin \sigma(f)$ iff $1/(f(x) - \lambda)$ exists a.e. and is bounded a.e.

Exercise 8.10. See Rickart [**35**, I.7].

Exercise 8.11. $\chi(e) = 1$ follows from $0 \neq \chi(a) = \chi(a)\chi(e)$.

Exercise 8.12. See Exercise 1.8.

Exercise 8.13. For every $f \in \mathcal{C}(K)$, $\widehat{f}(\delta_t) = f(t)$ is continuous with respect to the initial compact topology, which coincides with the coarser Gelfand topology. Moreover, $\widehat{f}(\delta_t) = f(t)$ shows that $\|\widehat{f}\|_\Delta = \|f\|_K$ and $\mathcal{C}(\Delta) = \mathcal{G}(\mathcal{C}(K))$.

Exercise 8.14. U is open in Δ because $U = \{\chi;\ |\widehat{z}(\chi)| < 1\}$, where z denotes the coordinate function. The rotation $\varrho : z \mapsto \xi z$ induces an isomorphism $f \mapsto f \circ \varrho$ of $H^\infty(U)$, and the adjoint of this isomorphism maps Δ_1 onto Δ_ξ.

Exercise 8.15. To prove that every $\chi \in \Delta(W)$ has the form $\chi = \delta_\tau$, let $u(t) = e^{it}$, so that $\|u\| = \|1/u\| = 1$ and $|\chi(u)| = 1$, since $|\chi(u)|, |\chi(1/u)| \leq 1$. There is some $\tau \in \mathbf{R}$ such that $\chi(u) = e^{i\tau} = u(\tau)$ and $\chi(u^k) = \chi(u)^k = u^k(\tau)$. Hence $\chi(P) = P(\tau)$ if $P(t) = \sum_{|k|\leq N} c_k e^{ikt}$, these trigonometric polynomials are dense in W, and $\chi(f) = \delta_t(f)$ for every $f \in W$. If f does not vanish at any point, then f has an inverse in W which must be $1/f$.

Exercise 8.16. $\delta_t = \delta_s$ if and only if $t - s = 2k\pi$, and $\delta_t f = \delta_z F$. Hence $\Delta = \mathbf{T}$ and $\widehat{f} = f$ (or F) as in Exercise 8.13. But not every function in $\mathcal{C}(\mathbf{T})$ is the sum of an absolutely convergent Fourier series, and $\widehat{W}$ is dense in $\mathcal{C}(\mathbf{T})$ (use the Stone-Weierstrass theorem). This implies that $\mathcal{G}$ cannot be an isometry.

Exercise 8.17. From the properties of the resolvent, there is a number $M > 0$ such that $\|(\lambda e - x)^{-1}\| < M$ for all $\lambda \notin U$, and

$$\lambda e - x - y = (\lambda e - x)(e - (\lambda e - x)^{-1}y)$$

is invertible, since $\|(\lambda e - x)^{-1}y\| < 1$. Choose $\delta = 1/M$.

Exercise 8.18. Note that $\|\widehat{a}\|_\Delta = r(a) = \lim_n \|a^{2^n}\|^{1/2^n}$. If $\mathcal{G}$ is an isometry, so that $\|\widehat{a}\|_\Delta = \|a\|$, it is clear that $\|a^2\| = \|\widehat{a}^2\|_\Delta = \|\widehat{a}\|_\Delta^2 = \|a\|^2$.

Exercise 8.19. $e^*x = (x^*e)^* = x^{**} = x$ and $(x^{-1})^*x^* = (xx^{-1})^* = e^* = e$.

Exercise 8.21. If $M_n = P_n(H)$ and $M = \bigoplus M_n$, then $y \in M$ if and only if $y = \sum_{n=1}^{\infty} y_n$, where $\sum_{n=1}^{\infty} \|y_n\|^2 < \infty$ and $y_n \in M_n$.

Consider $y_n \in M_n$ such that $\sum_{n=1}^{\infty} \|y_n\|^2 < \infty$, with $y_n \perp y_m$ when $n \neq m$. Then $S_N = \sum_{n=1}^{N} y_n$ is a Cauchy sequence, since $\|S_p - S_q\|^2 = \|\sum_{n=q+1}^{p} y_n\|^2 \leq \sum_{n=q+1}^{\infty} \|y_n\|^2 \to 0$ as $p \to \infty$.

Show that $M := \{\sum_{n=1}^{\infty} y_n, \lim_N S_N < \infty\}$ is the smallest closed subspace which contains every M_n. Then $Px := \sum_{n=1}^{\infty} P_n x \in M$, $Px = x$ if and only if $x \in M$, $P^2 = P$, and $(Px_1, x_2)_H = (x_1, Px_2)$.

Exercise 8.22. See Theorem 4.36.

Exercise 8.23. $B \subset K$, $(E(B)f, g)_H = \int_K \chi_B f \bar{g} = \int_K f \overline{\chi_B g} = (f, E(B)g)_H$ and $E(B)E(B)f = \chi_B \chi_B f = E(B)f$, so that $E(B)$ is an orthogonal projection. Also, $E(A \cap B)f = \chi_{A \cap B} f = \chi_A \chi_B f = E(A)E(B)f$. Moreover, $E(\biguplus_{n=1}^{\infty} B_n)\cdot = \chi_{\biguplus_{n=1}^{\infty} B_n}\cdot = \sum_{n=1}^{\infty} \chi_{B_n}\cdot$.

Exercise 8.24. $E(B) = 0$ if and only if $(E(B)x, x)_H = 0$ for every $x \in H$.

Exercise 8.25. Try $T = \begin{pmatrix} 0 & 1 \\ -1 & 0 \end{pmatrix}$.

Exercise 8.26. By Theorem 8.25, $T = T^*$ and $\sigma(T) \subset [0, \infty)$. Note that $0 \leq \widehat{T} \in \mathcal{C}(\sigma(T))$, $\widehat{T} = f^2$ for a unique $f \in \mathcal{C}(\sigma(T))$, $f \geq 0$, and there is a unique $S \in \langle T \rangle$ such that $\widehat{S} = f$ and $\widehat{S} \geq 0$, which is equivalent to $S \geq 0$.

Exercise 8.27. On $\sigma(T) = \sigma_{\langle T \rangle}(T)$, let $\lambda = p(\lambda)s(\lambda)$ with $p(\lambda) = |\lambda| \geq 0$ and $|s(\lambda)| = 1$ everywhere. Define $P = p(T)$ and $U = s(T)$.

Chapter 9

Exercise 9.1. If $F = \overline{\mathcal{D}(T)}$ is not the whole space H, write $H = F \oplus F^{\perp}$ and consider two different operators on $F^{\perp}$ to extend T from F to H.

Exercise 9.2. Since $|(Tx, y)_H| \leq \|x\|_H \|Ty\|$, $\{Tx; \|x\|_H \leq 1\}$ is weakly bounded and it is bounded by the uniform boundedness principle.

Exercise 9.3. A reduction to Example 9.2 is obtained by considering the Fourier transform of $\chi_{(n,n+1)}$.

Exercise 9.4. Note that $y \in (\operatorname{Im} T)^{\perp}$ if and only if $x \in \mathcal{D}(T) \mapsto (Tx, y)_H = 0$, so $y \in \mathcal{D}(T^*)$ and $T^* y = 0$.

Exercise 9.5. $\lambda \in \sigma(T)^c$ if $|\lambda - \lambda_0| < 1/R_T(\lambda_0)$; then $d(\lambda_0, \sigma(T)) \geq 1/R_T(\lambda_0)$.

Exercise 9.6. To show that $\mathcal{D}(T^*) \subset \mathcal{D}(T)$, let $y^* = T^*y$ for any $y \in \mathcal{D}(T^*)$, and choose $x \in \mathcal{D}(T)$ so that $Tx = y^*$. Then $y = x$ since, for every $z \in \mathcal{D}(T)$, $(Tz, y)_H = (z, y^*)_H = (Tz, x)_H$, that is, $(u, y)_H = (u, x)_H$ for all u.

Exercise 9.7. $\overline{\operatorname{Im} T} = \operatorname{Ker} T^\perp = H$. Let $y \in \mathcal{D}((T^{-1})^*)$; then $(A^{-1}x, y)_H = (x, y^*)_H$ for any $x \in \mathcal{D}(T^{-1})$ and $(z, y)_H = (Tz, y^*)_H$ if $z \in \mathcal{D}(T)$ ($z = T^{-1}x$), where $y^* \in \mathcal{D}(T)$ and $Ty^* = y$ since $T = T^*$. Then $y \in \operatorname{Im} T = \mathcal{D}(T^{-1})$ and $(T^{-1})^* y = y^* = T^{-1}y$, so that $(T^{-1})^* = T^{-1}$.

Exercise 9.8. Let $\lambda \notin \sigma_p(T)$, assume that $F = \mathcal{D}(T - \lambda I)$ is not dense, and choose $0 \neq y \in F^\perp$. But $(Tx - \lambda x, y)_H = 0$ for all $x \in H$ and, since $\lambda \in \mathbf{R}$ and $T^* = T$, also $(x, (T - \lambda I)y)_H = 0$, and for $x = (T - \lambda I)y$ we obtain $Ty = \lambda y$, a contradiction to $\lambda \notin \sigma_p(T)$.

Exercise 9.9. Show that $U(f) = f(A)z$ defines a bijective isometry (note that $U[t^n] = A^n z$) and check that $AUf = U[tf(t)]$.

Exercise 9.10. Cf. the constructions in Yosida [**44**, XI.5].

Exercise 9.11. Note that $\int_0^1 i\varphi'(t)\overline{\psi(t)}\,dt = \int_0^1 f(t)\overline{i\psi'(t)}\,dt$ to check that $S \subset S^*$. Denote $V(x) = \int_0^x v(t)\,dt$ ($v \in \mathcal{D}(S^*)$, and choose $u = 1$ in

$$\int_0^1 iu'(t)\overline{v(t)}\,dt = u(1)\overline{V(1)} - \int_0^1 u'(t)\overline{V(t)}\,dt \qquad (u \in \mathcal{D}(S))$$

to show that $V(1) = 0$. It follows that $iv - V \in (\operatorname{Im} S)^\perp = \{0\}$ and $\mathcal{D}(S^*) = H_0^1(0,1)$.

Exercise 9.12. See Exercise 9.11.

Exercise 9.13. Show that $-D^2 f = \lambda f$ with the conditions $f(0) = f(1) = 0$ has the solutions $\lambda_n = -\pi^2 n^2$, $f_n(x) = \sin(\pi n x)$ and prove that $\{f_n;\, n = 1, 2, 3, \ldots\}$ is an orthogonal total system in $L^2(0,1)$.

Exercise 9.14. Consider $-D^2 f = \lambda f$ with the conditions $f'(0) = f'(1) = 0$.

Exercise 9.15. Show that the operator has at least two different self-adjoint extensions, obtained in Exercises 9.13 and 9.14.

Exercise 9.16 $(T\psi, \psi)_2 = (\psi', \psi')_2 + \int_0^1 V(x)|\psi(x)|^2\,dx \geq \|\psi\|_2^2$. Indeed, $|\psi(t)| \leq \int_0^t |\psi'(x)|\,dx \leq t^{1/2}\|\psi'\|_2$; hence $\int_0^1 |\psi(t)|^2\,dt \leq \|\psi'\|_2^2$.

Exercise 9.17. Q_j and P_j are Q and $P = \frac{h}{2\pi i}D$ in the case $n = 1$.

Exercise 9.18. Prove that A_1 and A_2 commute if and only if A_2 commutes with every $E^1(B_1)$, and then A_2 will commutes with $E^1(B_1)$ if and only if

$E^1(B_1)$ commutes with every $E^2(B_2)$. Indeed,

$$(A_2A_1x, y)_H = \int t\, dE^1_{x,A_2y}, \quad (A_1A_2x, y)_H = \int t\, dE^1_{A_2x,y}$$

and also

$$(A_2E^1(B)x, y)_H = E^1_{x,A_2y}(B), \quad (E^1(B)A_2x, y)_H = E^1_{A_2x,y}(B).$$

It follows from $A_1A_2 = A_2A_1$ that $AA_2 = A_2A$ for all $A \in \langle A_1 \rangle = \{g(A_1);\, g \in \mathcal{C}(\sigma(A_1))\}$, and then $E^1_{x,A_2y} = E^1_{A_2x,y}$. If $A_2E^1(B) = E^1(B)A_2$, then also $E^1_{x,A_2y} = E^1_{A_2x,y}$.

Exercise 9.19. Show first that every $E(B_1 \times B_2)$ is an orthogonal projection and that E satisfies the conditions (1)–(3) of spectral measures on $\mathbf{R}^2$ (see Subsection 8.5.2) on Borel sets of type $B = B_1 \times B_2$. Then extend E to all Borel sets in $\mathbf{R}^2$ as in the construction of scalar product measures.

Exercise 9.20. Solution: $Au = -u'$ with $\mathcal{D}(A) = H^1(\mathbf{R})$. Note that $u \in H^1(\mathbf{R})$ when $u \in L^2(\mathbf{R})$ and the distributional limit $\lim_{h\to 0} h^{-1}[u(x-h) - u(x)]$ exists in $L^2(\mathbf{R})$.

Exercise 9.21. $Au = gu$, and $\mathcal{D}(A) = \{u \in L^2(\mathbf{R});\, gu \in L^2(\mathbf{R})\}$.

Bibliography

[1] R. A. Adams, *Sobolev Spaces,* Academic Press, Inc., 1975.

[2] N. I. Akhiezer and I. M. Glazman, *Theory of Linear Operators in Hilbert Space,* Dover, 1993.

[3] S. Banach, *Théorie des opérations linéaires,* Monografje Matamatyczne, Warsaw, 1932, and Chelsea Publishing Co., 1955.

[4] S. K. Berberian, *Lectures in Functional Analysis and Operator Theory,* Springer-Verlag, 1974.

[5] H. Brezis, *Analyse fonctionelle: Théorie et applications,* Masson, 1983.

[6] J. Cerdà, *Análisis Real,* Edicions de la Universitat de Barcelona, 1996.

[7] B. A. Conway, *A Course in Functional Analysis,* Springer-Verlag, 1985.

[8] R. Courant and D. Hilbert, *Methods of Mathematical Physics,* Oxford University Press, 1953.

[9] J. Dieudonné, *Foundations of Modern Analysis,* Academic Press, 1960.

[10] J. Dieudonné, *History of Functional Analysis,* North-Holland, 1981.

[11] P. A. M. Dirac, *The Principles of Quantum Mechanics,* Clarendon Press, 1930.

[12] N. Dunford and J. T. Schwartz, *Linear Operators: Part 1,* Interscience Publishers, Inc., 1958.

[13] R. E. Edwards, *Functional Analysis, Theory and Applications,* Hold, Rinehart and Windston, 1965.

[14] G. B. Folland, *Introduction to Partial Differential Equations,* Princeton University Press and University of Tokyo Press, 1976.

[15] I. M. Gelfand, and G. E. Chilov, *Generalized Functions,* Academic Press, New York, 1964 (translated from the 1960 Russian edition).

[16] I. M. Gelfand, D. A. Raikov and G. E. Chilov, *Commutative Normed Rings,* Chelsea Publishing Co., New York, 1964 .

[17] D. Gilbarg and N. S. Trudinger, *Elliptic Partial Differential Equations of Second Order,* Springer-Verlag, 2001 (revision of the 1983 edition).

[18] P. R. Halmos, *Measure Theory,* D. Van Nostrand Company, Inc., 1950.

[19] E. Hille and R. S. Phillips, *Functional Analysis and Semigroups,* Amer. Math. Soc. Colloquium Publ., vol. 31, 1957.

[20] L. Hörmander, *Linear Partial Differential Operators,* Springer-Verlag, 1963.

[21] J. Horvath, *Topological Vector Spaces and Distributions,* Addison Wesley, 1966.

[22] L. Kantorovitch and G. Akilov, *Analyse fonctionnelle,* "Mir", 1981 (translated from the 1977 Russian edition).

[23] T. Kato, *Perturbation Theory for Linear Operators,* Springer-Verlag, 1976.

[24] J. L. Kelley, *General Topology,* Van Nostrand, Princeton, NJ, 1963.

[25] A. N. Kolmogorov and S. V. Fomin, *Elements of the Theory of Functions and Functional Analysis,* Glaglock Press 1961 (translated from the 1960 Russian edition).

[26] G. Köthe, *Topological Vector Spaces I,* Springer-Verlag, 1969.

[27] P. D. Lax, *Functional Analysis,* Wiley, 2002.

[28] E. H. Lieb and M. Loss, *Analysis,* Graduate Studies in Mathematics, American Mathematical Society, 1997.

[29] V. Mazya, *Sobolev Spaces,* Springer-Verlag, 1985.

[30] R. Meise and D. Vogdt, *Introduction to Functional Analysis,* The Clarendon Press, Oxford University Press, 1997 (translated from the 1992 German edition).

[31] A. F. Monna, *Functional Analysis in Historical Perspective,* John Wiley & Sons, 1973.

[32] M. A. Naimark, *Normed Rings,* Erven P. Noordhoff, Ltd., 1960 (translation of the 1955 Rusian edition).

[33] E. Prugovečki, *Quantum Mechanics in Hilbert Space,* Academic Press, 1971.

[34] M. Reed and B. Simon, *Methods of Modern Mathematical Physics,* Academic Press, 1972.

[35] C. E. Rickart, *General Theory of Banach Algebras,* D. Van Nostrand Company, Inc., 1960.

[36] F. Riesz and B. Sz. Nagy, *Leçons d'analyse fonctionelle,* Akadémiai Kiadó, Budapest, 1952, and *Functional Analysis,* F. Ungar, 1955.

[37] H. L. Royden, *Real Analysis,* The Macmillan Company, 1968.

[38] W. Rudin, *Functional Analysis,* McGraw-Hill Book Company, 1973.

[39] W. Rudin, *Real and Complex Analysis,* McGraw-Hill Book Company, 1966.

[40] L. Schwartz, *Théorie des distributions,* Hermann & Cie, 1969 (new edition, 1973).

[41] E. M. Stein and G. Weiss, *Introduction to Fourier Analysis on Euclidean Spaces,* Princeton University Press, 1971.

[42] A. E. Taylor and D. C. Lay, *Introduction to Functional Analysis,* John Wiley & Sons, Inc., second edition, 1980.

[43] J. von Neumann, *Mathematical Foundations of Quantum Mechanics,* Princeton University Press, 1996 (translation of the 1932 German edition).

[44] K. Yosida, *Functional Analysis,* Springer-Verlag, 1968.

Index